NEUROTECHNOLOGY IN NATIONAL SECURITY AND DEFENSE

Practical Considerations, Neuroethical Concerns

Advances in Neurotechnology

Series Editor
James Giordano

Neurotechnology: Premises, Potential, and Problems
edited by James Giordano

Neurotechnology in National Security and Defense: Practical
Considerations, Neuroethical Concerns
edited by James Giordano

NEUROTECHNOLOGY IN NATIONAL SECURITY AND DEFENSE

Practical Considerations, Neuroethical Concerns

Edited by JAMES GIORDANO, PhD

Neuroethics Studies Program
Pellegrino Center for Clinical Bioethics
Georgetown University, Washington DC, USA

Center for Neurotechnology Studies
Potomac Institute for Policy Studies
Arlington, VA, USA

Human Science Center (HWZ)
Ludwig-Maximilians University, Munich, Germany

CRC Press
Taylor & Francis Group
Boca Raton London New York

CRC Press is an imprint of the
Taylor & Francis Group, an **informa** business

First published in paperback 2024

First published 2015
by CRC Press
2385 NW Executive Center Drive, Suite 320, Boca Raton FL 33431

and by CRC Press
4 Park Square, Milton Park, Abingdon, Oxon, OX14 4RN

CRC Press is an imprint of Taylor & Francis Group, an Informa business

© 2015, 2024 by Taylor & Francis Group, LLC

Publisher's Note
The publisher has gone to great lengths to ensure the quality of this reprint but points out that some imperfections in the original copies may be apparent.

ISBN: 978-1-4822-2833-5 (hbk)
ISBN: 978-1-03-292245-4 (pbk)
ISBN: 978-0-429-15973-2 (ebk)

DOI: 10.1201/b17454

Visit the Taylor & Francis Web site at
http://www.taylorandfrancis.com

and the CRC Press Web site at
http://www.crcpress.com

Dedication

To those who serve...

Contents

Series Preface

ADVANCES IN NEUROTECHNOLOGY: ETHICAL, LEGAL, AND SOCIAL ISSUES

Neuroscience and neurotechnology have progressed through a reciprocal relationship of tasks and tools: advances in technology have enabled discovery in the neurosciences, and neuroscientific questions and concepts have fueled development of new technologies with which to study and manipulate nervous systems and the brain. Neuroimaging, neurostimulatory and neuroprosthetic devices, neurogenetics, and unique pharmacological agents and approaches have shown considerable promise in improving the diagnosis and treatment of neurological injury and neuropsychiatric disorders and may expand the boundaries of human performance and human–machine interaction. Despite apparent limitations, these technologies increasingly are used to define, predict, and control cognition, emotions, and behaviors, and they have been instrumental in challenging long-held beliefs about the nature of the mind and the self, morality, and human nature. Thus, neuroscience and neurotechnology are implicitly and explicitly shaping our worldview—and our world.

On the one hand, neuroscience and neurotechnology may be regarded as a set of resources and tools with which to answer perdurable questions about humanity and nature, whereas, on the other hand, we must consider the profound questions that neurotechnology will foster. Can neurotechnology resolve the problem of consciousness—and if so, what will be the implications of such findings for social thought and conduct? Will these technologies take us to the point of being *beyond human*? What values drive our use—and potential misuses—of neurotechnologies? How will neurotechnology affect the international balance of economic, social, and political power? Should neurotechnology be employed by the military, and if so, how? Are there limits to neurotechnological research and use, and if so, what are they, who shall decide what they are, and what ethical criteria might be required to guide the use of neurotechnologies and the outcomes they achieve?

Such questions are not merely hypothetical or futuristic; rather, they reflect the current realities and realistic projections of neurotechnological progress, and they reveal the moral, ethical, and legal issues and problems that must be confronted when considering the ways that neurotechnology can and should be employed and how such utilization will affect various individuals, communities, and society at large. Confronting these issues necessitates discussion of the benefits, burdens, and harms that particular neurotechnologies could incur and the resources that are needed to govern neurotechnological progress. The discourse must be interdisciplinary in its constitution, delivery, and appeal, and therefore, any authentic effort in this regard must conjoin the natural, physical, and social sciences as well as the humanities.

This series seeks to contribute to and sustain such discourse by bringing together scholars from the aforementioned fields to focus on the issues that arise in and from studies and applications of neurotechnology. Each book will provide multidisciplinary perspectives on the ethico-legal and social issues spawned by the use of specific neurotechnologies and how more general developments in neuro-technology affect important domains of the human condition or social sphere. The insights afforded by this series are relevant to scientists, engineers, clinicians, social scientists, philosophers, and ethicists who are engaged in the academic, government, military, and commercial biotechnology sectors. It is my intention that this series will keep pace with and reflect advancements in this rapidly changing field and serve as both a nexus for interdisciplinary ideas and viewpoints and a forum for their expression.

James Giordano
Georgetown University
and
Ludwig–Maximilians University

Foreword

Jonathan D. Moreno

This volume captures the growing interest in the ethical, legal, and social aspects of neurotechnology in national security and defense uses through the writings of some of the best-informed and most provocative commentators in the field. Professor Giordano's ongoing work as a neuroscientist and neuroethicist, and his engagement with a number of international groups addressing the key issues important to the potential use and risks of misuse of neurotechnology to effect international relations, national security, and defense incentives, agendas, and operations, places him in good stead to steward this volume and provides keen insights to the field and brings together this group of scholars whose work focuses upon the timely and provocative topics it generates. This volume provides views and voices from multiple disciplines and perspectives, and as such, makes it an important addition to the literature.

My own interest in the relationship between ethics, neuroscience, and national defense began several years after the publication of my study of the history and ethics of human experiments for national security purposes (Moreno 2001) and an anthology on bioethics and bioterrorism (Moreno 2002). In the first years of the twenty-first century, the excitement about the potential of the new neuroscience, powered especially by imaging technologies, resulted in some public discussions of the social implications of emerging brain science. What struck me when I attended the early academic meetings about "neuroethics" was the lack of connection to the national security environment. Surely, there must be great interest in the applications of brain research among those responsible for national defense and counterintelligence, and surely they would raise important policy questions. Yet, there were virtually no attendees at neuroethics meetings a decade ago who had experience in the national security establishment, let alone discussions about neuroscience and defense policy.

While I was pondering these questions, Dana Press, part of the Dana Foundation, asked me if I was interested in publishing a book on the subject. The source of the invitation was especially intriguing, because the Foundation is a sponsor of much neuroscience education and research. As I began work on this book, it quickly became clear that some of the same challenges I faced in writing about the history and ethics of national security human experiments would apply to this new project. Among those challenges were some tough decisions about which topics fall precisely within the ambit of neuroscience and the need to reckon with the abundant conspiracy theories about clandestine experimental manipulations by government agencies. Although many outside the defense community assume that everything that might be interesting is "secret," this has been proven not to be true, since pretty much any science likely to be of national security interest is either explicit or implicit in the open literature.

The revised edition of my previous book is *Mind Wars: Brain Science and the Military in the 21st Century* (Moreno 2012). In *Mind Wars*, I argued that national security needs are sure to be at the cutting edge of neuroscience, both in terms of research for war fighting and intelligence and, over the long term, the introduction of new neurotechnologies in society. I also asserted that the encounter between national security and neuroscience will provide much fodder for the growing group of neuroethics scholars and that these developments can most fruitfully be seen in historical context. In that spirit, it is worth revisiting such cases as the Central Intelligence Agency and the U.S. Army's lysergic acid diethylamide (LSD) experiments of the 1950s and 1960s. Those projects were motivated by a number of concerns and goals, including interest in hallucinogens as a problem in counterintelligence (e.g., the fear that they could be used as a "truth serum" with a kidnapped U.S. nuclear scientist) and as a potential disruptor of combat units.

Another intriguing lesson of the LSD experience that might also apply to modern neuroscience was its unintended cultural consequences as the drug that came to symbolize a decidedly anti-establishment lifestyle. As we seem to be on the threshold of a new era of provocative, "enhancing" pharmacologics, more granular neuroscans, implantable devices, and other brain-related innovations, national security interest in neuroscience will surely be taken to a new level, leading as well to remarkable problems in ethics and public policy. As in the past, one may expect that there will be unanticipated social consequences of the embrace of new and more powerful neuroscience by organizations responsible for counterintelligence.

How things have changed in the past decade. Some measure of the increasing awareness of the ethical, legal, and social implications of neuroscience for national security can be derived from the fact that the National Academies have engaged in several projects directly related to neuroscience research policy since 2007, projects that were conducted under contract with national security agencies. The report "Emerging Cognitive Neuroscience and Related Technologies" was released in August 2008 for the Defense Intelligence Agency (National Research Council 2008). "Opportunities in Neuroscience for Future Army Applications" was published in May 2009 and conducted for the U.S. Army (National Research Council 2009). In addition, the Committee on Field Evaluation of Behavioral and Cognitive Sciences-Based Methods and Tools for Intelligence and Counter-Intelligence planned and hosted a public workshop in September 2009, at the request of the Defense Intelligence Agency and the Office of the Director of National Intelligence. Another report from the National Research Council appeared in 2014, one sponsored by the Defense Advanced Research Projects Agency (DARPA) titled "Emerging and Readily Available Technologies and National Security—A Framework for Addressing Ethical, Legal, and Societal Issues." (I was a member of the first, third, and fourth of these committees.) Other academies panels are also actively interested in neuroscience and security-related questions. And of course along with his announcement of the BRAIN Initiative in April 2013, President Obama included his request that the President's Commission for the Study of Bioethical Issues study ethical issues that that effort may entail, some of which will surely impinge on national security and defense.

International interest in the topic is also growing rapidly. Based on my personal experience alone, Australian, British, Dutch, German, and Japanese scholars and government officials or advisers are following these matters. In Great Britain, the Royal Society, the Nuffield Council, and the Scottish Parliament have held meetings and published reports. Controversies about dual use, remote-control weapons, and cyberwarfare are also driving the interest in neuroscience as a field that converges with bioengineering, artificial intelligence, and sensor technologies. Clearly, we are part of an international discussion that is only beginning and to which this volume is an important contribution; I commend Jim Giordano for his energy and commitment, and I am grateful to each and all of the contributing authors for both their fine chapters and their ongoing work in the field.

REFERENCES

Moreno, J. 2001. *Undue Risk: Secret State Experiments on Humans*. New York: Routledge.
Moreno, J. 2002. *In the Wake of Terror: Medicine and Morality in a Time of Crisis*. Cambridge, MA: MIT Press.
Moreno, J. 2012. *Mind Wars: Brain Science and the Military in the 21st Century*. New York: Bellevue Literary Press.
National Research Council. 2008. *Emerging Cognitive Neuroscience and Related Technologies (US)*. Washington, DC: National Academies Press.
National Research Council. 2009. *Opportunities in Neuroscience for Future Army Applications (US)*. Washington, DC: National Academies Press.
National Research Council. 2014. *Emerging and Readily Available Technologies and National Security—A Framework for Addressing Ethical, Legal, and Societal Issues (US)*. Washington, DC: National Academies Press.

REFERENCES

Acknowledgments

My long-standing involvement and interest in research, development, testing, and translation of neuroscience and neurotechnology (so-called neuro S/T) in national security initiatives have provided both strong impetus to develop this volume and fertile ground for its direction and content. Spanning from work initially conducted almost 30 years ago and extending to the present, I have had the opportunity to gain insight into the ways that neuro S/T was, is, and could be applied in military medicine, as well as tactical and strategic operations. Over the course of the past three decades, I have been fortunate to have worked with a number of outstanding individuals in both the civilian and military sectors, and indeed, they are far too numerous to mention here. On a more proximate scale and more recent timeframe, however, acknowledgment and gratitude must be extended to several colleagues who have been directly instrumental to the development and completion of this book, and I beg the reader to pardon me the indulgence to express my appreciation here.

First, please allow me to acknowledge the patience, support, and continuous encouragement of my acquisition editor Lance Wobus at CRC Press/Taylor & Francis. In taking the baton from Barbara Norwitz, Lance has been a coach, fan, administrator, and champion for each and every phase of this book and of the series to which it is part. As well, thanks go to Jill Jurgensen at CRC Press for her tireless assistance with all of the details of permissions, copyrights, and what seemed to be myriad forms and checklists. That this book came to fruition and has gotten into your hands is shining evidence of the excellence of their work in practice.

A deep nod of thanks goes to the late Professor Edmund Pellegrino, for supporting my work and activities in neuroethics at Georgetown University Medical Center, Washington, DC, over the past 10 years. Ed was a mentor, teacher, true scholar, gifted physician, and friend, and his enthusiasm for the blossoming field of neuroethics was stalwart. It was during a leave of absence from Georgetown, while working at the Potomac Institute for Policy Studies (PIPS) to direct the then nascent Center for Neurotechnology Studies (CNS), that the idea for a dedicated book focusing upon the potential viability, value, and contentious issues specific to the use and misuse of neuro S/T in national security and military contexts was developed. I thank the Institute's chief executive officer and president, Michael S. Swetnam, for his ongoing support and for providing me the latitude to pursue an in-depth address and analysis of the field and its implications. I am grateful to Tom O'Leary; Drs. Charles Herzfeld, Dennis McBride, Bob Hummel, Ashok Vaseashta, and John Retelle; Ambassador David Smith; Brigadier General David Reist, USMC (Ret.); and Rear Admiral James Barnett, USN (Ret.) at PIPS for their collegiality and collaboration. Special thanks must go to General Al Gray, USMC (Ret.), for his ebullience and promotion of my efforts in bringing these issues to the fore.

In returning to Georgetown in 2012, I have had the great fortune to rejoin my colleagues Drs. G. Kevin Donovan, Kevin FitzGerald, James Duffy, David Miller,

Rochelle Tractenberg, and Frank Ambrosio who each and all have provided support, insight, and ample good humor as this project was brought to completion. Many of the chapters in this volume grew from the authors' involvement and participation in a series of meetings and symposia held at CNS-PIPS and Georgetown, a special thematic issue of the journal *Synesis*, and the third national Neuroscience—Ethical, Legal, and Social Issues (NELSI-3) conference addressing the topic (for further information, see www.NELSI-3.com). For their support on various projects, symposia, and meetings, thanks go to Drs. Hriar Cabayan of the Strategic Multilevel Assessment Group; Diane DiEuliis at the Department of Health and Human Services; Kavita Berger and Heather Dean of the American Association for the Advancement of Science; Commander Demetri Ecomonos, MSC, USN; Tim Demy and Jeff Shaw of the U.S. Naval War College; Chris Forsythe and Wendy Shaneyfelt of Sandia National Laboratories; Captain Dylan Schmorrow, USN (Ret.); Commander Joseph Cohn, USN; and Major Jason Spitaletta, USMC.

Developing an analytic approach to issues of national security and defense and contexts and parameters of preparedness requires both an unembellished, unsentimental look at history and a view to the present and future. In this regard, I am grateful for prior and current collaborations with Drs. Bill Casebeer (formerly at the Defense Advanced Research Projects Agency [DARPA]), Jonathan Moreno at the University of Pennsylvania, and Carey Balaban of the Center for National Preparedness at the University of Pittsburgh, whom I am proud to call friends as well as much esteemed colleagues.

I acknowledge contributions of several international scholars with whom I have worked and who have provided insights to the ways that neuro S/T is engaged in a variety of global operations. I thank my longtime collaborator and friend, Dr. Roland Benedikter, for his contributions to our work on the *NeuroPolis* project assessing socioeconomic and political dynamics afforded by neuro S/T; Professors Ruth Chadwick and Malcolm Dando and my colleagues at the Human Science Center, Ludwig–Maximilians University, Munich, Germany: Drs. Niko Kohls (presently at the Coburg University of Applied Sciences), Herbert Plischke (presently at the Munich University of Applied Sciences), Ernst Pöppel, Albrecht von Müller, and Fred Zimmermann, for collaborations examining the potential for neuro S/T to fortify therapeutic approaches in military medicine.

I have been most fortunate to have had outstanding graduate students including (now Dr.) Rachel Wurzman, Elisabetta Lanzilao, M. Mercedes Sinnott, Rhiannon Bower, Anvita Kulkarni, and (now Dr.) Guillermo Palchik who have worked with me on several national security neuro S/T projects. Research scholars Liana Buniak, Dan Degerman, and Dan "A" Howlader worked diligently on various aspects and phases of this book and are afforded well-earned and much deserved kudos.

Research requires not only time, but financial support, and in this light, I acknowledge the funding of the J.W. Fulbright Foundation, Office of Naval Research, Lockheed Martin Corporation, Clark Fund, and the William H. and Ruth Crane Schaefer Endowment.

Finally, not so much a thanks, but a heartfelt expression of boundless appreciation for my wife, Sherry, who cheerfully and lovingly listened to hours of my talking about neuro S/T; came to every symposium, seminar, and conference; saw to all the little details of making things go smoothly; worked to coordinate meetings, conference calls, and editorial sessions with authors and the staff at CRC Press, and did it all with a smile and good humor; and at the end of the day, made our home a wonderful haven, retreat, and solace.

James Giordano

Editor

James Giordano, PhD, is chief of the Neuroethics Studies Program of the Edmund D. Pellegrino Center for Clinical Bioethics and is a professor on the faculties of the Division of Integrative Physiology/Department of Biochemistry; Department of Neurology; and Graduate Liberal Studies Program at Georgetown University, Washington, DC. He presently serves on the Neuroethics, Legal, and Social Issues Advisory Panel of the Defense Advanced Research Projects Agency (DARPA). As well, he is Clark Faculty Fellow at the Human Science Center of the Ludwig–Maximilians University, Munich, Germany, and a Senior Fellow of the Potomac Institute for Policy Studies, Arlington, Virginia.

Professor Giordano is editor-in-chief of the journal *Philosophy, Ethics, and Humanities in Medicine*; founding editor of *Synesis: A Journal of Science, Technology, Ethics, and Policy*; and associate editor of the international journal *Neuroethics*.

He is the author of over 200 publications in neuroscience, neurophilosophy, and neuroethics; his recent books include *Neurotechnology: Premises, Potential, and Problems* (CRC Press); *Scientific and Philosophical Perspectives in Neuroethics* (with Bert Gordijn, Cambridge University Press); and *Pain Medicine: Philosophy, Ethics, and Policy* (with Mark Boswell, Linton Atlantic Books).

His research addresses the neuroscience of pain; neuropsychiatric spectrum disorders and the neural bases of moral cognition and action; and the neuroethical issues arising in neuroscientific and neurotechnological research and its applications in medicine, public life, global relations, and national security. In recognition of his ongoing work, Giordano was awarded Germany's Klaus Reichert Prize in Medicine and Philosophy (with longtime collaborator Dr. Roland Benedikter); was named National Distinguished Lecturer of both Sigma Xi, the national research honor society, and the Institute of Electrical and Electronics; and was elected to the European Academy of Sciences and Arts.

He and his wife Sherry, an artist and naturalist, divide their time between Old Town Alexandria, Virginia, and Bad Tölz/Munich, Germany.

Contributors

Keith Abney
Philosophy Department
California Polytechnic State University
San Luis Obispo, California

Carey D. Balaban
Department of Otolaryngology
Eye & Ear Institute
University of Pittsburgh
Pittsburgh, Pennsylvania

Curtis Bell
Oregon Health and Science University
Portland, Oregon

Paolo Benanti
Pontificia Università Gregoriana
Rome, Italy

Chris Berka
Advanced Brain Monitoring
Carlsbad, California

Angela (Baskin) Carpenter
Cubic Advanced Learning Solutions, Inc.
Orlando, Florida

William D. Casebeer
Lockheed Martin Advanced
   Technologies Laboratories
Arlington, Virginia

James P. Farwell
Canada Centre for Global Security
   Studies
University of Toronto
Toronto, Ontario

and

The Farwell Group
New Orleans, Louisiana

Kevin T. FitzGerald
Department of Oncology
Georgetown University Medical
   Center
Washington, DC

Sven Fuchs
Design Interactive, Inc.
Oviedo, Florida

Lyn M. Gaudet
The Mind Research Network
Albuquerque, New Mexico

James Giordano
Neuroethics Studies Program
Georgetown University
Washington, DC

and

Human Science Center
Ludwig–Maximilians University
Munich, Germany

Lori Haase
University of California
San Diego, California

Kelly S. Hale
Design Interactive, Inc.
Oviedo, Florida

Douglas C. Johnson
University of California
and
Warfighter Performance Department
Naval Health Research Center
and
OptiBrain Consortium
San Diego, California

Patrick Lin
Philosophy Department
California Polytechnic State University
San Luis Obispo, California

Gary E. Marchant
Sandra Day O'Connor College of Law
Arizona State University
Tempe, Arizona

Jonathan H. Marks
Bioethics Program
The Pennsylvania State University
University Park, Pennsylvania

and

Edmond J. Safra Center for Ethics
Harvard University
Cambridge, Massachusetts

Robert McCreight
Penn State World Campus
The Pennsylvania State University
University Park, Pennsylvania

and

Institute for Crisis, Disaster & Risk
    Management
George Washington University
Washington, DC

Kaleb McDowell
U.S. Army Research Laboratory
Aberdeen Proving Ground, Maryland

Maxwell Mehlman
The Law-Medicine Center
Case Western Reserve University
    School of Law
and
Department of Bioethics
Case Western Reserve University
    School of Medicine
Cleveland, Ohio

Jonathan D. Moreno
Department of Medical Ethics and
    Healthy Policy
and
Department of History and Sociology
    of Science
University of Pennsylvania
Philadelphia, Pennsylvania

Steve Murray
University of San Diego
San Diego, California

Kelvin S. Oie
U.S. Army Research Laboratory
Aberdeen Proving Ground, Maryland

Martin P. Paulus
OptiBrain Consortium
and
Department of Psychiatry, Laboratory
    of Biological Dynamics and
    Theoretical Medicine
University of California
San Diego, California

Eric G. Potterat
Naval Special Warfare Center
and
OptiBrain Consortium
San Diego, California

Alan N. Simmons
University of California
San Diego, California

Kay M. Stanney
Design Interactive, Inc.
Oviedo, Florida

Judith L. Swain
OptiBrain Consortium
San Diego, California

James Tabery
Department of Philosophy
University of Utah
Salt Lake City, Utah

Kyle Thomsen
Department of Philosophy
Saint Francis University
Loretto, Pennsylvania

Rochelle E. Tractenberg
Departments of Neurology;
    Biostatistics, Bioinformatics &
    Biomathematics; and Psychiatry
Georgetown University
    Medical Center
Washington, DC

Karl Van Orden
Naval Health Research Center
and
OptiBrain Consortium
and
Department of Psychiatry, Laboratory
    of Biological Dynamics and
    Theoretical Medicine
University of California
San Diego, California

Rachel Wurzman
Interdisciplinary Program in
    Neuroscience
Georgetown University Medical Center
Washington, DC

Matthew A. Yanagi
Space and Naval Warfare Systems
    Center Pacific
San Diego, California

1 Neurotechnology, Global Relations, and National Security

Shifting Contexts and Neuroethical Demands

James Giordano

CONTENTS

NEUROSCIENCE AND NEUROTECHNOLOGY: ASSESSING—AND ACCESSING—THE BRAIN AND BEHAVIOR

This volume and book series address and reveal the reality that neuroscience and neurotechnology (neuro S/T) have become powerful forces that influence society, are influenced by various social forces, and incur a host of ethico-legal and social issues. Recent governmental and commercial investments in brain science and neuroengineering reflect growing interest and enthuse advancement(s) in neuro S/T and the information, products, and potential power these disciplines may yield. A dimension of this power is derived from the prospect of using neuro S/T to define—and affect—human nature. Current neuroscientific perspectives consider biological organisms to be complex internal environmental systems nested within complex external environmental systems (Schoner and Kelso 1985). Interactions within and among systems are based and depend upon numerous variables within these complex internal and external environments (Ridley 2003), framed by time, place, culture, and circumstance (Giordano 2011a, 2011b; Giordano et al. 2012).

This mandates appreciation of culture as an important force in determining the interactively biopsychosocial dimensions of human functioning. At the most basic level, *culture* refers to the development (i.e., cultivation) of living material, and hence, it becomes important, if not necessary to evaluate how "culture" engages and sustains the set of shared material traits, characteristic features, knowledge, attitudes, values, and behaviors of people in place and time. This definition rightly reveals that culture establishes and reflects particular biological characteristics (that develop and are preserved in response to environments) that can be expressed through cognitions and behaviors. In this way, culture is both a medium for biopsychosocial development and a forum and vector for its expression and manifestations (Ridley 2003; Giordano et al. 2008). Defining the neural bases of such biological–environmental interactions may yield important information about factors that dispose and foster various actions, including cooperation, altruism, conflict, and aggression (Casebeer, 2003; Cacioppo et al. 2006; Verplaetse et al. 2009).

Until rather recently, most efforts toward global relations, as well as national and international security and defense, have focused upon social factors influencing human behaviors, including hostility and patterned violence. Given that these behaviors are devised and articulated by human factors, and humans are most accurately defined as biopsychosocial organisms that are embedded within and responsive to geocultural environments, it is important to address and discern those (neuro)biological factors that are affected by and interact with psychosocial variables to dispose and instigate hostility and violence. Neuro S/T provides techniques and tools that are designed to assess, access, and target these neurological substrates, which could be employed to affect the putative cognitive, emotional, and behavioral bases of human aggression, conflict, and warfare.

NEURO S/T IN NATIONAL SECURITY AND DEFENSE

This establishes the viability and potential role, if not value of neuro S/T in programs of national security, intelligence, and defense (NSID). The challenges posed for and by using neuro S/T in these ways are as follows: (1) to develop a more complete understanding of mechanisms that precipitate aggression and patterned violence; (2) to provide practical means to assess, affect, alter, and/or impede these mechanisms; and (3) to base any such findings, options, and actions upon realistic appraisal of the capability, limitations, and ethico-legal and sociopolitical direction or constraint of this science, technology, and information.

As detailed throughout this book, a number of neuroscientific techniques and technologies are being utilized in NSID efforts, including:

1. Neural systems modeling and human–machine interactive networks in intelligence, training, and operational systems (see Chapters 3 and 4).
2. Neuro S/T approaches to optimizing performance and resilience in military personnel (see Chapter 5).
3. Neuro S/T in operational medicine (see Chapter 6).

4. Neuro S/T approaches to intelligence gathering and articulation (i.e., NEURINT) and direct weaponization of neuro S/T (e.g., a variety of neuropharmacologic agents and neurotechnologic devices such as certain types of neuroimaging, and forms of interventional neurotechnologies; see Chapter 7).

While the use of neuro S/T in military medicine might be viewed in ways that are similar to civilian applications, such uses certainly foster neuroethico-legal and social issues—a number of which are addressed in the first volume of this series, *Neurotechnology: Premises, Potential and Problems* (Giordano 2012b). Yet, issues such as (biomedical) augmentation, enablement, optimization, and enhancement often produce particular unease and anxieties if and when construed in contexts and terms of the cognitive, emotional, and behavioral performance of warfighters and other NSID personnel. Concerns and fears increase when neuro S/T is more explicitly engaged in the development of intelligence and weapons' systems (see Chapters 11 through 14, and 16). As noted in a 2008 report conducted by the *ad hoc* Committee on Military and Intelligence Methodology for Emergent Neurophysiological and Cognitive/Neural Science Research, in the Next Two Decades "... for good or for ill, an ability to better understand the capabilities of the body and brain will require new research that could be exploited for gathering intelligence, military operations, information management, public safety and forensics" (National Research Council 2008). These efforts are not limited to the West, but will be undertaken on an international scale (Bitzinger 2004; Benedikter and Giordano 2012), and the extent and directions that such research can and likely will assume remain vague, as "... military and intelligence planners are uncertain about the ... scale, scope, and timing of advances in neurophysiological research and technologies that might affect future ... warfighting capabilities" (National Research Council 2008).

In light of this, I assert that it would be pragmatic and prudent to develop a stance that is based upon preparation, resilience, and in some cases intervention to guide and govern the advancement of certain research, development, testing, evaluation (RDTE) and translational trajectories of neuro S/T in NSID. Surveillance of international RDTE would be necessary to guard against potentially negative and harmful uses of neuro S/T. However, surveillance alone will not be sufficient; rather, true preparedness will entail greater insights into (1) those possible ways that neuro S/T could be used in NSID, (2) the effects and manifestations that such use might evoke, and (3) the means and readiness to counter or mitigate any such effects (Lederberg 1999; Dando 2007; Bousquet 2009). Such an agenda will require the coordinated efforts of professionals (in the military and civilian sectors—inclusive of politics) as well as the public (Moreno 2006; see also Chapters 11 and 14 through 17), although any public discourse would necessitate diligent stewardship of information so as to delicately balance relative transparency of the issues with vulnerability of sensitive details relevant to the integrity of national security efforts (Giordano 2010). The discourse should conjoin academic, corporate, and governmental sectors (the so-called triple helix of the scientific estate; Etzkowitz 2008) at a variety of levels and stages in this enterprise (Wurzman 2010). This is not new; we need only to look at the Manhattan Project and the "space race" for examples of this estate in practice. But this framework did not restrict escalation, nor did it prevent continued threat and/or far-reaching social effects (e.g., the Cold War; "Star Wars"-type defense concerns; persistent danger of nuclear/radiological weapons), and so the triple-helix approach may require modification(s) to facilitate stronger collaboration between its constituents, on a scale that extends beyond national boundaries (Anderson et al. 2012; Giordano 2012a). I opine that an important aspect of this expanded approach will entail increased involvement of both the humanities and the public (at least to some reasonable extent), because real effect can only be leveraged through guidelines, laws, and policies that are sensitive to ethical and social ramifications, issues, and problems that are relevant to those affected.

PARADOXES AND QUESTIONS

Here we encounter a number of paradoxical realities, which foster fundamental questions, tensions, and even conflict about the ways that neuro S/T should be regarded, studied, overseen, guided, employed, and/or restricted in NSID agendas. For example, it could be claimed that certain forms of neuro S/T can—and perhaps should—be utilized to define, predict, and, thereby, prompt intervention(s) to prevent or minimize individual and/or group aggression, violence, and combativeness, and in this way afford public protection (Farahany 2009; Greely 2013; Giordano et al. 2014). How might such "protection" be balanced with individual and public privacy? To protect against possible harms that might be incurred by use or misuse of neuro S/T in NSID, it would be important to acquire and/or develop knowledge of real and potential threats posed by the use, misuse, and/or unintended consequences of neuro S/T in NSID and squelch events before they escalate into scenarios that place the population at risk of large-scale harm. How might the need for such research be balanced with attempts at restricting its direct or dual use in NSID operations? It could be argued that extant and perhaps newly developed international policies and treaties that affect the conduct of neuro S/T research are viably important to preventing or limiting possible military use; yet, it is equally important to recognize that international policies and signatory treaties do not guarantee cooperation (Gregg 2010) and may establish imbalances in capability and power that can subsequently be exploited (Giordano et al. 2010; Benedikter and Giordano 2012; Brindley and Giordano 2014).

To address such questions, we have advocated that neuro S/T continued to be studied for its potential viability—specifically to *decrease* harms necessary to preserve national security and defense. In this light, we are developing a proposed set of criteria for the consideration and possible use of neuro S/T in NSID settings (Bower and Giordano 2012; Giordano et al. 2014), these include the following:

1. There is less harm done by using the neuro S/T in question.
2. If an individual(s) pose(s) a realistic and immediate threat of severe harm to others, the most effective science and technology (including neuro S/T)—and least harmful among these—should be utilized toward mitigating these threats.

3. The use of neuro S/T must be admissible in a court of law under the most current and stringent legal standards (Orofino 1996). As well, it will be important to examine other ethico-legal frameworks and standards to enable a more internationally relevant approach to using neuro S/T in NSID agendas (Farahany 2009; Freeman and Goodenough 2009; Eagleman 2011; Spranger 2012; Morse and Roskies 2013).

4. If neuro S/T is employed for intelligence purposes, only information pertinent to an ongoing investigation or a specific issue of security and/or deterrence should be obtained and used, and this should be stored in official police and/or government records.

5. There must be other corroborating evidence to substantiate interventive action(s)—outside of evidence gathered by neuro S/T—as is necessary based upon maturity and reliability of techniques (see criterion 3 in this list; also Chapter 9).

6. There must be a valid precedent and/or exigency to incur the use of neuro S/T in these circumstances (see criteria 2 and 3 in this list; also Chapters 9 through 11, 15, and 17).

7. Applying these technologies in a predictive or preventive manner is still practically problematic and should not be implemented until further scientific research and technical development has been undertaken and more detailed and effective ethico-legal frameworks are generated (Farahany 2009; Greely 2013; Giordano et al. 2014).

Of course, this necessitates (1) questioning the ecological validity and reliability of the ways that neuro S/T is used, as well as (2) concern about the ethico-legal probity and value of neuro S/T-based assessment and intervention in specific NSID applications. To be sure, such questions and concerns are challenging; the challenge reflects and must address the fundamental questions in neuroscience. That is, what are the nature and type of neurobiological characteristics that affect cognition, emotion, and behavior? Can these characteristics be accurately assessed, and what types and combinations of techniques, technologies, and metrics are required in this task? Can these techniques and tools be used to (1) describe and perhaps predict bio-psychosocial factors of group violence; (2) provide putative targets for multidisciplinary intervention to deter or mitigate aggression and violence; and (3) if so, in what ways, to what extent—and through what process and method(s)—can and should these approaches be utilized (and conversely, in what situations and circumstances should they be constrained or prohibited)?

It is unwise—and inapt—to overestimate (or underestimate) the capability of neuro S/T, and it is equally foolish to misjudge the power conferred by this science or the tendency for certain groups to misdirect and misuse these technologies and the power they yield (Giordano 2012c). Granted, there is robust political—and thus national security—power that can be gained and leveraged through neuro S/T. Although contributory to new dimensions of military capability, such power is not limited to war-fighting capacity. The changing political "power shift from the West towards the East ... [as] consequence of the latest ... financial crises" (as asserted by France's Premier Francois Fillon on November 6, 2011; Evans-Pritchard 2011), and the growing prominence of non-Western nations in neuro S/T research and production (Bitzinger 2004; Sobocki et al. 2006; Lynch and McCann 2009) afford a number of heretofore ineffectual nations a greater economic presence upon the world stage. Taken together with the relative accessibility of neuro S/T to nonstate actors (Forsythe and Giordano 2011), this creates new scenarios of socioeconomic and political dependencies, and rebalances power equations of global politics, influence, and defense needs and capabilities.

The issue of how to apply neuro S/T also prompts the human question in the more strict philosophical and ethical sense, as reflective of, and inherited by Western post-Enlightenment constructs, which are now situated among increasingly prevalent and influentially pluralist ideals and ethics (Anderson et al. 2012; Giordano and Benedikter 2012). How might neuro S/T be used to establish definitions, norms, mores, and attitudes toward particular individuals, groups, or communities? To reiterate, neuro S/T—like any form of science and technology—can be used to effect good and harm. And while the tendency to use science and technology inaptly or toward malevolent ends is certainly not novel, the extent and profundity of what neuroscientific information implies (i.e., about the nature of the mind, self-control, identity, and morality) and what neuro S/T can exert over these aspects of the human being, condition, predicament, and relationships mandates thorough review and discernment.

Therefore, a particularly high level of scrutiny is needed when looking to, and relying upon neuro S/T to describe, evaluate, predict, or control human thought, emotions, and behavior. It will be crucial to develop measurements for such scrutiny and the means—and ethico-legal lenses, voices, parameters, and paradigms—that will be required to translate these metrics to nationally, regionally, and internationally relevant standards (Farahany 2009; Dando 2007; Nuffield Council Report 2013). Extant criteria, as provided by ethics, laws, policies, and treaties, while viable to some degree, can reflect—and are often contributory to—the scientific, social, and economic "climate" in which various techniques and technologies are regarded, embraced, and utilized. Thus, in most cases they are only temporary philosophical and/or political agreements and represent the dominant socially accepted viewpoint(s). How can these be evaluated and weighed when sociocultural (and political) perspectives, needs, and values differ?

To wit, I hold that any and all analyses and guidelines for the use of neuro S/T must be based upon pragmatic assessment of technological and human dimensions of science and technology, the capabilities and limits of scientific and technological endeavor, and the manifestations incurred by studying and using such science and technology in the public sphere(s) in which NSID exerts influence and effect.

ADDRESSING CHALLENGES AND OPPORTUNITIES: A PATH FORWARD

Toward such ends, it becomes essential to appropriately address these "deep" questions, both separately and in their interrelatedness. A first step is to more fully recognize the rapid development and use of neuro S/T, the variety of new fields of application and transformation generated in the mid-to-long term by neuroscientific techniques and tools, and the information and capability they yield. To date, there has been what appears to be a somewhat indecisive posture—a "waiting game"—toward the unavoidable increase in contextual (indirect) and classical (direct) political power connected to the use of neuro S/T in the coming decades. Jeannotte et al. (2010) and Sarewitz and Karas (2012) warn of the inadequacy, if not real danger, of such hesitancy.

I concur and believe that what is needed—at present—is the formulation and articulation of a cosmopolitan, yet contextually receptive neuroethics framework (Giordano 2010, 2011a, 2011b; Giordano and Benedikter 2012; Lanzilao et al. 2013; Shook and Giordano 2014). This type of neuroethics does not yet exist on a level that affords international relevance and application; thus, it might represent a watershed development that could (1) strategically enable transnational neuroscientific innovation; (2) remain cross-culturally sensitive and responsive; and (3) be instrumental to negotiating more stable economic, political, and national security relationships.

Neuro S/T is and remains a field in evolution. The questions generated by the field and its applications are complicated and arguably more numerous than the certainties achieved thus far. The common ground of these questions is not whether "deep-reaching" scientific and technological shifts will occur, but in which ways these shifts will be expressed by and/or affect global relations, political forces, and postures of national and international security and defense. Could some uniform regulations for research and use be viable in any and all situations? And if so, by which mechanisms might these codes be developed and articulated and enforced? Or, will progress in neuro S/T incur more isolationist leanings? And, in the event, how would various nations then maneuver neuroscientific efforts to retain viable presence on the global technological, economic, security, and defense map(s)? Might this trend toward pervasive use of neuro S/T in these silos of power be regarded as a form of "neuro-politics"?

It is exactly this scientific-to-social span of neuro S/T-related and neuro S/T-derived effects that necessitates a stronger focus and investment in both the science and a meaningful neuroethics. It will be necessary to (1) analyze problems borne of neuroscientific and neurotechnological progress, and (2) develop recommendations and guidelines that direct the scope and tenor of current neuroscientific research and applications, so as to (3) ensure preparedness for the consequences of neuro S/T advancements at present and in the future (both as specifically relevant to NSID, and in other, more socially broad-based contexts). Toward these ends, it may be that existing ethical and legal concepts and systems will need to be amended, or even newly developed, to sufficiently account for the changes and challenges that neuro S/T is evoking in an evermore pluralized society and the NSID environment defined by the international dynamics of the twenty-first century (Giordano 2010; Levy 2010; Racine 2010; Lanzilao et al. 2013; Shook and Giordano 2014).

CONCLUSIONS

Apropos the aforementioned exigencies and contingencies, the intent of this volume is not to explicitly advocate the use of any particular neuro S/T in NSID, but rather to describe the practical capabilities and limitations of these approaches, toward a fuller consideration of the neuroethico-legal issues, questions, and problems—and perhaps paths to resolution—that they incur. Given revivified "big science" incentives in neuro S/T, such as the Brain Research through Advancing Innovative Neurotechnolgies (BRAIN) agenda in the United States, the European Union's Human Brain Project, and Decade of the Mind enterprises in Asia, it may be that the real "grand challenge" will be dedicating effective investments of time, effort, and funding required to meet the urgent neuroethical demands spawned by current and future iterations of neuro S/T. It is my hope that this volume, and the series at large, may provide perspectives, information, and insights that are useful and helpful to this crucial endeavor.

ACKNOWLEDGMENTS

This work was supported, in part, by funding from the J.W. Fulbright Foundation; Office of Naval Research; and the Neuroethics Studies Program of the Edmund D. Pellegrino Center for Clinical Bioethics, Georgetown University Medical Center, Washington, DC. The author gratefully acknowledges the assistance of Sherry Loveless in the preparation of this manuscript.

REFERENCES

Anderson, M.A., N. Fitz, and D. Howlader. 2012. "Neurotechnology research and the world stage: Ethics, biopower and policy." In *Neurotechnology: Premises, Potential and Problems*, ed. J. Giordano. Boca Raton, FL: CRC Press, 287-301.
Benedikter, R. and J. Giordano. 2012. "Neurotechnology: New frontiers for European policy." *Pan Euro Network Sci Tech* 3: 204–208.
Bitzinger, R.A. 2004. "Civil-military integration and Chinese military modernization." *Asia-Pacific Center for Security Studies*. http://www.apcss.org/Publications/APSSS/Civil-MilitaryIntegration.pdf. Accessed January 30, 2014.
Bower, R. and J. Giordano. 2012. "The use of neuroscience and neurotechnology in interrogations: Practical considerations and neuroethical concepts." *AJOB Neurosci* (Suppl 3): 3.
Brindley, T. and J. Giordano. 2014. "International standards for intellectual property protection of neuroscience and neurotechnology: Neuroethical, legal and social (NELS) considerations in light of globalization." *Stanf J Law Sci Policy* 7:33.
Bousquet, A. 2009. *The Scientific Way of Warfare: Order and Chaos on the Battlefields of Modernity*. New York: Columbia University Press.
Cacioppo, J.T., P.S. Visser, and C.L. Pickett. 2006. *Social Neuroscience: People Thinking about Thinking People*. Cambridge, MA: MIT Press.
Casebeer, W. 2003. "Moral cognition and its neural constituents." *Nat Rev Neurosci* 4: 840–847.
Dando, M. 2007. *Preventing the Future Military Misuse of Neuroscience*. New York: Palgrave Macmillan.
Eagleman, D. 2011. *Incognito: The Secret Lives of the Brain*. New York: Pantheon.
Etzkowitz, H. 2008. *The Triple Helix: University–Industry–Government Innovation in Action*. New York: Routledge.
Evans-Pritchard, A. 2011. "France cuts frantically as Italy nears debt spiral." *The Telegraph*, November 8, 2011, http://www.telegraph.co.uk/finance/financialcrisis/8875444/France-cuts-frantically-as-Italy-nears-debt-spiral.html. Accessed December 30, 2013.
Farahany, N. (ed.). 2009. *The Impact of Behavioral Science on Criminal Law*. New York: Oxford University Press.

Forsythe, C. and J. Giordano. 2011. "On the need for neurotechnology in national intelligence and defense agenda: scope and trajectory." *Synesis J Sci Technol Ethics Policy* 2(1): 88–91.

Freeman, M. and O. Goodenough. 2009. *Law, Mind and Brain*. London: Ashgate.

Giordano, J. 2010. "Neuroethics: Coming of age and facing the future." In *Scientific and Philosophical Perspectives in Neuroethics*, eds. J. Giordano and B. Gordijn. Cambridge: Cambridge University Press, pp. xxv–xxix.

Giordano, J. 2011a. "Neuroethics: Traditions, tasks and values." *Human Prospect* 1: 2–8.

Giordano, J. 2011b. "Neuroethics: Two interacting traditions as a viable meta-ethics?" *AJOB Neurosci* 3: 23–25.

Giordano, J. 2012a. "Integrative convergence in neuroscience: Trajectories, problems and the need for a progressive neurobioethics." In *Technological Innovation in Sensing and Detecting Chemical, Biological, Radiological, Nuclear Threats and Ecological Terrorism (NATO Science for Peace and Security Series)*, eds. A. Vaseashta, E. Braman, and P. Sussman. New York: Springer.

Giordano, J. (ed.). 2012b. *Neurotechnology: Premises, Potential and Problems*. Boca Raton, FL: CRC Press.

Giordano, J. (ed.). 2012c. "Neurotechnology as demiurgical force: Avoiding Icarus' folly." In *Neurotechnology: Premises, Potential and Problems*. Boca Raton, FL: CRC Press, pp. 1–14.

Giordano J. and R. Benedikter. 2012. "Neurotechnology, culture, and the need for a cosmopolitan neuroethics." In *Neurotechnology: Premises, Potential and Problems*, ed. J. Giordano. Boca Raton, FL: CRC Press, pp. 233–242.

Giordano, J., R. Benedikter, and N.B. Kohls. 2012. "Neuroscience and the importance of a neurobioethics: A reflection upon Fritz Jahr." In *Fritz Jahr and the Foundations of Global Bioethics: The Future of Integrative Bioethics*, eds. A. Muzur and H.-M. Sass. Berlin, Germany: LIT Verlag.

Giordano, J., J. Engebretson, and R. Benedikter. 2008. "Pain and culture: Considerations for meaning and context." *Camb Q Rev Healthc Ethic* 77: 45–59.

Giordano, J., C. Forsythe, and J. Olds. 2010. "Neuroscience, neurotechnology, and national security: The need for preparedness and an ethics of responsible action." *AJOB Neurosci* 1: 35–36.

Giordano, J., A. Kulkarni, and J. Farwell. 2014. "Deliver us from evil? The temptation, realities, and neuroethico-legal issues of employing assessment neurotechnologies in public safety initiatives." *Theor Med Bioeth* 1: 1–17.

Greely, H.T. 2013. Mindreading, neuroscience and the law. In *A Primer on Law and Neuroscience*, eds. S. Morse and A. Roskies. New York: Oxford University Press.

Gregg, K. 2010. *Compliance with the Biological and Toxin Weapons Convention (BWC)*. Biosecurity, FAS in a Nutshell, 2010, http://www.fas.org/blog/nutshell/2010/08/compliance-with-the-biological-and-toxin-weapons-convention-bwc/. Accessed January 5, 2014.

Jeannotte, A.M., K.N. Schiller, L.M. Reeves, E.G. Derenzo, and D.K. McBride. 2010. "Neurotechnology as a public good." In *Scientific and Philosophical Perspectives in Neuroethics*, eds. J. Giordano and B. Gordijn. Cambridge: Cambridge University Press, pp. 302–320.

Lanzilao, E., J. Shook, R. Benedikter, and J. Giordano. 2013. "Advancing neuroscience on the 21st century world stage: The need for—and proposed structure of—an internationally-relevant neuroethics." *Ethics Biol Eng Med* 4(3): 211–229.

Lederberg, J. (ed.). 1999. *Biological Weapons: Limiting the Threat*. Cambridge, MA: MIT Press.

Levy, N. 2010. "Neuroethics—A new way of doing ethics." *AJOB Neurosci* 2(2): 3–9.

Lynch, Z. and C.M. McCann. 2009. "Neurotech clusters 2010: Leading regions in the global neurotechnology industry 2010–2020." NeuroInsights Report. http//www.neuroinsights.com. Accessed January 4, 2014.

Moreno, J. 2006. *Mind Wars*. New York: Dana Press. Reprint, New York: Bellevue Press, 2012.

Morse, S. and A. Roskiers, eds. 2013. *A Primer on Law and Neuroscience*. New York: Oxford University Press.

National Research Council. 2008. *Emerging Cognitive Neuroscience and Related Technologies*. Washington, DC: National Academies Press.

Nuffield Council Report. 2013. *Novel Neurotechnologies: Intervening in the Brain*. London: Nuffield Council on Bioethics.

Orofino, S. 1996. "*Daubert v. Merrell Dow Pharmaceuticals, Inc*.: The battle over admissibility standards for scientific evidence in court." *J Undergrad Sci* 3: 109–111.

Racine, E. 2010. *Pragmatic Neuroethics*. Cambridge, MA: MIT Press.

Ridley, M. 2003. *Nature via Nurture: Genes, Experience and What Makes Us Human*. London: HarperCollins.

Sarewitz, D. and T.H. Karas. 2012. "Policy implications of technologies for cognitive enhancement." In *Neurotechnology: Premises, Potential and Problems*, ed. J. Giordano. Boca Raton, FL: CRC Press, pp. 267–286.

Schoner, G. and J.A.S. Kelso. 1985. "Dynamic pattern generation in behavioral and neural systems." *Science* 239: 1513–1520.

Shook, J.R. and J. Giordano. 2014. "Toward a principled and cosmopolitan neuroethics: Considerations for international relevance." *Philos Ethics Humanit Med* 9(1).

Sobocki, P., I. Lekander, S. Berwick, J. Oleson, and B. Jonsson. 2006. "Resource allocation to brain research in Europe—A full report." *Eur J Neurosci* 24(10): 1–24.

Spranger, T. (ed.) 2012. *International Neurolaw: A Comparative Analysis*. Dordrecht, Germany: Springer.

Verplaetse, J., J. DeSchrijver, S. Vanneste, and J. Braeckman, (eds.) 2009. *The Moral Brain*. Berlin, Germany: Springer.

Wurzman, R. 2010. "Inter-disciplinarity and constructs for STEM education: At the edge of the rabbit hole." *Synesis J Sci Technol Ethics Policy* 1: G32–G35.

2 Transitioning Brain Research
*From Bench to Battlefield**

Steve Murray and Matthew A. Yanagi

CONTENTS

BACKGROUND

Contributions of brain research to human knowledge have flourished in recent years, owing in large measure to the increasing sophistication of direct brain monitoring, imaging, and interventional technologies and cognitive modeling tools (National Research Council 2008). Focused research collaborations, such as the Decade of the Mind (Albus et al. 2007), sought to ensure the rapid emergence of further insights into human cognition, emotion, and behavior. While neuroscience research (NR) is yielding important benefits to mental health, education, and computational science (the motivating aspects for the Decade of the Mind), the rapid pace of such work can also offer other capabilities to enhance national security, with knowledge and capability to improve the following:

- Human cognitive performance—Through better understanding of basic processes involved with memory, emotion, and reasoning, including the formation of biases and heuristics (Canton 2004). Such knowledge can provide improved task design, information structuring and presentation, and decision support to enhance human analysis, planning, and forecasting capabilities.
- Training efficiency—Enabling rapid mastery of knowledge and skills, with longer retention times (Giordano 2009), through individualized, real-time

---

* This chapter is adapted with permission from Murray, S. and M. A. Yanagi. 2011. Transitioning brain research: From the laboratory to the field. *Synesis: A Journal of Science, Technology, Ethics, and Policy* 2(1):T:17–T:25. The author of this manuscript (Steve Murray) was a U.S. Government employee, and this manuscript was written as part of his official duties as an employee of the U.S. Government.

tailoring of instructional material. Such capability could provide more flexible job assignment and more effective employment of available manpower.

- Medical treatment and rehabilitation—Providing more rapid and complete recovery from injury and enhanced resilience to the stresses and hazards of military operations (Miller 2008) so as to prevent or ameliorate the human costs of military service and enhance post-military health.
- Team process performance—By sensing, modeling, and supporting the dynamic social cognition processes needed to bridge organizational, cultural, and expertise gaps across team members (Fiore et al. 2003), NR can enhance the productivity of heterogeneous groups.
- System engineering—Including technologies to support shared-initiative problem solving between humans and machines (Smith and Gevins 2005), thereby enhancing the information-processing capabilities of both individuals and organizations.

The knowledge and tools generated by and from cognitive neuroscience activities have the potential to fundamentally alter many national security processes. Optimization of human–system performance capabilities, better employment (and protection) of available manpower, and reduced operational costs can dramatically expand the options available to government and military leaders in implementing national security policy. The value of brain-mind research to national security needs has been recognized among government, military, and science agencies, and investment in human cultural, cognitive, behavioral, and neural sciences has steadily emerged as a national budgetary priority (Defense Research and Engineering 2007).

Some of the potential benefits of cognition research (CR) and NR activities are detailed throughout this chapter. Realization of scientific potential in the practical world, however, is the result of labor required to fashion new knowledge and technologies into forms suitable for operational use, with accompanying acceptance and policy change. Therefore, the goal of this chapter is to explore the requirements necessary to evaluate and implement brain-mind-related research products in some of the settings important to national security, particularly the Department of Defense (DoD). Furthermore, because the insights generated by neuroscience promise to change many current assumptions about both human and machine capabilities, transition will likely face unique, additional challenges. This chapter introduces some of these challenges to highlight the discussion and planning needed to anticipate and avoid them.

THE SCIENCE AND TECHNOLOGY TRANSITION PROCESS

All DoD transition programs are designed to shepherd new products into acquisition programs, where they are purchased for operational use. Although the science and technology (S&T) community sponsors research projects through funding support, it is the acquisition community that handles the major tasks of transition. This expands the range of people and issues that researchers must accommodate to guide their work into operational practice. Among other responsibilities, acquisition agencies ensure that products contain sufficient clarity of purpose (i.e., fit a need and will be used), reliability (i.e., will perform "as advertised"), and sustainability (i.e., can be maintained and supported, and operators can be trained, throughout the product's operational life).

The current DoD S&T acquisition strategy—the Joint Capabilities Integration and Development System (JCIDS)—emphasizes incremental development and evaluation, to allow projects to mature and improve as a result of new discoveries and lessons learned (Defense Acquisition University 2005b). Two forms of transition are used, they are as follows:

1. Incremental development, in which the technology is essentially known from the beginning till maturation, occurs over a linear process
2. Spiral development, in which the technology is less defined, refers to the process wherein the transition is resolved through iterative user experience and feedback

Transition is a process involving continual performance measurement, with product performance reflected as a technology readiness level (TRL). The TRL model provides an operational description of both the S&T performance goals and the evaluation environments in which they are demonstrated (Deputy Under Secretary of Defense for Science and Technology 2005); success in a more challenging environment results in a higher TRL.

TRLs are based on a variety of factors (Dobbins 2004), including the following:

- Effectiveness—Does the technology work?
- Suitability—Does the technology work for the intended user, in the intended environment?
- Cost—Is the technology affordable?
- Schedule—Can the technology be delivered when needed?
- Quality—Does the technology represent the best available S&T?
- Reliability—Will the technology work consistently?
- Producibility—Can the technology be produced in quantity?
- Supportability—Can the technology be maintained in operational use?

DoD manages a variety of processes to accelerate the transition process by channeling attention and resources upon the S&T product by all of the government communities required for evaluation (Defense Acquisition University 2005b). Primary among these processes are the Advanced Technology Demonstration (ATD), focused on technology development, and the Advanced Concept Technology Demonstration (ACTD), focused on technology integration. Although each military service has its own transition programs that address unique user community needs, all of their methods are aligned with the objectives of these DoD-wide processes. Thus, while each service also supports rapid transition opportunities for high potential products (Kubricky 2008), selected products must still negotiate critical maturity and performance benchmarks before transition is concluded.

These processes can structure, but not ensure, viable transition. While much of the brain-mind-based technologies developed over the last decade have demonstrated enormous potential, few of these have transitioned into operational use, due in large part to the considerable challenges of demonstrating the higher maturity levels of the TRL model or of addressing each of the acquisition factors (reliability, producibility, etc.).

The results of new CR/NR efforts will be similarly at risk unless these requirements are addressed early and explicitly in the S&T planning process. Two classes of issues require consideration: (1) general challenges of the transition process and (2) potentially unique challenges resulting from the unknown impact of the science itself.

BRAIN-MIND RESEARCH TRANSITION—GENERAL CHALLENGES

Regardless of the cognizant agency, or the specifics of the S&T product, certain themes appear across all transition processes. The first of these is that transition is fundamentally a needs-driven process. While S&T sponsors always welcome and encourage disruptive "breakthroughs," in a climate of conflicting budget priorities and transition schedules, needs-based S&T products will almost always have priority. A needs-driven product directly addresses an existing or anticipated capability shortfall, and S&T is harnessed to solve a recognized problem; its value is apparent. An opportunity-driven product, however, emerges from new discoveries and applications must be identified; its ultimate value may be understood only as the product evolves. Although S&T programs typically include a mix of needs-driven (tech pull) and opportunity-driven (tech push) efforts and because CR-NR generates a high level of fundamental new knowledge, its applications are almost certain to be opportunity-driven. The most critical outcomes may not be those that were anticipated when research began, and fitting to a needs-driven process can therefore be extremely challenging (MacGillicuddy 2007).

All of the tasks required to validate a product for acquisition—for example, producibility, cost, and supportability—imply that transition is also an engineering process. That is, DoD acquisition processes conform to a system engineering model, which involves predictable steps of evaluation and gradual improvement toward a robust, understood, and supportable outcome (Deputy Under Secretary of Defense for Science and Technology 2005). Because the typical artifacts of such outcomes are tangible hardware and software products with perceivable, measurable performance effects, it can be difficult to cast the artifacts of CR-NR into such engineering forms (with the exception of certain medical applications, where this research may simply inform operational practices and technologies).

To realize the practical benefits of brain research for national security, the products of that research must relate to perceivable user needs and must be defined with engineering constructs. These requirements can be illustrated with a representative example—a human–machine interface system that uses direct physiological sensing to determine a human operator's cognitive state in real time and adaptively modifies system operation to enhance mission performance.

Measures of brain activity and psychological signals can reveal human states of workload, comprehension, and fatigue that could be used by system processing routines to adjust information presentation rates or dynamically move manual tasks to automated execution (Scerbo 2008). DoD interest in this form of neuroscience has been both intense and long (Scott 1994). The brain-based interface application represents a candidate for near-term S&T transition, in that it

- Addresses known operational performance problems. Many military systems can tax the information-processing capabilities of human operators.

These capabilities, furthermore, fluctuate with changes in operator fatigue, motivation, ability, and attention. These factors can have major impact on decision quality and mission performance.

- Involves current hardware and software systems used for surveillance and weapons control. While sensing algorithms and control logic are still the focus of active research, even first-generation interface applications could be treated as front-end modules to the existing military technologies.
- Is grounded in a wealth of foundational psychophysiological research and principles. The science needed to construct a physiological sensing system is sufficiently understood to establish at least basic working software models with useful operational impact (although other technological hurdles still exist).
- Has been discussed and socialized within DoD communities via S&T demonstration efforts (St. John et al. 2004).

To initiate transition, the originator of such an advanced human–machine interface, and affiliated S&T sponsors, would need to

1. *Define, or map to, an operational need.* This requires engagement with one or more prospective user communities to establish identifiable capability gaps. Although military user communities are comfortable with technology and typically have a good familiarity with emerging science, they necessarily think in terms of mission needs. It is the responsibility of the S&T community to engage with users to reach a consensus on operational needs. Because many S&T products represent new discoveries and often reveal new capabilities, such gaps may not be perceived. If the need is not apparent, however, then further development is necessary and the product is not yet a transition candidate. For the interface example, there must be a destination platform, such as a command center surveillance system, a shipboard radar console, or an aircraft navigation suite. Next, a case must be made that the current interface is inadequate. Finally, because NR can yield deep understanding of cognitive processes, systems that depend on this research will likely be more sophisticated; a case must be made that any anticipated performance improvement is worth the price in complexity.

2. *Establish performance metrics in engineering-relevant terms.* The S&T sponsor and acquisition team must develop specific answers to relevant questions that are often deferred until very late in the research process. For example, will the advanced interface reduce operator workload? If so, by how much and at what times in the mission profile? Will the interface elicit better decision making from the operator? Under what circumstances? In fact, how will workload and decision quality be defined so that user communities can understand and accept such metrics? What current metrics like this can be used as baselines to evaluate improvements from the technology? Will the advanced interface relax the need for strict operator selection or lengthy training (e.g., due to enhanced automation support)? And what data on current performance are being used as a baseline, given that brain and other physiological signals are not currently

measured in operations? Even dramatic laboratory success may not translate into operational environments, so metrics must be carefully chosen and agreed upon early in the transition effort. This task is made easier if capability gaps (step 1) are first clearly defined, which can facilitate the early definition of performance metrics as the research effort proceeds.

3. *Address each of the topics required for acquisition planning.* As previously addressed (and detailed further in Defense Acquisition University [2005a]), each of the topics in acquisition planning is essential for successful S&T transition. Systems must have plans in place for production, operator training, maintenance, and logistics before they are fielded. Although these topics are not commonly considered during the research and development process (which has more fundamental technical issues to contend with), such delay has ended countless business ventures; the "launching" of a new S&T product into military use is no less vulnerable to such failure. If the example human–machine interface system is to be transitioned into operational use, the acquisition community must identify and resolve a multitude of practical issues associated with operational introduction. How many systems will be installed, and where? Will the new interface be implemented as a front end to existing systems or completely integrated with existing technology? Who will fund installations, repair, and technical support? What existing systems, possibly unrelated to the new S&T, might be impacted? Who will be trained to work with the new interface, and will they be available prior to system installations? Will different installations (i.e., units that have the new interface and units that do not) affect personnel assignability? Are failure models and maintenance procedures sufficiently developed to maintain operational readiness, at least through initial evaluation periods? S&T research sponsors can be of immense help in such situations, as the bridge between fundamental research, potential user applications, and transition requirements is their working domain.

These are essential but tractable tasks that are necessary to negotiate the transition process. The advanced interface was chosen because it fits the current transition requirements better than some alternate technologies. Nevertheless, while it appears that most general research domains can directly navigate government acquisition processes to realize new capabilities, certain characteristics of CR-NR will still likely stretch the current transition model, as discussed in the section Brain Research Transition—Unique Challenges.

BRAIN RESEARCH TRANSITION—UNIQUE CHALLENGES

While aggressive government funding of CR-NR has enabled rapid scientific progress (Moreno 2006), significant and profound gaps still exist in our understanding of brain-mind function at many levels (Halpern 2000). Further consideration of current research in this area will highlight additional transition issues that emerge from the nature of the science itself. Among these issues are the disruptive impact of brain-based technologies and the additional analyses required to account for agency, responsibility, and transparency (described below) connected with their use. This gives a pause to any effort to insert such products into national security applications, where reliability is essential, and those involved in revolutionary S&T must respect the operational tension between innovation and conservatism.

Although system design based on human reasoning (e.g., expert systems, artificial intelligence) is not a new field, and the science of human–machine interaction based on real-time physiological states has shown great promise, the ontological implications of human–machine relations have not been resolved to any degree and will likely have an initially disruptive impact on planning and practice of military operations. Advances in CR-NR could, for example, enable dramatic improvements in mission performance of both human operators and autonomous machines (National Research Council 2009). Such capabilities will require a rethinking of military operating doctrine at several levels. How much information will operators need to reveal about their cognitive functioning (i.e., by allowing their physiological status to be monitored) to obtain improved mission performance? How can the decision processes of brain-based human–machine systems or autonomous systems be evaluated when the underlying algorithms are dynamic and may differ from mission to mission, from person to person, and at different times within a mission? How will brain-based systems interact with non-brain-based systems in distributed networks? Initial answers to such questions, and assessments of their impact on military doctrine, must accompany any effort to introduce these technologies into operational use. While government transition processes provide for graduated testing of technologies (see Figure 2.1), the impact of the issues described here may not be manifested until operational experience and exposure with such systems is accumulated.

Brain-based systems used in national security applications will almost certainly reflect a combination of autonomous initiative and original problem solving by both human and machine. This means shared agency (who or what acts) and responsibility (who or what is accountable for the result) in military decisions. Although shared agency between humans and computers lies at the core of many combat tasks, such sharing is largely based on predetermined decision models that persist across operational conditions, and the machine's role is an instantiation of one or more rule sets. The issue of agency and responsibility is expanded, when machine intelligence is more powerful and based on real-time exercise of humanlike faculties, even if those faculties are used to support human decisions. Recognizing the new status of such advanced machine capabilities will require a large adjustment in military and societal thinking about what constitutes a legitimate "mind" in military operations. The effective proliferation of these technologies into any arena of human activity will depend on how much attention and debate is offered to resolving such issues sooner, rather than later, in the transition process.

System operation based on either autonomous or shared information processing must be visible. Much of the difficulty encountered during early attempts to introduce intelligent (e.g., expert) systems into organizational settings was the lack of explanatory capabilities (Woods 1996) or transparency; systems could not make their reasoning explicit and understandable to operators, and therefore, system output was often not trusted. Because systems based on cognition and neuroscience principles

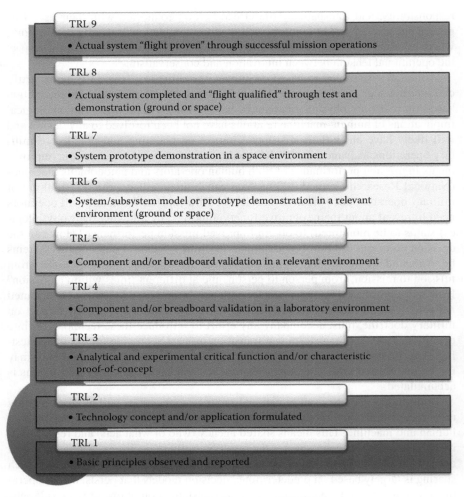

FIGURE 2.1 Technology readiness levels. (Courtesy of NASA.)

contain many of the defining features of artificial intelligence (e.g., shared-initiative decision making based on real-time conditions), they necessarily contain many of the same potential problems, including the need to understand operating state and to recognize degraded conditions. Although system engineering models address these issues, their manifestation in brain-based technologies may not be easy to characterize.

New capabilities typically lead to new consequences that ripple through organizations, so the issues discussed here will necessarily influence the reactions of complex organizations to the introduction of CR-NR products. Certainly, the existing traditions and values of national security organizations, including military communities, act to stabilize their activities and limit the pace of change (Pasmore 1988). While government agencies strive to make S&T transition expeditious, the process exercises an important restraining influence by imposing a structure and sequence to its component steps. Transition involves factors independent of the value of a new technology, and researchers must respect the sociotechnical context of the agencies with which they work. The advanced human–machine interface, used earlier, can again serve to illustrate some of the strategies that may be required to transition such fundamentally new S&T capabilities in the context of existing operational and technical traditions, which are as follows:

1. *Minimize disruption.* A stepwise introduction of disruptive capabilities, involving deployment of modest but well-understood applications, may become the desirable model for introduction to the effects of brain-based technologies. The human–machine interface might, for example, be tested using only operator workload or fatigue as a performance parameter and might only provide information feedback (instead of dynamic task support), delaying more advanced capabilities until initial performance has been operationally documented and user communities are satisfied with the results.

2. *Begin the dialogue to define agency and responsibility.* Who—human or machine—is ultimately responsible for decisions made or actions taken during mission execution with a brain-based interface? If a decision is wrong, which entity is responsible (i.e., legally or politically liable)? If these advanced interfaces are truly interactive, is the human operator sufficiently knowledgeable about their role in the task process to accept responsibility? And, understanding these issues, is the government or the larger society willing to accept the consequences involved in shared cognition? These questions require engagement with communities beyond those typically involved in S&T transition. Because these issues are likely consequences of the use of such systems (Illes 2006), however, the brain research community (and their sponsors) should lead the way in establishing early discussion and debate with all of the operational communities that will deal with their impact.

3. *Design for transparency.* The advanced interface example is grounded in the real-time measurement of the operator's cognitive state, which fluctuates according to mission conditions and individual factors. Because these measurements and the algorithms that operate on them are imperfect, some means of making these operations visible to the operator are essential. Although this is similar to the design of many artificial intelligence and expert systems, operations based on neural and physiological sensing may be accessing very personal information about the individual. What methods will be used to gather and reflect individual cognitive state, and who will have access to those data? There are currently no standards for collecting, displaying, and using such information in daily practice, highlighting a new requirement for S&T and user community engagement to develop such methods in advance of successful transition.

4. *Respect sociotechnical contexts.* The transition effectiveness of the advanced interface considered here—or any other brain-based technology—could be

enhanced through understanding and accommodating the conditions, values, and limitations of prospective user communities; some examples are as follows:

a. Developing operational employment concepts in parallel with funda-mental research. The ideas that generate scientific hypotheses should also serve to motivate thinking about applications, even if new discov-eries during the research process result in revisions to initial applica-tion concepts. How might the interface system be used? Under what conditions? For what missions? By which operators? The objective is to prepare solutions early to questions that are sure to be asked by military planners and users.

b. Soliciting discussion and debate from all user communities. The gen-eration fundamental precedent and foundation ethics regarding brain-based systems should necessarily find a place across agencies and applications. Key individuals and groups should be sought and engaged to develop early policy for interface employment and its consequences.

c. Establishing, elaborating, or leveraging government transition tools to reduce risk early in the science process. Military or other agency facili-ties—having consistent involvement with prospective users—could provide a persistent (standing) exploration and development arena that could provide both the socialization of users to new interface science concepts and the long-term validation testing required to develop con-fidence in the system. Such technology "nurseries" might also serve to allow time and experience for operating doctrine to catch up with the potential of the science.

d. Exploring applications across a wide front. The cultural traditions and operating conditions that govern the acceptance of new concepts may differ within and across user communities. A wide engagement can avoid the seduction of success based on only limited or specialized S&T introduction. Conversely, the choice of which technology, or which parts of a technology, to introduce first can impact transition success; early introduction and shrewd selection in one arena could counter resistance elsewhere. S&T transition is a tactical, as well as a strategic, effort.

The most relevant analog for transition of CR-NR into operational systems may be human factors engineering (HFE) and human system integration (HSI) products (Booher 2003), which directly address human-centric sciences and technologies. These engineering activities possess methodologies to ensure the effective function-ing of human–machine systems at all levels and address many of the issues described here, for example, identifying the need, engaging user communities, defining perfor-mance metrics, and addressing the full panoply of operational employment issues (e.g., maintenance, training, etc.). Two relevant lessons from HFE experience are that

- The user community can often identify novel applications for technologies before the developers themselves; early engagement pays dividends.
- Effective transition requires persistent involvement with the operational environment; success depends on iteration.

The technological implications of CR-NR represent advances not just of degree but of kind and will therefore reshape how we think about both humans and machines. Such understanding, involving individual and group cognition, is far more personal and opens many more unexplored issues than most other topics of technology transition. Additional steps are needed, therefore, to tune transition processes to better address the consequences of such new science in operational practice.

SUMMARY

Brain research offers extraordinary potential for expanding human performance in a wide range of national security endeavors, and new, as-yet unforeseen capabilities will emerge as new knowledge of neural function is gained. It is because knowledge about the human mind is growing so rapidly, however, that additional efforts at mutual education—among the research, S&T, acquisition, and user communities—are so essential; converting brain research results into operational capabilities requires the contributions of many agents. The need for a broad, collaborative approach to transitioning such science from the laboratory to national security capabilities is apparent when matching the potentially disruptive products of the research enterprise to hard engineering and acquisition requirements and to current operational demands.

Brain research holds significant potential to advance national security in original, fundamental ways. The applications discussed here were selected to bring a subset of issues into focus; certainly, other applications such as training, cognitive enhancement, improved social interaction, and health care present additional issues that must also be debated and resolved before useful products are realized. The common points of all these applications are that cognition and neuroscience researchers must navigate a practical and structured transition process if the products of the laboratory are to be realized as tangible human capabilities and that brain research discoveries may require elaboration of the transition process itself, to anticipate potentially disruptive consequences to operations. Early engagement around these topics among researchers, government transition communities, and users will, however, develop the conceptual foundation needed for significant advances in national security capabilities.

REFERENCES

Albus, J., G. Bakey, J. Holland et al. 2007. "A proposal for a decade of the mind initiative." *Science* 317(5843):1321.
Booher, H. 2003. *Handbook of Human Systems Integration*. Hoboken, NY: Wiley.
Canton, J. 2004. "Designing the future: NBIC technologies and human performance enhancement." *Annals of the New York Academy of Sciences* 1013:186–198.
Defense Acquisition University. 2005a. *Introduction to Defense Acquisition Management*, 7th ed. Washington, DC: Defense Acquisition University.
Defense Acquisition University. 2005b. *Manager's Guide to Technology Transition in an Evolutionary Acquisition Environment*, 2nd ed. Washington, DC: Defense Acquisition University.
Defense Research and Engineering. 2007. *Strategic Plan*. Washington, DC: Office of the Director of Defense Research and Engineering.

Deputy Under Secretary of Defense for Science and Technology. 2005. *Technology Readiness Assessment Deskbook*. Washington, DC: Defense Acquisition University.

Dobbins, J. 2004. "Planning for technology transition." *Defense AT&L*, March–April, pp. 14–18.

Fiore, S., E. Salas, H. Cuevas et al. 2003. "Distributed coordination space: Toward a theory of distributed team process and performance." *Theoretical Issues in Ergonomics Science* 4(3/4):340–364.

Giordano, J. 2009. "Education applications from understanding the human mind." *Paper Presented at the Decade of the Mind Conference IV*, Santa Ana Pueblo, NM, January 13–15.

Halpern, D. 2000. "Mapping cognitive processes onto the brain: Mind the gap." *Brain and Cognition* 42(1):128–130.

Illes, J. 2006. *Neuroethics*. New York: Oxford University Press.

Kubricky, J. 2008. "The rapid insertion of technology in defense." *Defense AT&L*, July–August, pp. 12–17.

MacGillicuddy, J. 2007. "Some thoughts on technology transition." *Transition Point* 2:8–10.

Miller, G. 2008. "Neurotechnology: Engineering a fix for broken nervous systems." *Science* 322(5903):847.

Moreno, J. D. 2006. *Mind Wars: Brain Research and National Defense*. Washington, DC: Dana Press.

National Research Council. 2008. *Emerging Cognitive Neuroscience and Related Technologies*. Washington, DC: National Academies Press.

National Research Council. 2009. *Opportunities in Neuroscience for Future Army Applications*. Washington, DC: National Academies Press.

Pasmore, W. 1988. *Designing Effective Organizations: The Sociotechnical Systems Perspective*. New York: John Wiley & Sons.

Scerbo, M. 2008. "Adaptive automation." In *Neuroergonomics: The Brain at Work*, eds. R. Parasuraman and M. Rizzo. New York: Oxford University Press, pp. 239–252.

Scott, W. 1994. "Neurotechnologies linked to performance gains." *Aviation Week & Space Technology* 141(7):55.

Smith, M. and A. Gevins. 2005. Neurophysiologic monitoring of mental workload and fatigue during operation of a flight simulator. *Proceedings of SPIE* 5797(18):127–138.

St. John, M., D. A. Kobus, J. G. Morrison et al. 2004. "Overview of the DARPA augmented cognition technical integration experiment." *International Journal of Human-Computer Interaction* 17(2):131–149.

Woods, D. 1996. "Decomposing automation: Apparent simplicity, real complexity." In *Automation and Human Performance: Theory and Applications*, eds. R. Parasuraman and M. Mouloua. Boca Raton, FL: CRC Press, pp. 3–18.

3 Neural Systems in Intelligence and Training Applications*

*Kay M. Stanney, Kelly S. Hale, Sven Fuchs,
Angela (Baskin) Carpenter, and Chris Berka*

CONTENTS

Functional neuroimaging is progressing rapidly and is likely to produce important findings over the next two decades. For the intelligence community and
the Department of Defense, two areas in which such progress could be of great
interest are enhancing cognition and facilitating training.

National Research Council (2008, 6)

INTRODUCTION

Lahneman et al. (2006) suggested that the intelligence community (IC) will soon
"… experience an imbalance between the demand for effective overall intelligence
analysis and the outputs of the individually oriented elements and outlooks of its
various analytic communities." The IC is producing analysts tailored to engage

---

* This chapter is adapted with permission from Stanney, K.M., K.S. Hale, S. Fuchs, A. Baskin, and
C. Berka. 2011. "Training: Neural systems and intelligence applications." *Synesis: A Journal of
Science, Technology, Ethics, and Policy* 2(1):T38–T44.

specific, focused missions. There is a need to train analysts to perform in a flexible manner that is conducive to supporting both kinetic—especially asymmetric—warfare and nonkinetic operations. This need can be supported with current training technologies that enable trainees "... to practice intelligence functions at all levels of war, from unconventional, low-intensity, tactical engagements to conventional, high-intensity, force-on-force conflicts" (George and Ehlers 2008). Advances in modeling and simulation can now integrate realistic, dynamic, and unpredictable virtual training environments with real-world mission data (e.g., unmanned aerial system feeds and satellite-orbit displays) and substantially improve intelligence training. While these simulated environments deliver realistic training opportunities, we posit that, to maximize learning, senior analysts and trainees must be equipped with tools that both enable the measurement of learning outcomes and the evaluation of training effectiveness. Thus, an unmet challenge is how to best measure, diagnose, and mediate intelligence operations training exercises so as to ensure that learning is maximized. To address this need, neurotechnology could be used for such assessment. Specifically, neurotechnology could be employed in both the bottom-up and top-down processing cycles of information analysis.

TRAINING INTELLIGENCE OPERATIONS

In training intelligence operations, it is critical to address a variety of interrelated activities that comprise the nested top-down and bottom-up cycles through two reciprocal processes (Pirolli and Card 2005):

- Bottom-up processing is used to integrate series of events and data into evidence, schemas, and hypotheses that explain *how* an event occurred and/or is likely to occur in the future.
- Top-down processing seeks additional evidence to support previously created hypotheses/propositions to explain or predict *why* an event occurred and/or is likely to occur.

Top-down and bottom-up processes constitute the "think-loop cycle" and are mutually engaged to create and validate accurate hypotheses that are important to accurate information analysis (Bodnar 2005). The complexity of this process is evident in the nested cycles presented in Figure 3.1. At the highest level, the cycle is broken down into two subprocesses with distinctly different goals, the foraging loop and the sense-making loop (Pirolli and Card 2005). The foraging loop gathers information. At this stage, the analyst will perform high-level searches of data repositories for information related to a specific question or topic of interest (e.g., biotechnological capabilities of a target country). Once the information that is gathered begins to evolve into a "story," a more comprehensive analysis of the selected sources is performed to extract more detailed information to address gaps related to the question/area being evaluated. When enough nuggets of information have been extracted to afford a relative understanding of a given question/area, the sense-making cycle is entered. At this stage, the analyst uses schemas to create propositions/hypotheses based on the evidence extracted during foraging. After developing hypotheses, analysis may progress

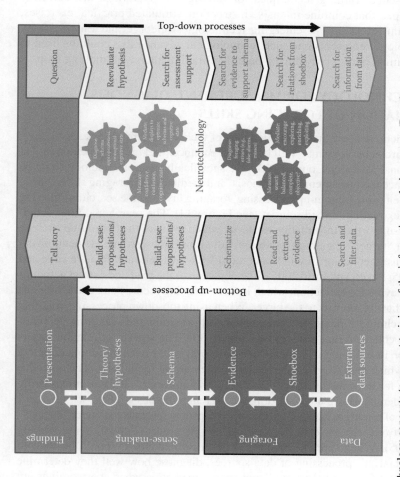

FIGURE 3.1 Neurotechnology concepts to support training of the information analysis process. (Based upon the work of Bodnar, J. "Making sense of massive data by hypothesis testing." *Proceedings of the 2005 International Conference on Intelligence Analysis*, McLean, VA, May 2–6, 2005; Pirolli, P. and S. Card. "The sensemaking process and leverage points for analyst technology as identified through cognitive task analysis." *Presentation at the 2005 International Conference on Intelligence Analysis*, McLean, VA, May 2–4, 2005.)

to a more top-down process to link extracted data to support or refute the hypotheses. If at any point additional data are required to validate the evolving "story," the foraging cycle is reentered to further gather supportive or contrary information.

Due to the high ambiguity and limited reliability of intelligence sources, along with the effort of data interpretation, analysts represent a key component in any authentic processing of intelligence sources. It is therefore important to effectively train and support intelligence personnel in those skills required in high-level analysis. Leading-edge training techniques for intelligence operations are currently being developed that combine intensive instruction, simulated practical exercises, and maximum leveraging of emerging technologies (George and Ehlers 2008). Neurotechnology represents one such technological advance that could be used to enhance training, particularly during both the foraging and sense-making loops.

NEUROTECHNOLOGY FOR MEASUREMENT, DIAGNOSIS, AND MEDIATION OF FORAGING SKILLS

"While predictions about future applications of technology are always speculative, emergent neurotechnology may well help to ... enhance [the] training techniques" of the IC (National Research Council 2008, 6). Specifically, neurotechnology could be used to enhance the measurement, diagnosis, and mediation of foraging skills during training. When using bottom-up processing, foraging involves the searching/filtering and reading/extracting of data sources (see Figure 3.1). The search and filter activities focus on supporting information retrieval, which involves the bottom-up strategy of defining a "target" (e.g., high-value individual, infiltrator, rogue element, physical system, insurgent camp, area of operation) for which relevant data are collected (Bodnar 2005; Pirolli and Card 2005). The objective of the read and extract activities is to support evidence extraction, which involves reading records and extracting bits of evidence that are relevant to the target. When using top-down processing, foraging involves searching for relations from collected information and searching for information from collected data sources (see Figure 3.1). The search for relations activity focuses on researching documents that have previously been accumulated into a "shoebox," which are thought to be related to the current hypothesis. The objective of the search for information activity is to seek available external data sources to dig deeper and identify new leads regarding the current hypothesis.

The heart of foraging, whether bottom-up or top-down, is thus to collect "nuggets" of evidence from relevant data sources that can be used to support sense-making (Bodnar 2005). When training foraging skills, neurotechnology could be used to monitor an analyst's processing of data sources, diagnose how well they determine "relevance" and formulate "nuggets," and trigger mediation when shortcomings are found. Specifically, during training of foraging skills, neurotechnology could be used for the following purposes:

- Measure:
  - Determine when and what data elements analysts are reading/viewing and which they are discarding to assess if the search is balanced, complete, and objective.

- Capture confidence, confusion, and/or interest via neurophysiological indicators that could be associated with each evidence "nugget" gathered to quantify the quality of each evidence item and, eventually, the analysis as a whole.
- Diagnose:
  - Identify which specific data elements cause analysts to become more or less interested, confused, or confident and compare to pre-identified areas of interest (AoIs).
  - Detect errors (e.g., false alarms, misses) during foraging that negatively impact the extraction of relevant data elements.
  - Detect narrow search, which indicates potentially relevant information is being discarded based on top-down processing or "explaining away."
- Mediate:
  - Collect, highlight, and/or force review of data elements to which analysts showed "interest" subconsciously (i.e., detected via neurophysiological indicators), but did not include in their analysis procedure (e.g., because they were "explained away" by prior or tacit knowledge).
  - Encourage exploring or monitoring more of the information space, enriching the data elements that have been collected for analysis by creating smaller, higher-precision data sets, and exploiting gathered items through more thorough review if narrow search is detected during the foraging process.

NEUROTECHNOLOGY FOR MEASUREMENT, DIAGNOSIS, AND MEDIATION OF SENSE-MAKING SKILLS

Neurotechnology could also be used to enhance the measurement, diagnosis, and mediation of sense-making skills during training. During bottom-up processing, sense-making involves fitting evidence into schemas (i.e., representations from which conclusions can easily be drawn), defining hypotheses/propositions and building supporting cases, and telling the story that is laid out by the evidence (see Figure 3.1). The schematizing activity focuses on supporting evidence marshaling, where the analyst begins to build a case by assembling individual pieces of evidence into a simple schema (Bodnar 2005; Pirolli and Card 2005). The objective of the hypothesis generation/case building activity is to support theory formulation, which synthesizes a number of interlocking schemas into a proposed theory. The storytelling activity focuses on developing a presentation (i.e., present hypotheses with supporting arguments: the ideas, facts, experimental data, intelligence reports, etc. that support/refute it) through which to convey and disseminate the analysis results. During top-down processing, sense-making involves questioning, reevaluating hypotheses, searching for assessment support, and searching for evidence (see Figure 3.1). The objective of the questioning activity is to reexamine the current story to identify if a new/refined theory must be considered. The objective of the reevaluating activity is to explicitly state a new hypothesis to enable identification of the kinds of evidence to search for that will support it or refute it. The objective of the search for assessment support activity is to look for available propositions that support or refute the working hypothesis and rebuild or modify these past propositions to address the new hypothesis. The objective of the search for evidence activity is to uncover available "nuggets" of evidence within the "shoebox" that support or refute the working hypothesis, or to identify that new evidence needs to be found to test it.

The heart of sense-making, whether bottom-up or top-down, is thus to build a story by synthesizing "nuggets" of evidence into schemas and testing/refining hypotheses (Bodnar 2005). When training sense-making skills, neurotechnology could be used to assess the appropriateness of schemas and hypothesis-testing quality and correctness during training, and assist in further unraveling the cognitive nuances of the analytic sense-making process. Various candidate areas exist where implementation of neurotechnology could lead to substantial improvements during the training of sense-making skills; examples follow:

- Measure:
  - Capture confidence and/or confusion via neurophysiological indicators during "nugget handling" that could be associated with the appropriateness of utilized schemas (e.g., "strong" versus "weak" linkages between evidence "nuggets" and propositions/hypotheses) to identify when sense-making links do not fit with current mental models.
  - Capture the cognitive state of the analyst (e.g., sensory memory, working memory, attention, executive function) via neurophysiological indicators (Schmorrow and Stanney 2008).
- Diagnose:
  - Identify when cognitive state is nonoptimal (e.g., working memory load is high, attention bottlenecks are being experienced), which may indicate limited sense-making capacity.
  - Identify when mental models are no longer appropriate for assessing evidence (e.g., confusion), which may indicate a potential bias or inaccurate analysis.
  - Evaluate in real time the effectiveness of information analysis (Joint Chiefs of Staff 2004), specifically its timeliness (e.g., during a training scenario, is evidence-marshaling and theory-building accomplished in a timely manner such that it supports planning, influences decisions, and prevents surprise?), relevance (e.g., is the intelligence being gathered relevant to answering the instructors' objectives?), and accuracy (e.g., does the intelligence being gathered provide a balanced, complete, and objective theory of the target threat—including addressing uncertainties, as well as alternative, contradictory assessments that present a balanced and bias-free analysis?).
- Mediate:
  - Trigger mediation techniques aimed at optimizing cognitive state (e.g., offload information patterns onto multimodal displays, such as visual mediators; Stanney and Salvendy 1995; Pirolli and Card 2005).
  - Trigger mediation techniques that encourage reassessment of mental models/schemas when deemed no longer relevant (e.g., highlight weak evidence links).

Thus, the application of neurotechnology to monitor the analytic processes within the foraging and sense-making loops during training has the potential to provide objective measures in a highly subjective process, as it could be used to enhance the analysis outcome through measurement, diagnosis, and mediation of analysis shortcomings.

Recent work conducted by Paul Sajda and colleagues has established the utility of a neurotechnology-centric platform for analysts engaged in high-volume analysis of satellite images (Sajda et al. 2010; Pohlmeyer et al. 2011). These studies have shown that single-trial event-related potentials (ERPs) can provide evidence of the neural signatures of target detection as analysts view rapid presentations of image fragments; relevant images can be sorted automatically, significantly improving analyst throughput capacity (Parra et al. 2008; Sajda et al. 2009, 2010). In addition, Curran and colleagues (2009) developed electroencephalography (EEG)-based change detection, which offers the prospect of an automated platform to facilitate analysis of changes when repeated images of a specific region are studied to identify relevant alterations of vehicles, personnel, or other resources. Another application designed by this group employs EEG-based facial recognition to allow rapid search of facial databases to identify and classify recognized faces. These examples demonstrate the growing use of neurotechnology to support analytic processes. One particular application will now be discussed in detail.

CASE STUDY: NEUROTECHNOLOGY SUPPORT OF FORAGING DURING IMAGE ANALYSIS

Neurotechnology solutions have been developed that target improving the speed and quality of evidence gathering (i.e., foraging) in intelligence operations through individualized training. One such solution suite is the Auto-Diagnostic Adaptive Precision Training (ADAPT) system that is designed to substantially enhance anomaly detection by incorporating neurophysiology measurement techniques into a closed-loop system that tracks the visual search process and automatically identifies skill deficiencies/inefficiencies in scanning, detecting, and recognizing anomalies (Hale et al. 2012). ADAPT utilizes two distinct neurophysiology instruments, specifically eye tracking and EEG technology. Eye tracking technology offers a unique methodology for cognitive assessment in that these systems can determine exactly what a person has visually perceived, and it "has become one of the most important and productive ways for investigating aspects of mind and brain across a wide variety of topic areas" (van Gompel et al. 2007, p. 3). Further, eye tracking technology has shown behavioral differences between novices and experts in search patterns, percentage of time looking at AoIs, and fixations (Kurland et al. 2005). EEG technology has excellent temporal resolution and tracking of neural activity, representing the flow of information from sensory processing and analysis to initiation of a response (i.e., EEG signals of humans searching for a specific target item in rapidly viewed images can reveal the perception of a specific item 130–150 ms post-stimulus, thus before conscious recognition occurs; Hopf et al. 2002). Past research also indicates the feasibility of using EEG/ERP to differentiate between correct responses (i.e., hits and correct rejections) and highly biased responses (e.g., false alarms and misses; Vogel and Luck 2000; Yamaguchi et al. 2000; Sun et al. 2004;

Hale et al. 2007, 2008), thus supporting the potential for using neurotechnology to enhance training of decision accuracy. Using these technologies, ADAPT allows trainees to review images naturally, while using eye fixations, EEG, and behavioral responses to determine whether or not AoIs are scanned, detected, and recognized in a timely fashion. Using behavioral responses alone, a system would know whether the outcome decision was correct or incorrect (i.e., whether a threat present was correctly perceived or not). If incorrect, however, the system would have no knowledge about why the error occurred—it may have been that critical cues were missed completely, or such areas were viewed, but were dismissed via top-down processing during the sense-making loop. Incorporating both eye tracking and EEG, ADAPT can conduct root cause analysis to discover error patterns present within a trainee's behavior and appropriately drive future training to target diagnosed deficiencies/inefficiencies.

The ADAPT system can support training on both static (Hale et al. 2012) and dynamic stimuli, and incorporates two main instructional methods: exposure and discrimination training. Exposure training presents trainees with a series of images and requires trainees to identify the presence or absence of a threat; discrimination training involves pairs of targets presented in two separate side-by-side images and requires trainees to identify whether the images are the same or different (Fiore et al. 2006). Appropriately aligning training methods with identified deficiencies/inefficiencies is expected to optimize the training cycle, enhancing training effectiveness and efficiency through individualized, tailored training. For example, a pattern of misses would lead to discrimination training, as trainees should benefit from focused attention on distinguishing details of anomalies between two images. These benefits are expected because it is theorized that learning results from strategic skills through comparison making, as well as stimulus-specific knowledge through repeat exposure (Hale et al. 2012).

In summary, ADAPT is a neurotechnology solution that could be used to support the measurement, diagnosis, and mediation of intelligence operations training exercises, specifically the foraging loop, as follows:

- Measure:
  - It can be used to determine when and what visual data elements to which analysts are attending, and the appropriateness of their assessment of interest/relevance for each area reviewed (e.g., identify hits, correct rejections, false alarms, misses, areas within the image that were not visually fixated upon).
- Diagnose:
  - It can be used to identify errors in foraging, such as visual data elements that distract analysts (i.e., false alarms) and which visual data elements are missed (i.e., AoIs that were discounted or not appropriately considered).
- Mediate:
  - It can be used to focus training on observed error patterns via real-time scenario adaptation (e.g., incorporate subsequent images that highlight missed areas) or after-action review (e.g., provide remediation training to identifying a consistently missed visual element in a particular orientation or location).

CONCLUSIONS

Neurotechnology has the potential to enhance the training of intelligence analysis, as it can provide objective measures of the highly subjective analytic process. During foraging in particular, identifying and remediating nonoptimal cognitive states (e.g., low engagement, high distraction), the detection of a lack of critical evaluation (e.g., misses), or discerning of continual attentional focus on distracting data (e.g., false alarms) could enhance information gathering. Further, neurophysiological monitoring during the sense-making process could evaluate critical thinking while connections and relationships between information snippets are created, encourage the consideration of alternative views when critical thinking is found to be insufficient based on neurophysiological indicators, and identify potential bias when creating argument chains. Taken together, these applications of neurotechnology could lead to a level of training effectiveness evaluation of foraging and sense-making skills that is well beyond the capability of current assessment techniques.

ACKNOWLEDGMENTS

This material is based upon the work supported in part by the Intelligence Advanced Research Projects Activity (IARPA)/Air Force Research Laboratory (AFRL) under BAA contract FA8750-06-C-0197.

DISCLAIMER

The authors affirm that any claims opinions, findings, and conclusions or recommendations expressed in this material are those of the authors and do not necessarily reflect the views or the endorsement of the IARPA/AFRL.

REFERENCES

Bodnar, J. 2005. "Making sense of massive data by hypothesis testing." *Proceedings of the 2005 International Conference on Intelligence Analysis*, McLean, VA, May 2–6.

Curran, T., L. Gibson, J.H. Horne, B. Young, and A.P. Bozell. 2009. "Expert image analysts show enhanced visual processing in change detection." *Psychonomic Bulletin & Review* 16(2):390–397.

Fiore, S.M., Scielzo, S., Jentsch, F. and Howard, M.L. (2006). Understanding performance and cognitive efficiency when training for X-ray security screening. *Proceedings of the Human Factors and Ergonomics Society 50th Annual Meeting*, pp. 2610–2614.

George, D.S. and R. Ehlers. 2008. "Air-intelligence operations and training: The decisive edge for effective airpower employment." *Air & Space Power Journal* 22(2): 61–67.

Hale, K.S., A. Carpenter, M. Johnston, J. Costello, J. Flint, and S.M. Fiore. 2012. Adaptive training for visual search. *Paper presented at the 2012 Interservice/Industry Training, Simulation, and Education Conference*, Orlando, FL, December 3–6.

Hale, K.S., S. Fuchs, P. Axelsson, A. Baskin, and D. Jones. 2007. "Determining gaze parameters to guide EEG/ERP evaluation of imagery analysis." In *Foundations of Augmented Cognition*, eds. D.D. Schmorrow, D.M. Nicholson, J.M. Drexler, and L.M. Reeves. Arlington, VA: Strategic Analysis, pp. 33–40.

Hale, K.S., S. Fuchs, P. Axelsson, C. Berka, and A.J. Cowell. 2008. "Using physiological measures to discriminate signal detection outcome during imagery analysis." *Human Factors and Ergonomics Society Annual Meeting Proceedings* 52(3):182–186.

Hopf, J.-M., E. Vogel, G. Woodman, H.J. Heinze, and S.J. Luck. 2002. "Localizing visual discrimination processes in time and space." *Journal of Neurophysiology* 88(4):2088–2095.

Joint Chiefs of Staff. 2004. *Joint and National Intelligence Support to Military Operations.* Washington, DC: Defense Technical Information Center.

Kurland, L., A. Gertner, T. Bartee, M. Chisholm, and S. McQuade. 2005. *Using Cognitive Task Analysis and Eye Tracking to Understand Imagery Analysis.* Bedford, MA: MITRE Corporation.

Lahneman, W.L., J.S. Gansler, J.D. Steinbruner, and E.J. Wilson. 2006. *The Future of Intelligence Analysis: Volume I Final Report.* College Park, MD: Center for International and Security Studies.

National Research Council. 2008. *Emerging Cognitive Neuroscience and Related Technologies.* Washington, DC: National Academies Press.

Parra, L.C., C. Christoforou, A. Gerson, M. Dyrholm, A. Luo, M. Wagner, M.G. Philiastides, and P. Sajda. 2008. "Spatio-temporal linear decoding of brain state: Application to performance augmentation in high-throughput tasks." *IEEE Signal Processing Magazine* 25(1):95–115.

Pirolli, P. and S. Card. 2005. "The sensemaking process and leverage points for analyst technology as identified through cognitive task analysis." *Presentation at the 2005 International Conference on Intelligence Analysis*, McLean, VA, May 2–4.

Pohlmeyer, E.A., J. Wang, D.C. Jangraw, B. Lou, S.-F. Chang, and P. Sajda 2011. "Closing the loop in cortically-coupled computer vision: A BCI for searching image databases." *Journal of Neural Engineering* 8(3):036025.

Sajda, P., L.C. Parra, C. Christoforou, B. Hanna, C. Bahlmann, J. Wang, E. Pohlmeyer, J. Dmochowski, and S.F. Chang. 2010. "In a blink of an eye and a switch of a transistor: Cortically-coupled computer vision." *Proceedings of the IEEE* 98(3):462–478.

Sajda, P., M.G. Philiastides, and L.C. Parra. 2009. "Single-trial analysis of neuroimaging data: Inferring neural networks underlying perceptual decision-making in the human brain." *IEEE Reviews in Biomedical Engineering* 2:97–109.

Schmorrow, D.D. and K.M. Stanney. 2008. *"Augmented Cognition: A Practitioner's Guide."* Santa Monica, CA: Human Factors and Ergonomics Society.

Stanney, K.M. and G. Salvendy. 1995. "Information visualization: Assisting low spatial individuals with information access tasks through the use of visual mediators." *Ergonomics* 38(6):1184–1198.

Sun, Y., H. Wang, Y. Yang, J. Zhang, and J.W. Smith. 2004. "Probabilistic judgment by a coarser scale: Behavioral and ERP evidence." *Paper presented at the 26th Annual Conference of the Cognitive Science Society*, Chicago, IL, August 5–7.

Van Gompel, R.P.G., M.H. Fischer, W.S. Murray, and R.L. Hill. 2007. "Eye-movement research: An overview of current and past developments." In *Eye Movements: A Window on Mind and Brain*, eds. R.P.G. van Gompel, M.H. Fischer, and W.S. Murray. New York: Elsevier Science & Technology Books, pp. 1–28.

Vogel, E.K. and S.J. Luck. 2000. "The visual NI component as an index of a discrimination process." *Psychophysiology* 37:190–203.

Yamaguchi, S., S. Yamagata, and S. Kobayashi. 2000. "Cerebral asymmetry of the 'top-down' allocation of attention to global and local features." *Journal of Neuroscience* 20(9):RC72.

4 Neurocognitive Engineering for Systems' Development[*]

Kelvin S. Oie and Kaleb McDowell

CONTENTS

Recent analyses by the Office of the Chairman of the Joint Chiefs of Staff have pointed to three key aspects of the security environment that will drive the development of operational capabilities and concepts needed to ensure success on the battlefield now and into the future: a wider variety of adversaries, a more complex and distributed battle space, and increased technology diffusion and access (Office of the Chairman of the Joint Chiefs of Staff 2004). Successful future human–system materiel development will depend on an approach that is able to account for the complex interactions of these critical aspects of the environment, as well as the numerous other environmental, task, and personnel factors that impact performance. For example, consider the case of a military commander in charge of a platoon or a transportation officer in charge of a security team at an airport. What factors will affect their performance? Some factors will be external to them and will be out of their control, such as the size of the enemy force, the time of day, or the effectiveness of their

---

[*] This chapter is adapted with permission from Oie, K.S. and K. McDowell. 2011. "Neurocognitive engineering for systems development." *Synesis: A Journal of Science, Technology, Ethics, and Policy* 2(1):T:26–T:37. The authors of this manuscript were U.S. Governmental employees, and this manuscript was written as part of their official duties as an employee of the U.S. Government.

security systems. Other factors will be internal, such as their ability to communicate and lead their personnel, their personalities, and their fatigue levels. An individual's cognitive functioning, or how they think about the situation and the information presented to them and how they translate that thinking into effective behaviors, will be critical to their performance. However, as will be discussed, ensuring adequate levels and sustainment of cognitive performance needed for mission success is nontrivial and will depend on the development and integration of advanced technologies and understandings of human neurocognitive behavior that lead to the effective design of socio-technical systems (i.e., complex systems accounting for both people and technology).

The potential impact of environmental complexity on cognitive function can be seen in the analysis of those military and industrial disasters where decision makers need to interact with equipment and personnel in a stressful, dynamic, and uncertain environment. Analysis of the shooting down of Iran Air flight 655 by the USS *Vincennes* in 1988 and the partial core meltdown of the nuclear reactor on Three Mile Island in 1979 revealed that cognitive aspects of complex human–system interactions can have dramatic and unexpected consequences (Cooke and Durso 2007). One of the primary contributors in these and similar incidents was the highly dynamic and information-rich environment enabled by advances in computer and information technologies (see below for further discussion). Similarly, contributors to the complexity of future sociotechnical interactions are likely to include the following: the increasingly dynamic and nonlinear nature of the battle space; the adoption by the adversary of advanced information technologies, such as the Internet, cellular telephones, GPS devices; nontraditional approaches to warfare, such as the widespread use of improvised explosive devices and suicide bombings; the high level of interactions between our forces and the local populations and political leaders; and the envisioned nature and demands of future warfare, which will involve reduced manpower; greater availability of information; greater reliance on technology, including robotic assets; and full functionality under suboptimal conditions (McDowell et al. 2007). These challenges will fundamentally alter the balance and nature of the sociotechnical interactions in the emerging operating environment such that meeting the *cognitive demands* posed by these environments will necessitate the change from a model that primarily relies on personnel to one that involves a balance between personnel and system. While such a shift may be necessary to provide the capabilities needed on the future battlefield, it can also lead to new patterns of errors (Wiener and Curry 1980) and impose new demands on systems developers.

From the materiel development perspective, the complexity of the aforementioned security environment presents significant difficulties. It is widely believed that the profound advances in computing, information and communications technologies, will provide a path forward toward meeting those demands. Realizing such capabilities, however, will require the research and development community to better understand the impact that the complexity of the operational environment has on behavior and to develop and implement systems that will best provide the capabilities required to work in harmony with our personnel, while disrupting the capabilities of the opposition. More specifically, we believe that systems should be designed to work in ways that are consistent with the function of the human brain, augmenting its capabilities to compensate for and overcome its limitations and capitalizing on inherent neurocognitive strengths in those domains where effective technological solutions cannot be attained. In this way, human–system performance can be maximized to meet the challenges of a complex, dynamic, and ever-changing security environment.

In this chapter, we discuss an approach to materiel development utilizing cognitive engineering supported by neuroscience, that is, *neurocognitive engineering*. We use as an illustrative example the problem space of the information-intensive security environment and the widely accepted approach to addressing its challenges, namely, decision superiority and information dominance that can be enabled through advanced information networks. We argue that traditional approaches to addressing the cognitive needs of systems development will not be met by traditional methods and that adopting tools and approaches from neuroscience provides opportunities to enable neurocognitive engineering to demonstrably improve systems designs, both within this context and more generally across human-technology interactions. Finally, we discuss several challenges, wherein a neurocognitive engineering approach has the potential for improving soldier, system, and integrative soldier–system performance.

THE INFORMATION-INTENSIVE SECURITY ENVIRONMENT

As information technologies have proliferated over the past several decades, it has become widely believed that information and its use on the battlefield is vital to the success of our armed forces; that is, "superiority in the generation, manipulation, and use of information," or "information dominance," is critical to enabling military dominance (Libicki 1997). Winters and Giffin (1997) assume an even more aggressive position, defining information dominance as a qualitative, rather than simply quantitative, superiority that provides "overmatch" for all operational possibilities, while at the same time denying adversaries equivalent capabilities. As reported in Endsley and Jones (1997), each of the major branches of the armed forces has embraced the critical importance of information dominance on the future battlefield.

A more recent elaboration on the concept of information dominance is the notion of "decision superiority": the process of making decisions better and faster than adversaries (Office of the Chairman of the Joint Chiefs of Staff 2004). Decision superiority is one of the seven critical characteristics of the future joint force (Office of the Chairman of the Joint Chiefs of Staff 2004). It rests upon a paradigm of information dominance to provide the capabilities to acquire, process, display, and disseminate information to decision makers at every echelon across the force.

THE CAPABILITIES OF THE INFORMATION AGE

Technological capabilities to process, store, transmit, and produce information have increased remarkably. As reported by Chandrasekhar and Ghosh (2001), progress in information and communications technologies over the past 40 years has been truly remarkable. For example, between the early 1970s and late 1990s, there was a greater than 10,000-fold increase in the number of transistors that could be placed on a computer chip; a 5,000-fold decrease in the cost of computing power; a 4,000-fold decrease in the cost of data storage, and a 1,000,000-fold decrease in the cost to transmit information. Other authors have produced similar, but varying, estimates of the growth and impact of information and communications technologies (Bond 1997; Nordhaus 2007), as well as predictions of the continued growth in such technologies in the near and mid-range future (Moravec 1998; Nordhaus 2001). This growth has, in turn, stimulated the development and availability of devices that have revolutionized the ability to produce, acquire, organize, retrieve, display, manipulate, and disseminate information at levels that have been historically unprecedented. It has been estimated that worldwide production of original information in 2002 was between 3.4 and 5.6 exabytes (1 exabyte = 10^{18} bytes of information). They further provide some context: 5 exabytes of information is equivalent to about 37,000 times the size of the 17-million-book collection of the U.S. Library of Congress, or about 2,500 times the size of all of the U.S. academic research libraries combined. Instead of focusing on the production of new information, more recently, Bohn and Short (2012) estimated that in 2008, Americans consumed approximately 3.6 zettabytes (1021 bytes), with the consumption of information growing at an annual rate of 5.4% from 1980 to 2008.

It is widely maintained that the complexity of the current and future battlespace can be addressed through the development and use of information and related computer technologies, and thus, these technologies are considered vital for national security at the highest political and scientific levels (National Research Council 1994; National Science and Technology Council 1994). Specifically, it is envisioned that decision superiority and information dominance can be realized through the development and effective utilization of advanced information networks (Cotton 2005). Indeed, the development of a Global Information Grid (GIG) was mandated in 2002 by Department of Defense Directive 8100.1, "Global Information Grid Overarching Policy" (Zimet and Barry 2009).

In 2004, the National Military Strategy discussed the development of the GIG, which would facilitate "information sharing, effective synergistic planning, and execution of simultaneous, overlapping operations" (Office of the Chairman of the Joint Chiefs of Staff 2004). According to that analysis, the GIG "has the potential to be the single most important enabler of information and decision superiority." Similarly, proposals for the Army's future forces also rely heavily upon an advanced battlefield network to provide superior battlespace awareness and strategic and tactical advantages by providing precise and timely information of enemy and friendly positions, capabilities, activities, and intentions (Office of the Chairman of the Joint Chiefs of Staff 2004). Such information is intended, in turn, to make flexible, adaptive planning possible in the face of a complex, dynamic security environment. The belief that information and communications technologies can support increased operational capabilities appears to be both clear and pervasive, although alternative perspectives have been expressed (Gentry 2002). By 2012, the GIG had become the "principal common network backbone" for the implementation of network-centric operations for the military (Zimet and Barry 2009).

INFORMATION INTENSITY AND CONSEQUENCES FOR HUMAN PERFORMANCE

While the technological capabilities for collecting, processing, displaying, and disseminating information have dramatically increased over the past several decades, human information processing capabilities have not increased in the same manner. The cognitive capacities of the human brain, despite its vast complexity, have been shown to be limited, and such limitations are widely noted (Arnell and Jolicœur 1999; Cowan 2001; Marois and Ivanov 2005; Miller 1956; Ramsey et al. 2004; Shiffrin 1976; Vogel et al. 2001). These limitations of human cognitive capabilities will have obvious and significant consequences for performance in the face of an increasingly complex and information-intensive operational environment when considering a paradigm of information dominance. This may be especially true when the demands of a task (or set of tasks) exceed an operator's capacity (i.e., under conditions of mental or cognitive overload) (Kantowitz 1988).

Performance suffers under overload conditions. Numerous studies have shown the negative performance effects of increased information load on task performance across a range of human performance, including driving (Wood et al. 2004), simulated flight control (Veltman and Gaillard 1996), production management and scheduling (Speier et al. 1999), and business and consumer decision making (Ariely 2003). For example, Jentsch and colleagues (1999) have found that increased task and information load leads to losses in situational awareness (SA) among flight crew members, resulting in poor task performance. These researchers analyzed 300 civilian air traffic incidents and found that pilots were more likely to lose SA when at the controls of the aircraft than when their copilot was at the controls; a finding that was valid regardless of aircraft type, flight segment (e.g., takeoff, approach), or weather conditions.

The significance of the detrimental effects of information load and the potential for substantial deficits of soldier–system performance has also been well acknowledged within the defense community (Cotton 2005; Fuller 2000; Sanders and Carlton 2002). Leahy (1994) discusses two examples in which information overload has had serious, and sometimes disastrous, effects. During Operation Desert Storm, a 1000-page, computerized listing of all coalition air operations, the Air Tasking Order (ATO), was produced every day. With limited time to read and process this amount of data, the result was that tactical air planning staffs focused only on information pertaining to their specific missions and were "often unaware of other missions in the same area," though that information may have been available in the ATO. In 1988, the USS *Vincennes* mistakenly classified Iran Air Flight 655 as an enemy F-14 fighter jet, shooting it down and killing 290 civilian passengers and crew. In his analysis, in "No Time for Decision Making," Gruner (1990) pointed out that investigators concluded that the ship's information systems "functioned as designed," but that bad decisions on the part of the captain and crew were due to information overload, among other factors, during a time-critical operation. He states: "Simply put, the rate at which the brain can comprehend information is too slow under fast-paced action. It has neither the time to understand all the inputs it receives, nor the ability to effectively perform all the other function [sic] it would be capable of in a less harried environment." Acknowledging these issues, several conceptual and empirical efforts have examined the issues of information and cognitive task load and its effects on human performance in military and defense-related domains (Kerick et al. 2005; Sanders and Carlton 2002; Svensson et al. 1997) and have explored potential solutions that could mitigate these effects (Dumer et al. 2001; Lintern 2006; Walrath 2005).

The negative impact of cognitive and information load incurs additional effects beyond task performance. As Kirsh (2002) reported, a survey of middle and senior managers in the United Kingdom, Australia, Hong Kong, and Singapore revealed delays and deficits in making important decisions, but also showed loss of job satisfaction, ill health, and negative effects on personal relationships, as a consequence of the stress associated with information overload. In another study, Kinman (1998) conducted a survey of 2000 academic and academic-related staff in the United Kingdom and found that 61% of the respondents cited information overload as a cause for stress related to time management and 66% reported that time management pressures forced them to compromise on the quality of their work. This result is consistent with Cooper and Jackson's (1997) contention that the increased prevalence of information technology has resulted in information overload and an accelerated pace of work, and more recent and general findings that perceived information overload is associated with increased stress and poorer health (Misra and Stoklos 2011). Similarly, Cotton (2005) has argued that the proliferation of information on the battlefield would increase stress on warfighters across the joint force, with the need for "faster access to information, quicker decision cycles, increased productivity, and measurable improvements," while at the same time producing unintended, but significant negative psychological, cardiovascular, and other health-relevant consequences.

SYSTEMS DESIGN AND COGNITIVE PERFORMANCE

In the earlier sections, we have discussed the challenges imposed by the increased complexity of the current and future security environment, the belief that solutions based on information and communications technologies can meet those challenges, and the detrimental effects of information and cognitive load on human performance. Given that the human brain's finite cognitive capacities and limited information processing capabilities are a major limitation to soldier–system performance, one of the major goals of technology developers should be to design systems that can work in ways that are consistent with human brain function(s). Such an approach would exploit the unique capabilities of the human nervous system, while accounting for its limitations, to maximize soldier–system performance.

Unfortunately, the general model for technological development has not taken this approach. Instead, the standard has been to allow technologies to advance essentially unfettered and to depend upon the capabilities of the human operator to adapt to the latest innovations. Consider, for example, the current prevalence of navigation and route-guidance systems and/or information and entertainment systems in automobiles. These systems are intended to improve safety and convenience for drivers, but they also add additional tasks—some of which can be information intensive (e.g., searching through a list of restaurants or songs)—to the primary driving task (Körner 2006). Green (2004) has reported that the use of such in-vehicle systems (1) is a contributing factor in accidents, (2) causes drivers to lose awareness of their primary driving task, and (3) is associated with accidents that happen during good driving conditions, suggesting that such accidents are distraction-related occurrences (rather than alcohol-related or fatigue-related). And while many states have already limited cell phone use during driving—for example, 12 states prohibit hand-held phone use and 43 states have banned texting while driving (Governers Highway Safety Association 2014), the development of guidelines and regulation of the installation and use of most in-vehicle information systems is still lacking (National Highway Traffic Safety Administration 2014). So, while systems-design approaches that rely upon the adaptive capacities of the human nervous system have been generally successful, it is important to note that as technologies are increasingly inserted into the systems in use, new approaches to integrating these systems and mitigating information overload will be necessary.

NEUROCOGNITIVE APPROACHES TO SYSTEM DESIGN

Designing systems that can work in ways that are consistent with human brain function is nontrivial when considering the factors that influence human neural activity. For example, substantial evidence points to interindividual differences in neural function; adaptation of neural function as a function of training, experience, and transfer effects; and changes in brain state due to stress, fatigue, and the use of various pharmacological and even nutritional agents. These factors point to a systems engineering approach that examines not only the use of the system itself, but (1) the impact of environmental stressors, training and experience with the system, and other related technologies and (2) the capabilities of users at various levels of skill, experience, and internal state. Perhaps the most critical aspect of such an approach is to first enhance our understanding of cognitive function in operationally-relevant contexts.

ASSESSMENT IN OPERATIONAL ENVIRONMENTS

Previously, traditional cognitive psychology, human factors, and engineering approaches have often been successful in addressing some of the cognitive-based needs of technology development. However, increased information intensity of the current and future battlefield, as discussed, is likely to challenge soldiers in ways not previously considered. Given the importance of cognitive performance in facing these challenges, we believe that systems that are not harmonized to human neural information processing will diminish the potential impact of our investments in technology. More importantly, this would lead to deficits in soldier–system performance on the battlefield, endangering soldier sustainability, survivability, and mission success.

Understanding the impact of an increasingly complex and information-intense operational environment on cognitive performance is a fundamental step toward developing approaches to systems design that can mitigate the negative consequences of cognitive and information overload. We contend that, to provide systems developers with the knowledge of human (i.e., warfighter) cognition needed to make critical design and development decisions, such understandings must be objective, nonintrusive, high resolution, and operationally relevant. Real-time assessments of warfighter cognitive capabilities and limitations would provide further potential for systems research, development, test, and evaluation that can integrate online knowledge of soldier functional state and adapt system behavior to suit the operator's current operational needs and abilities.

Unfortunately, traditional methods of cognitive performance evaluation alone cannot provide an understanding of the mechanisms and technical approaches required. Here again, we consider the concept of cognitive or information workload. Generally, there are four traditional techniques for assessing workload: performance measures, subjective ratings, physiological measurements, and subject matter expert opinion. Performance measures such as reaction time or response time are used extensively in psychological and human factors research on simple tasks. However, as Veltman and Galliard (1996) suggest, performance measures often cannot be used to index workload in complex task environments. This is especially true when assessing workload for subtasks, as changing task priorities make it impossible to determine whether such measures accurately reflect specific subtask performance. Even if such subtask evaluation(s) were possible, there is not a formalized methodology for combining scores on different tasks into a single score to adequately reflect overall task performance. Operators will also adapt to increasing task demands by "exerting additional effort" (Sarter et al. 2006), which may lead to equivalent assessments of task and cognitive performance when assessed through task outcome measures alone, even though cognitive workload has increased. This means that performance-based measures can only provide information on workload when some estimate of the operator's effort can also be indexed.

Rating scales, which are based upon post hoc, subjective reports of perceived workload, might provide such estimates. Several instruments (e.g., National Aeronautics and Space Administration [NASA] Task Load Index [TLX], Subjective Workload Assessment Technique [SWAT], and Workload Profile) have been extensively used in previous research (Fréard et al. 2007; Hart 1988; Rubio et al. 2004; Scallen et al. 1995; Tsang and Velazquez 1996; Veltman and Gaillard 1996; Verwey and Veltman 1996; Wu and Liu 2007) and have been shown to be effective in assessing subjective workload associated with performance on routine laboratory tasks (Rubio et al. 2004). However, it has also been argued to the contrary that individuals do not always report their current psychological, mental, or emotional status accurately (Zak 2004). Veltman and Gaillard (1996) posit that rating scales are limited by the effect of participants' memory, perception, and biases. For example, participants appear to be unable to discriminate between task demands and their effort invested in task performance. As well, subjective rating scales are not well suited for online estimation of workload, as they often require significant task interruptions, imposing at least some cost(s) to performance due to task switching (Monsell 2003; Speier et al. 1999).

Measurement of physiological function and state offers a third approach to assessing cognitive processing. Central and peripheral physiological measures can provide a more objective means of assessment than can be obtained via traditional performance and rating scale methods, and numerous different measures have been related to cognitive performance, including heart rate, heart rate variability, blood pressure, respiration rate, skin temperature, pupillary responses, and galvanic skin response (Boehm-Davis et al. 2003; Jackson 1982; Verwey and Veltman 1996). Unfortunately, physiological measures taken in isolation from central nervous system activity do not seem to have a high degree of sensitivity to cognitive performance across different task and environmental conditions.

Barandiaran and Moreno (2006) conceptualize this problem from an evolutionary perspective: biological systems are intrinsically purposeful in terms of their self-sustaining nature, which is the result of their internal, metabolic organization. At the most fundamental level, this is the source of what we refer to as intentionality. The evolution of the nervous system enabled organisms to actively modify their relationship with the external environment (e.g., by enabling the organism to move to different locations within the geographic space) to satisfy biologically defined constraints. In the case of systems that are distinctly cognitive, however, constraint satisfaction and metabolically driven intentionality do not seem to be able to fully explain the phenomenology of cognition (e.g., behavior that does not seem to be solely in response to metabolic needs). The authors suggest that the nervous system can be considered to be "decoupled" from the metabolic (and constructive) processes of the organism such that the interactions of the nervous system that underlie cognitive state are no longer explicitly governed by the metabolic organization that supports the nervous system's architecture. A significant implication of such a perspective is that the local states of metabolic systems (i.e., the physiological states of the heart, lungs, and kidneys) alone will not be able to predict the dynamic behavior and states of the nervous system.

Given the shortcomings of traditional approaches to cognitive assessment, it is unlikely that incremental improvements in our knowledge based upon these approaches alone can (or will) provide the necessary understandings of cognitive function that would be needed to address the challenges of systems design for the current and future security environment. However, recent progress in the neurosciences has expanded knowledge of how brain function underlies human cognitive performance. Increasingly, the connection between human experience and its bases in nervous system function is considered to be the foundation for understanding how we sense, perceive, and interact with the external world. In particular, the advancement of noninvasive neuroimaging technologies has provided new venues toward understanding the brain (National Research Council 2009).

However, much of the recent knowledge of human brain function has been gained from the highly controlled environments of the laboratory, with tasks that often are not representative of those that humans perform in real-world scenarios. Such experimental conditions are required both to minimize motion as much as possible and to maximize measurement fidelity (e.g., in functional magnetic resonance imaging [fMRI] and magnetoencephalography) and to control for the effects of potentially confounding variables that could affect the interpretation of experimental data. The dynamic, complex nature of military operational tasks and environments, by contrast, is likely to affect the human nervous system, and its functioning, in ways that are significantly different than the tasks and environments traditionally employed in laboratory studies. It is clear that not only do different individuals process information differently (see below), but the same individuals may engage different brain regions to cognitively process information in ways that are dependent on context (Hasson et al. 2010; Edelman and Gally 2001; National Research Council 2009). Thus, assessing the cognitive demands of human operators during the performance of real-world tasks in real-world environments will be critical for understanding how we really process information, integrate neural function, and behave (Gevins et al. 1995) (i.e., ecological validity). Such an understanding is vital for generalizing results of laboratory studies to more naturalistic behaviors and environments (i.e., external validity).

Toward this end, several research groups have advanced the use of electroencephalography (EEG) within environments previously thought to be unapproachable (Castermans et al. 2011; Huang et al. 2007; Kerick et al. 2009; Matthews et al. 2008). EEG, as a direct measure of the electrical activity of the brain detected at the scalp, provides an objective measure that is more closely associated with cognitive function than other psychophysiological measures (e.g., heart rate or respiration). EEG also provides measurement at very high temporal resolution, enabling observation and analysis at timescales (~1 ms) that are relevant to the dynamic behavior of the brain, unlike performance measures or rating scales. And while current technologies are still fairly cumbersome to use (e.g., requiring significant setup time and the application of electrolytic gels), technological advances hold the promise of nearly noninvasive, zero-preparation EEG recording (Lin et al. 2008; Matthews et al. 2007; McDowell et al. 2013; Sellers et al. 2009).

Progress in computational power and data analytic techniques has also enabled the development and application of novel signal analysis and decomposition methods (Gramann et al. 2011; Jung et al. 2001; Makeig et al. 1996), as well as advanced data mining techniques (Garrett et al. 2003) for data processing and knowledge discovery in highly multidimensional data in ways that have clearly surpassed our previous capabilities. These advances have great potential to improve EEG technology, enhancing its spatial resolution relative to the current state-of-the-art in neuroimaging technologies (e.g., fMRI) and moving neuroscience-based cognitive assessment into the operational realm.

While further technological advances and methodological developments are still needed, the current tools of neuroscience, when integrated with complementary approaches of more traditional methods, can provide more complete characterizations and understandings of cognitive (or somewhat more specifically, neurocognitive) performance in operational environments (for recent discussions, see Gordon et al. 2012). A recently developed system that implements this "multi-aspect" approach was described in Doty et al. (2013). Though that implementation was optimized for the study of stress, the advances embodied in the system could be applied to study other naturally occurring phenomena. The multiaspect approach will be critical not only for systems engineers who are developing systems to meet the challenges of the current and future security environment, but also for cognitive systems engineers who aim to facilitate performance by focusing on the "thinking" aspects within such socio-technical systems (McDowell et al. 2009). In the following two sections, we discuss an important issue—individual differences—in which a neurocognitive engineering approach may have significant potential for enhancing systems design.

DIFFERENCES IN OPERATOR CAPABILITIES

One of the most common, yet more difficult systems engineering issues is the need to account for the individual difference in operator capabilities. Cognitive research has revealed that people not only differ in classical categories of mental function, such as intelligence, skill set, or relating to past experience, but they also differ on a more fundamental level in how they think (i.e., cognitive styles, abilities, and strategies). These differences arise from many factors, including inherent characteristics of the operators and how operators are affected by stressors, such as emotionality and fatigue. A growing body of evidence suggests that individual differences in cognition, behavior, and performance of skilled tasks are rooted, at least to some extent, in differences in neural function and/or structure (National Research Council 2009). This has been supported by the association of genetic markers with variability in brain size, shape, and regional structure (Tisserand et al. 2004); elucidation of differences in nervous system connectivity that relate to different patterns of cognitive activity (Baird et al. 2005; Ben-Shachar et al. 2007); and demonstrating variability in individual patterns of brain activity (Chuah et al. 2006; Miller and van Horn 2007; Miller et al. 2002). These findings suggest the need for, and perhaps the basis of, plausible engineering solutions that are directed at developing integrated systems that accommodate and maximize individual structure–function relationships in the brain (for a recent discussion, see Vettel et al. 2014).

TRAINING, EXPERTISE, AND EXPOSURE

Individual differences between operators, such as those associated with the related factors of training, expertise, and exposure, can change how an individual processes information and makes decisions. An example of this is how people "naturally" envision force and motion. Research has indicated that people who have formal education in Newtonian physics can understand motion differently than neophyte physics students, who generally have a naïve "impetus" view of motion (Clement 1982; Mestre 1991). The concept appears to be related to differences in the neural processing involved with learned knowledge versus simple "beliefs" about physics (Dunbar et al. 2007). From this example, one can see the potential for different system designs to alter the cognitive processing associated with performance; if the system is inconsistent with the operator's view of the world, different and perhaps increased neural resources will be required to complete tasks. This possibility has been supported by research showing distinct neural and time factors involved in skill acquisition (Chein and Schneider 2005; Kerick et al. 2004; Poldrack et al. 2005). Further, these studies provide insights into the ways that future systems might employ neuroscience-based technologies to assess and adapt to how an operator "naturally" interacts with the system.

Both training and expertise are related, in part, to exposure. Many national security technologies are envisioned that may have unique aspects to which operators have not been previously exposed. Thus, while they may have been trained on related technologies, even with extensive exposure, many operators may never achieve expert levels of performance with new technology. It is known that exposure to an enriched environment produces changes in synaptic growth, brain morphology, and neurochemistry, as well as behavior (van Praag et al. 2000). Studies have shown positive effects of exposure to multiple channels of stimuli (e.g., audio, visual, and tactile, as compared to unimodal stimuli) in the performance of single tasks (Seitz et al. 2006, 2007). However, unintended and interference effects of multimodal stimuli have also been shown (Shams et al. 2002). This latter finding highlights a possible negative effect of learning: the strength of past events may influence future perceptions when conditions are sufficiently similar. This illustrates that an operator's expected exposure to a given technology must be considered in system design and gives insights into potential system designs that might be utilized to predict operator perceptual biases over time and to adapt to and eliminate these potentially negative effects.

FUTURE APPLICATIONS AND CONSIDERATIONS

To be sure, recent advances in neurotechnology are enabling understandings of neurocognitive functioning in ways, and within environments, that are highly relevant to national security. This is prompting neurocognitive engineering approaches to materiel development that has the potential to revolutionize human–system design(s). One of the primary capabilities afforded by these advances is the leveraging of insights into nervous system function, with particular attention to individual differences, so as to design systems that are consistent with "natural" patterns of information processing in the human brain. In this light, one could imagine designs that allow the presentation of information in a manner that limits the neural resources required for processing and thereby increases the speed of perception and performance by accessing, facilitating, and/or augmenting the cognitive style and abilities of an individual operator. Insights into the neural basis of performance also allow detection of real-time, moment-to-moment changes in neural activity that can be fed back into an adaptive system (Thorpe et al. 1996). Such information have been used to develop proof-of-principle systems that merge EEG classification technologies to interpret when an operator has seen a significant target within natural environments (Curran et al. 2009) with automated target recognition systems to improve overall accuracy and speed of target detection (see Gerson et al. 2006, for an early example). Current efforts are underway to further improve the measurement and classification of perceptual states, signal-to-noise ratio, and detection accuracy of EEG-based approaches (Lawhern et al. 2012; Marathe et al. 2014). Ultimately, it is envisioned that this type of technology could be integrated into operationally relevant systems where target detection is a component of complex operator tasking (see Touryan et al. 2013, for an early proof-of-concept example).

Of equal importance, insight into the neural basis of performance is leading to an ability to predict future operator capability. Recently, applications of neural decoding techniques to spatial patterns of activity measured with fMRI (Haynes and Rees 2005; Kamitani and Tong 2005) and high-resolution temporal patterns of neural activity within EEG (Giesbrecht et al. 2009) have been shown to predict performance in a dual-task target detection paradigm. Such results, when taken together with advanced neurophysiological measurement technologies, suggest the potential not only to monitor ongoing neurocognitive activity but also to use such measurements to predict possible performance failures, giving systems engineers an opportunity to design systems that can mitigate the detrimental effect(s) of such errors and thereby enhance soldier survivability and mission success. Such concepts illustrate one potential type of application that could emerge from technologies that effectively integrate operator neural activity in real-time into sociotechnical systems. These brain–computer interaction technologies are envisioned to push applications beyond human–computer interfaces and to change the very nature of how people interact with technology and their environment across a broad range of domains, including operational performance, education and training, medicine, recreation, and technology development (Lance et al. 2012). For example, brain-based technologies will allow computers, for the first time, to leverage sophisticated analyses of the emotional and cognitive states and processes of the people using them, revolutionizing the basic interactions people have, not only with the systems they use, but also with each other (see Lance et al. 2012 for recent discussion).

In summary, rapid advancements in technology coupled with the dynamic, complex nature of the national security environment create novel challenges for the materiel developer. The information-intensive environment and widely accepted approach of decision superiority and information dominance force the creation of sociotechnical systems that share the cognitive burden between personnel and the systems with which they interact. A neurocognitive engineering approach is posited to offer insights into developing such systems, from designing more effective displays to systems that adapt to the state of the operator. Any such approach must take into account traditional cognitive engineering issues such as the changing capabilities of the operator, the environments under which the systems will be used, and the different potential tasks the operator system may attempt to undertake. As neuroscience and its constituent and allied fields rapidly advance, it is expected that the neurocognitive engineering approach will advance, as well. In this way, future progress not only involves the direct employment of neurotechnology (e.g., moment-to-moment brain–computer interface; Serruya et al. 2002), but will likely be fortified by the use of nutriceuticals and pharmaceuticals that work in tandem with any such technology to enhance individual capabilities (Kosfeld et al. 2005; McCabe et al. 2001; Zak et al. 2005).

ACKNOWLEDGMENTS

This work was conducted under the U.S. Army Research Laboratory's Army Technology Objective (Research), "High-Definition Cognition in Operational Environments."

DISCLAIMER

The findings in this report are those of the authors and are not to be construed as an official Department of the Army position unless so designated by other authorized documents. Kelvin Oie and Kaleb McDowell are U.S. Government employees and this work was written as part of their official duties as employees of the U.S. Government.

REFERENCES

Ariely, D. 2003. "Controlling the information flow: effects on consumers' decision making and preferences." *Journal of Consumer Research* 27(2):233–248.

Arnell, K.M. and O. Jolicœur. 1999. "The attentional blink across stimulus modalities: Evidence for central processing limitations." *Journal of Experimental Psychology: Human Perception and Performance* 25(3):630–648.

Baird, A.A., M.K. Colvin, J.D. VanHorn, S. Inati, and M.S. Gazzaniga. 2005. "Functional connectivity: Integrating behavioral, diffusion tensor imaging, and functional magnetic resonance imaging data sets." *Journal of Cognitive Neuroscience* 17(4):687–693.

Barandiaran, X. and A. Moreno. 2006. "On what makes certain dynamical systems cognitive: A minimally cognitive organization program." *Adaptive Behavior* 14(2):171–185.

Ben-Shachar, M., R.F. Dougherty, and B.A. Wandell. 2007. "White matter pathways in reading." *Current Opinion in Neurobiology* 17(2):258–270.

Boehm-Davis, D.A., W.D., Gray, L. Adelman, S. Marshall, and R. Pozos. 2003. *Understanding and Measuring Cognitive Workload: A Coordinated Multidisciplinary Approach.* Dayton, OH: Air Force Office of Scientific Research.

Bohn, R. and J. Short. 2012. Measuring consumer information. *International Journal of Communication* 6:980–1000.

Bond, J. 1997. "The drivers of the information revolution–cost, computing power, and convergence." In *The Information Revolution and the Future of Telecommunications*, ed. J. Bond, Washington, DC: The World Bank Group, pp. 7–10.

Castermans, T., M. Duvinage, M. Petieau, T. Hoellinger, C.D. Saedeller, K. Seetharaman et al. 2011. "Optimizing the Performances of a P300-Based Brain-Computer Interface in Ambulatory Conditions." *IEEE Journal on Emerging and Selected Topics in Circuits and Systems* 1(4):566–577.

Chandrasekhar, C.P. and J. Ghosh. 2001. "Information and communication technologies and health in low income countries: The potential and the constraints." *Bulletin of the World Health Organization* 79(9):850–855.

Chein, J.M. and W. Schneider. 2005. "Neuroimaging studies of practice-related change: fMRI and Meta-analytic evidence of a domain-general control network for learning." *Cognitive Brain Research* 25(3):607–623.

Chuah, Y.M.L., V. Venkatraman, D.F. Dinges, and M.W.L. Chee. 2006. "The neural basis of interindividual variability in inhibitory efficiency after sleep deprivation." *Journal of Neuroscience* 26(27):7156–7162.

Clement, J. 1982. "Students' preconceptions in introductory mechanics." *American Journal of Physics* 50(1):66–71.

Cooke, N.J. and F.T. Durso. 2007. *Stories of Modern Technology Failures and Cognitive Engineering Successes.* Washington, DC: CRC Press.

Cooper, C. and S.E. Jackson. 1997. *Creating Tomorrow's Organizations: A Handbook for Future Research in Organizational Behavior.* Chichester: Wiley.

Cotton, A.J. 2005. *Information Technology—Information Overload for Strategic Leaders.* Carlisle Barracks, PA: U.S. Army War College.

Cowan, N. 2001. "The magical number 4 in short-term memory: A reconsideration of mental storage capacity." *The Behavioral and Brain Sciences* 24(1):87–114.

Curran, T., L. Gibson, J.H. Horne, B. Young, and A.P. Bozell. 2009. "Expert image analysts show enhanced visual processing in change detection." *Psychonomic Bulletin & Review* 16(2):390–397.

Doty, T.J., W.D. Hairston, B. Kellihan, J. Canady, K.S. Oie, and K. McDowell. 2013. "Developing a Wearable Real-World Neuroimaging System to Study Stress." *Proceedings of the 6th International IEEE/EMBS Conference on Neural Engineering (NER),* November 6–8, San Diego, CA.

Dumer, J.C., T.P. Hanratty, J. Yen, D. Widyantoro, J. Ernst, and T.J. Rogers. 2001. "Collaborative agents for an integrated battlespace." *Proceedings of the 5th World Multi-Conference on Systemics, Cybernetics, and Informatics,* Orlando, FL, July 22–25.

Dunbar, K.N., J.A. Fugelsang, and C. Stein. 2007. "Do naïve theories ever go away? Using brain and behavior to understand changes in concepts." In *Thinking with Data,* ed. M.C. Lovett and P. Shah. Hillsdale, NJ: Lawrence Erlbaum Associates, pp. 411–450.

Edelman, G.M. and J.A. Gally. 2001. "Degeneracy and Complexity in Biological Systems." *Proceedings of the National Academy of Sciences of the United States of America* 98(24):13763–13768.

Endsley, M.R., and W.M. Jones. 1997. *Situation Awareness, Information Dominance, and Information Warfare.* Dayton, OH: U.S. Air Force Armstrong Laboratory.

Fréard, D., E. Jamet, O. Le Bohec, G. Poulain, and V. Botherel. 2007. "Subjective measurement of workload related to a multimodal interaction task: NASA-TLX vs. workload profile." *Human Computer Interaction* 4552:60–69.

Fuller, J.V. 2000. *Information Overload and the Operational Commander.* Newport, RI: Joint Military Operations Department.

Garrett, D., D.A. Peterson, C.W. Anderson, and M.H. Thaut. 2003. "Comparison of linear and nonlinear methods for EEG signal classification." *IEEE Transactions on Neural Systems and Rehabilitation Engineering* 11(2):141–144.

Gentry, J.A. 2002. "Doomed to fail: America's blind faith in military technology." *Parameters* 32(4):88–103.

Gerson, A.D., L.C. Parra, and P. Sajda. 2006. "Cortically Coupled Computer Vision for Rapid Image Search." *IEEE Transactions on Neural Systems and Rehabilitation Engineering* 14(2):174–179.

Gevins, A., H. Leong, R. Du, M.E. Smith, J. Le, D. DuRousseau et al. 1995. "Towards measurement of brain function in operational environments." *Biological Psychology* 40(1/2):169–186.

Giesbrecht, B., M.P. Eckstein, and C.K. Abbey. 2009. "Neural decoding of semantic processing during the attentional blink." *Journal of Vision* 9(8):124.

Governors Highway Safety Association. "Distracted Driving Laws." 2014. www.ghsa.org/html/stateinfo/laws/cellphone_laws.html.

Gramann, K., J.T. Gwin, D.P. Ferris, K. Oie, T.-P. Jung, C.-T. Lin et al. 2011. "Cognition in action: imaging brain/body dynamics in mobile humans." *Reviews in the Neurosciences* 22(6):593–608.

Green, P. 2004. *Driver Distraction, Telematics Design, and Workload Managers: Safety Issues and Solutions*. Warrendale, PA: Society of Automotive Engineers.

Gruner, W.P. 1990. "No time for decision making." *Proceedings of the U.S. Naval Institute* 116:31–41.

Hart, S.G. and L.E. Staveland. 1988. "Development of NASA-TLX (task load index): Results of empirical and theoretical research." In *Human Mental Workload*, ed. P.A. Hancock and N. Meshkati. Amsterdam, the Netherlands: North Holland Press, pp. 239–250.

Hasson, U., R. Malach, and D.J. Heeger. 2010. "Reliability of cortical activity during natural simulation." *Trends in Cognitive Science* 14(1):40–48.

Haynes, J.D. and G. Rees. 2005. "Predicting the orientation of invisible stimuli from activity in human primary visual cortex." *Nature Neuroscience* 8(5):686–691.

Huang, R.S., T.P. Jung, and S. Makeig. 2007. "Event-related brain dynamics in continuous sustained-attention tasks." *Lecture Notes in Computer Science* 4565:65–74.

Jackson, B. 1982. "Task-evoked pupillary responses, processing load, and the structure of processing resources." *Psychological Bulletin* 91(2):276–292.

Jentsch, F., J. Barnett, C.A. Bowers, and S.E. Eduardo. 1999. "Who is flying this plane anyway? What mishaps tell us about crew member role assignment and air crew station awareness." *Human Factors* 41(1):1–14.

Jung, T.-P., S. Makeig, M.J. McKeown, A.J. Bell, T.-W. Lee, and T.J. Sejnowski. 2001. "Imaging brain dynamics using independent component analysis." *Proceedings of the IEEE* 89(7):1107–1122.

Kamitani, Y. and F. Tong. 2005. "Decoding the visual and subjective contents of the human brain." *Nature Neuroscience* 8(5):679–685.

Kantowitz, B.H. 1988. "Mental workload." In *Human Factors Psychology*, ed. P.A. Hancock. Amsterdam, the Netherlands: Elsevier, pp. 81–121.

Kerick, S.E. 2005. "Cortical activity of soldiers during shooting as a function of varied task demand." In *Foundations of Augmented Cognition*. ed. D. Schmorrow, Mahwah, NJ: Lawrence Erlbaum Associates, pp. 252–260.

Kerick, S.E., L.W. Douglass, and B.D. Hatfield. 2004. "Cerebral cortical adaptations associated with visuomotor practice." *Medicine and Science in Sports and Exercise* 36(1):118–29.

Kerick, S.E., S.E. Oie, and K. McDowell. 2009. *Assessment of EEG Signal Quality in Motion Environments*, Report No. ARL-TN-355. Aberdeen, MD: U.S. Army Research Laboratory.

Kinman, G. 1998. *Pressure Points: A Survey into the Causes and Consequences of Occupational Stress in UK Academic and Related Staff*. London: Association of University Teachers.

Kirsh, D. 2002. "A few thoughts on cognitive overload." *Intellectica* 30(1):19–51.

Körner, J. 2006. "Searching in lists while driving: Identification of factors contributing to driver workload." PhD thesis. Ludwig-Maximilians-Universität, Munich, Germany.

Kosfeld, M., M. Heinrichs, P.J. Zak, U. Fischbacher, and E. Fehr. 2005. "Oxytocin increases trust in humans." *Nature* 435(2):673–676.

Lance, B., S.E. Kerick, A.J. Ries, K.S. Oie, and K. McDowell. 2012. "Brain-computer interface technologies in the coming decades." *Proceedings of the IEEE* 100:1585–1599.

Lawhern, V., W.D. Hairston, K. McDowell, M. Westerfield, K. Robbins. 2012. "Detection and classification of subject-generated artifacts in EEG signals using autoregressive models." *Journal of Neuroscience Methods* 208(2):181–189.

Leahy, K.B. 1994. "Can computers penetrate the fog of war?" Newport, RI: U.S. Naval War College.

Libicki, M.C. 1997. "Information dominance." National Defense University Strategic Forum, No. 132. https://digitalndulibrary.ndu.edu.

Lin, C.-T., L.-I. Ko, J.-C. Chiou, J.-R. Duann, R.-S. Huang, S.-F. Liang et al. 2008. "Noninvasive neural prostheses using mobile and wireless EEG." *Proceedings of the IEEE* 96(7):1167–1183.

Lintern, G. 2006. "A functional workspace for military analysis of insurgent operations." *International Journal of Industrial Ergonomics* 36(5):409–422.

Lyman, P. and H.R. Varian. 2003. "How much information." http://www.sims.berkeley.edu/how-much-info-2003.

Makeig, S., A.J. Bell, T.-P. Jung, and T.J. Sejnowski. 1996. "Independent component analysis of electroencephalographic data." In *Advances in Neural Information Processing Systems*, ed. D.S. Touoretzky, M.C. Mozer, and M.E. Hasselmo. Cambridge, MA: MIT Press, pp. 145–151.

Marathe, A.R., A.J. Ries, and K. McDowell. 2014. "Sliding HDCS: Single-trial EEG classification to overcome and quantify temporal variability." *IEEE Transactions on Neural Systems and Rehabilitation Engineering* 22(2):201–211.

Marois, R. and J. Ivanov. 2005. "Capacity limits of information processing in the brain." *Trends in Cognitive Sciences* 9(6):296–305.

Matthews, R., N.J. McDonald, H. Anumula, J. Woodward, P.J. Turner, M.A. Steindorf et al. 2007. "Novel hybrid bioelectrodes for ambulatory zero-prep EEG measurements using multi-channel wireless EEG system." *Lecture Notes in Computer Science* 4565: 137–146.

Matthews, R., P.J. Turner, N.J. McDonald, J. Ermolaev, T. McManus, R.A. Shelby et al. 2008. "Real time workload classification from an ambulatory wireless EEG system using hybrid EEG electrodes." *Proceedings of the 30th Annual International IEEE Engineering in Medicine and Biology Society Conference*, Vancouver, BC, August 20–24.

McCabe, K., D. Houser, L. Ryan, V. Smith, and T. Trouard. 2001. "A functional imaging study of cooperation in two-person reciprocal exchange." *Proceedings of the National Academies of Science* 98(20):11832–11835.

McDowell, K., K.S. Oie, B.T. Crabb, V. Paul, and T.T. Brunye. 2009. "The need for cognitive engineering in the United States army." *Insight* 12(1): 7–10.

McDowell, K., K.S. Oie, T.M. Tierney, and O.M. Flascher. 2007. "Addressing human factors issues for future manned ground vehicles (MGVs)." *Army Acquisition, Logistics and Technology Magazine*, January–March, pp. 20–23.

Mestre, J.P. 1991. "Learning and instruction in pre-college physical science." *Physics Today* 44(9): 56–62.

Miller, G. 1956. "The magical number seven plus or minus two: Some limits on our capacity for processing information." *Psychological Review* 63(2):343–352.

Miller, M.B. and J.D. Van Horn. 2007. "Individual variability in brain activations associated with episodic retrieval: A role for large-scale databases." *International Journal of Psychophysiology* 63(2):205–213.

Miller, M.B., J.D. Van Horn, G.L. Wolford, T.C. Handy, M. Valsangkar-Smyth, S. Inati et al. 2002. "Extensive individual differences in brain activations associated with episodic retrieval are reliable over time." *Journal of Cognitive Neuroscience* 14(8):1200–1214.

Misra, S. and D. Stoklos. 2012. "Psychological and health outcomes of perceived information overload." *Environment and Behavior* 44(6):737–759.

Monsell, S. 2003. "Task switching." *Trends in Cognitive Sciences* 7(3):134–140.

Moravec, H. 1998. "When will computer hardware match the human brain?" *Journal of Evolution and Technology* 1(1):1–12.

National Highway Traffic Safety Administration, Department of Transportation (U.S.). "Visual-Manual NHTSA Driver Distraction Guidelines: Portable and Aftermarket Electronic Devices" 2014. NHTSA-2013-0137-0001. www.regulations.gov.

National Research Council, Committee on Human Factors (U.S.). 1994. *Emerging Needs and Opportunities for Human Factors Research*. Washington, DC: National Academies Press.

National Research Council, Committee on Opportunities in Neuroscience for Future Army Applications (U.S.). 2009. *Opportunities in Neuroscience for Future Army Applications*. Washington, DC: National Academies Press.

National Science and Technology Council (U.S.). 1994. *Fact Booklet*. Office of Science and Technology Policy, The White House Washington, DC.

Nordhaus, W.D. 2001. "The progress of computing." Yale Cowles Foundation for Research in Economics, Research Paper Series, Discussion Paper No. 1324.

Nordhaus, W.D. 2007. "Two centuries of productivity growth in computing." *Journal of Economic History* 67(1):128–159.

Office of the Chairman of the Joint Chiefs of Staff (U.S.). 2004. *The National Military Strategy of the United States of America: A Strategy for Today; A Vision for Tomorrow (Unclassified Version)*. Washington, DC: United States Department of Defense.

Oie, K.S., S. Gordon, and K. McDowell. 2012. "The multi-aspect measurement approach: Rational, technologies, tools, and challenges for systems design." In *Designing Soldier Systems: Current Issues in Human Factors*, eds. P. Savage-Knepshield, J. Martin, J. Lockett III, and L. Allender, pp. 217–248. Burlington, VT: Ashgate.

Poldrack, R.A., F.W. Sabb, K. Foerde, S.M. Tom, R.F. Asarnow, S.Y. Bookheimer et al. 2005. "The neural correlates of motor skill automaticity." *The Journal of Neuroscience* 25(22):5356–5364.

Ramsey, N.F., J.M. Jansma, G. Jager, T.R. van Raalten, and R.S. Kahn. 2004. "Neurophysiological factors in human information processing capacity." *Brain* 127(3):517–525.

Rubio, S., E. Diaz, J. Martin, and J.M. Puente. 2004. "Evaluation of subjective mental workload: A comparison of SWAT, NASA-TLX, and workload profile methods." *Applied Psychology* 53(1):61–86.

Sanders, D.M. and W.B. Carlton. 2002. Information Overload at the Tactical Level: An Application of Agent Based Modeling and Complexity Theory in Combat Modeling. West Point, NY: United States Military Academy.

Sarter, M., W.J. Gehring, and R. Kozak. 2006. "More attention must be paid: The neurobiology of attentional effort." *Brain Research Reviews* 51(2):145–160.

Scallen, S.F., P.A. Hancock, and J.A. Duley. 1995. "Pilot performance and preference for short cycles of automation in adaptive function allocation." *Applied Ergonomics* 26(6):387–403.

Seitz, A.R., R. Kim, and L. Shams. 2006. "Sound facilitates visual learning." *Current Biology* 16(14):1422–1427.

Seitz, A.R., R. Kim, V. van Wassenhove, and L. Shams. 2007. Simultaneous and independent acquisition of multisensory and unisensory associations. *Perception* 36(10):1445–1453.

Sellers, E.W., P. Turner, W.A. Sarnacki, T. McManus, T.M. Vaughn, and R. Matthews. 2009. "A novel dry electrode for brain-computer interface." *Lecture Notes in Computer Science* 5611:623–631.

Serruya, M.D., N.G. Hatsopoulos, L. Paninski, M.R. Fellows, and J.P. Donoghue. 2002. "Brain-machine interface: Instant neural control of a movement signal." *Nature* 416:141–142.

Shams, L., Y. Kamitani, and S. Shimojo. 2002. "Visual illusion induced by sound." *Cognitive Brain Research* 14(1):147–152.

Shiffrin, R.M. 1976. "Capacity limitations in information processing, attention, and memory."
    In *Handbook of Learning and Cognitive Processes/Vol. 4, Attention and Memory*, ed.
    W.K. Estes. Mahwah, NJ: Lawrence Erlbaum Associates, pp. 177–236.
Speier, C., J.S. Valacich, and I. Vessey. 1999. "The influence of task interruption on individual
    decision making: An information overload perspective." *Decision Sciences* 30(2):337–360.
Svensson, E., M. Angelborg-Thanderz, L. Sjöberg, and S. Olsson. 1997. "Information
    complexity—mental workload and performance in combat aircraft." *Ergonomics*
    40(3):362–380.
Thorpe, S., D. Fize, and C. Marlot. 1996. "Speed of processing in the human visual system."
    Nature 381(6):520–522.
Tisserand, D.J., M.P.J. van Boxtel, J.C. Pruessner, P. Hofman, A.C. Evans, J. Jolles. 2004.
    "A voxel-based morphometric study to determine individual differences in gray matter den-
    sity associated with age and cognitive change over time." *Cerebral Cortex* 14(9):966–973.
Touryan, J., A.J., Ries, P. Weber, and L. Gibson. 2013. "Integration of automated neural pro-
    cessing into an army-relevant multitasking simulation environment." In *Foundations of
    Augmented Cognition*, eds. D.D. Schmorrow and C.M. Fidopiastis, pp. 774–782. Berlin;
    Heidelberg, Germany: Springer.
Tsang, P.S. and V.L. Velazquez. 1996. "Diagnosticity and multidimensional subjective work-
    load ratings." *Ergonomics* 39(3):358–381.
Van Praag, H., G. Kempermann, and F.H. Gage. 2000. "Neural consequences of environmental
    enrichment." *Nature Reviews Neuroscience* 1(3):191–198.
Veltman, H. and A.W.K. Gaillard. 1996. "Physiological indices of workload in a simulated
    flight task." *Biological Psychology* 42(3):323–342.
Verwey, W.B. and H.A. Veltman. 1996. "Detecting short periods of elevated workload: A com-
    parison of nine workload assessment techniques." *Journal of Experimental Psychology:
    Applied* 2(3):270–285.
Vettel, J.M., P.J. McKee, A. Dagro, M. Vindiola, A. Yu, K. McDowell, and P. Franaszczuk.
    2014. Scientific accomplishments for ARL brain structure-function couplings research
    on large-scale brain networks from FY11-FY13. *DSI Final Report* ARL-TR-6871.
    Aberdeen Proving Ground, MD: U.S. Army Research Laboratory.
Vogel, E.K., G.F. Woodman, and S.J. Luck. 2001. "Storage of features, conjunction and objects
    in visual working memory." *Journal of Experimental Psychology: Human Perception
    and Performance* 27(1):92–114.
Walrath, J.D. 2005. *Information Technology for the Soldier: The Human Factor*. Aberdeen,
    MD: U.S. Army Research Laboratory.
Wiener, E.L. and R.E. Curry. 1980. "Flight-deck automation: Promises and problems."
    Ergonomics 23(10):995–1011.
Winters, J. and J. Giffin. 1997. Issue paper: Information dominance vs. information superi-
    ority. Fort Monroe, VA: Information Operations Division, U.S. Training and Doctrine
    Command. http://www.iwar.org.uk/iwar/resources/info-dominance/issue-paper.htm.
Wood, C., K. Torkkola, and S. Kundalkar. "Using driver's speech to detect cognitive work-
    load." *Proceedings of the 9th Conference of Speech and Computer*, September 20–22,
    St. Petersburg, Russia.
Wu, C. and Y. Liu. 2007. "Queuing network model of driver workload and performance."
    IEEE Transactions on Intelligent Transportation Systems 8(3):528–537.
Zak, P.J. 2004. "Neuroeconomics." *Philosophical Transactions of the Royal Society of London,
    Series B: Biological Sciences* 359(1451):1737–1748.
Zak, P.J., R. Kurzban, and W.T. Matzner. 2005. "Oxytocin is associated with human trustwor-
    thiness." *Hormones and Behavior* 48(5):522–527.
Zimet, E. and C.L. Barry. 2009. "Military service overview." In *Cyberpower and National
    Security,* eds. F.D. Karmer, S.H. Starr, and L.K. Wentz, pp. 285–308. Washington, DC:
    National Defense University Press.

5 Neural Mechanisms as Putative Targets for Warfighter Resilience and Optimal Performance

Martin P. Paulus, Lori Haase, Douglas C. Johnson, Alan N. Simmons, Eric G. Potterat, Karl Van Orden, and Judith L. Swain

CONTENTS

EXPOSURE TO EXTREME ENVIRONMENTS, PERFORMANCE, AND BRAIN FUNCTIONING

Military deployment and, in particular, combat deployment (Smith et al. 2008) can have profound effects on mental health (Thomas et al. 2010) and lead to increased prevalence rates of psychiatric disorders (Larson et al. 2008) as well as alcohol and substance use (Seal et al. 2011); subsequently, these effects negatively impact social functioning and employability and result in increased utilization of health care services (Hoge et al. 2006). However, much less understood and studied are the effects of recovery from deployment stress as they relate to soldier's performance. Others have stressed that there is a clear need for a quantitative approach to assess cognitive functioning of soldiers exposed to high-stress environments, to provide early detection of individual health and military performance impairments and management of occupational and deployment health risks (Friedl et al. 2007). The need for an objective assessment of performance is underscored by a recent study which found that self-reported cognitive functioning was not correlated with objective cognitive abilities. Instead, perceived cognitive deficits were associated with depression, anxiety, and posttraumatic stress disorder (PTSD; Spencer et al. 2010). Moreover, longitudinal work is required to examine the trajectory of performance changes and recovery (Vasterling et al. 2009).

There is some evidence that military deployment may compromise performance on tasks of sustained attention, verbal learning, and visual-spatial memory, whereas it enhances performance on simple reaction time (Vasterling et al. 2006). Others have found that following acute stress of combat simulation, soldiers showed a shift in attention away from threat (Wald et al. 2010). The assessment of neurocognitive performances is particularly important because there is evidence suggesting that performance decrements are associated with the development of postdeployment PTSD symptoms (Marx et al. 2009), which is consistent with the idea that domains of cognitive function may serve as risk or protective factors for PTSD (Gilbertson et al. 2006). The effect of combat exposure on brain function is still very much under study. While some have suggested that no consistent findings have emerged across advanced brain imaging techniques (Zhang et al. 2010). More recently, there is converging evidence to suggest involvement of the frontal and limbic systems. Using both electroencephalography (EEG) and (structural) magnetic resonance imaging (MRI), alteration of EEG phase synchrony was associated with the changes in structural integrity of white matter tracts of the frontal lobe (Sponheim et al. 2011). Moreover, there is evidence that combat stress increases amygdala and insula reactivity to biologically salient stimuli, which has been taken as evidence that threat appraisal affects interoceptive awareness and amygdala regulation (van Wingen et al. 2011). Research has also found that combat exposure affects the balance between activation in ventral frontolimbic regions, which are important for processing emotional information, and dorsolateral prefrontal regions, which are important for executive functioning (Morey et al. 2008). Additionally, increased activation to combat-related stimuli has been reported in the visual cortex (Hendler et al. 2003). Taken together, these and other studies show that cognitive and brain function is significantly affected by deployment. However, what is much less clear is how cognitive and brain function recovers as soldiers return home from combat deployment.

RESILIENCE

Resilience is a complex and possibly multidimensional construct (Davidson 2000). It includes trait variables such as temperament and personality, as well as cognitive functions such as problem solving that may work together for an individual to adequately cope with traumatic events (Campbell-Sills et al. 2006). Here, we focus on resilience in terms of a process through which individuals successfully cope with (and bounce back from) stress. For instance, after being fired from a job, a resilient individual adopts a proactive style in improving his job hunting and work performance. In contrast, a less resilient individual may adopt a simple recovery from insult where job loss causes a period of initial depressive mood followed by a return to affective baseline without an attempt to modify habitual coping mechanisms. Further studies are needed to show that resilience, which is a critical characteristic of optimal performance in extreme environments, has significant effects on brain structures that are thought to be important for such performance. Our approach has been to use tasks of emotion face assessment that we have previously shown to be sensitive to levels of trait anxiety (Stein et al. 2007), can be modulated by antianxiety drugs (Paulus et al. 2005), and are well known to be sensitive to genetic differences across individuals (Hariri et al. 2002). As elaborated below, we hypothesize that limbic and paralimbic structures play an important role in helping individuals adjust to extreme conditions. Thus, we hypothesize that the level of resilience critically modulates activation of the amygdala and insular cortex. In particular, if the anterior insula plays an important role in helping to predict perturbations in the internal body state, one would hypothesize that greater activation in this structure is associated with better resilience. Moreover, if one assumes that the amygdala is important in assessing salience in general, and the potential of an aversive impact in particular, one would hypothesize that greater resilience is associated with relatively less activation in the amygdala during emotion face processing.

INTEROCEPTION AND ITS NEURAL SUBSTRATES: A NOVEL APPROACH TO RESILIENCE AND STRESS ASSESSMENT

Interoception comprises receiving, processing, and integrating body-relevant signals together with external stimuli to affect motivated behavior (Craig 2002, 2009). Different conceptualizations of interoception have included its definition as the state of the individual at a particular point in time (Craig 2010), or as the sensing of body-related information in terms of awareness (Pollatos et al. 2005), sensitivity (Holzl et al. 1996), or accuracy of the sensing process (Vaitl 1996). Interoception provides an anatomical framework for identifying pathways focused on modulating the internal state of the individual. This framework comprises peripheral receptors (Vaitl 1996), c-fiber afferents, spinothalamic projections, specific thalamic nuclei, posterior and anterior insula as the limbic sensory cortex, and anterior cingulate cortex (ACC) as the limbic motor cortex (Augustine 1996; Craig 2007). Central to the concept of interoception is that body-state relevant signals comprise a rich and highly organized source of information that affects how an individual engages in motivated behavior. Importantly, interoception is linked to homeostasis (Craig 2003), which implies that an individual's motivated approach or avoidance behavior toward stimuli and resources in the outside world is aimed at maintaining an equilibrium. For example, a person will approach a heat source in a cold environment but avoid it when the ambient temperature is high.

Interoception is an important process for optimal performance because it links the perturbation of internal state, as a result of external demands, to goal-directed action that maintain a homeostatic balance (Paulus et al. 2009). In particular, the interoceptive system provides information about the internal state to neural systems that monitor value and salience and are critical for cognitive control processes. We recently proposed that maintaining an interoceptive balance, by generating body prediction errors in the presence of significant perturbations, may be a neural marker of optimal performance (Paulus et al. 2009). This notion is consistent with findings that elite athletes pay close attention to bodily signals (Philippe and Seiler 2005) and may be particularly adept in generating anticipatory prediction errors (Aglioti et al. 2008). Moreover, Tucker (2009) has proposed that individuals regulate performance via perceived exertion through a "teleoanticipation" process, which is the combination of afferent and efferent brain processes that attempt to effectively adjust between the metabolic and biomechanical limits of the body and the demands of the exercise task (Hampson et al. 2001).

The insular cortex is a complex brain structure, which is organized macroscopically along an anterior–posterior (Craig 2002) and superior–inferior axis (Kurth et al. 2010) and microscopically as granular, dysgranular, and agranular from posterior to anterior insula, respectively (Chikama et al. 1997; Shipp 2005). The anterior cluster is predominately activated by effortful cognitive processing, whereas the posterior region is mostly activated by interoception, perception, and emotion (Cauda et al. 2012). Moreover, the anterior insula, potentially together with the ACC, appears to pivotally influence the dynamics between default mode and executive control networks (Sutherland et al. 2012). The insula is thought to be the central nervous system hub for interoceptive processing, such that somatosensory relevant afferents enter the posterior insula and are integrated with the internal state in the mid-insula, and re-represented as a complex feeling state within the anterior insular cortex. Although there has been some debate, a recent meta-analysis suggests that the anterior insula is critical and necessary for emotional awareness (Gu et al. 2013).

Functional neuroimaging studies using resting state, task-related, and structural connectivity measures have shown that individual brain structures are organized in functional networks (Bellec et al. 2006). Therefore, interoception cannot be reduced to simply a function of the insula. In particular, functional neuroimaging studies have delineated a medial default mode network, a frontal control network, and a limbic salience network (Spreng et al. 2013). Depending on the approach, the insular cortex has been divided into two (Taylor et al. 2009) or three (Deen et al. 2011) compartments, which may serve different functions in these large-scale networks. Specifically, the dorso anterior insula is consistently associated with the frontal control network (Chang et al. 2012), whereas others have suggested that the anterior insula is critical for the saliency network and is functionally connected with frontal, cingulate, parietal, and cerebellar brain areas (Sullivan et al. 2013). In comparison, the posterior insula is closely connected to sensorimotor, temporal, and posterior cingulate areas (Cauda et al. 2012). Some have proposed that the right frontoinsular cortex together with the ACC plays a causal role in switching between the frontal control network and the default mode network (Sridharan et al. 2008) and is involved in switching during a variety of perceptual, memory, and problem-solving tasks (Tang et al. 2012). Consistent with this notion is the observation that the anterior insula is involved in the processing of temporal predictions (Limongi et al. 2013) as well as the influence of self-regulation on functional connectivity (Haller et al. 2013). These connectivity patterns suggest that the anterior insula is important for translating emotional salience into activation of the cognitive control network to implement goal-directed behavior (Cloutman et al. 2012).

The ACC has been labeled the limbic motor cortex by some (Craig 2007) (Allman et al. 2001) and is thought to be the critical interface between cognitive and emotion processing (Vogt et al. 2003). In particular, Von Economo neurons, which are projection neurons located in layer V within the ACC and frontoinsular cortex, have been implicated in integrative function of the ACC (Butti et al. 2013). However, whether different parts of the ACC are involved in distinct processes and whether these processes are segregated for different functions is still highly debated. On the one hand, several investigators have proposed an anatomically based topography of the ACC consisting of subgenual, pregenual, and anterior midcingulate cortices, which are cytoarchitecturally distinct and have different connectivity with other brain structures (Vogt 2005). In particular, whereas the rostral ACC comprising both subgenual and pregenual ACC is important for emotional processing, the dorsal or midcingulate cortex is thought to implement cognitive control and emotion regulation (Mohanty et al. 2007). However, there is also considerable overlap between the "cognitive" division of the ACC and the midcingulate area that processes pain and fear (Vogt 2005). This overlap is consistent with the idea that dorsal–caudal regions of the ACC and medial prefrontal cortex are involved in appraisal and expression of negative emotion (Etkin et al. 2011). On the other hand, based on a meta-analysis of imaging studies, some investigators have proposed that negative affect, pain, and cognitive control activate an overlapping region within the anterior midcingulate cortex, which can be thought of as a hub that links reinforcers to motor centers responsible for expressing affect and executing goal-directed behavior (Shackman et al. 2011). In particular, it has been proposed that the ACC supports the selection and maintenance of extended, context-specific sequences of behavior directed toward particular goals that are learned through a process of hierarchical reinforcement learning (Holroyd and Yeung 2012). This generalized view of ACC functioning is consistent with the proposal that this structure, among other functions, orchestrates approach or avoidance behaviors in response to particular internal body states that involve homeostatic perturbations (Weston 2012). This function of the ACC is supported by the strong functional (Taylor et al. 2009) and anatomical (Ongur and Price 2000) connections between the anterior insula and the ACC. This systemic view is also aligned with a prediction error-based conceptualization of the specific computational processes that may be carried out within this structure. Taken together, the ACC receives information about the individual's current state, the expected state, and computes various types of error signals that help to establish the selection of an action which is optimally adapted to the higher order goal state.

OPTIMAL PERFORMANCE: A LINK TO EFFICIENT INTEROCEPTIVE PROCESSING

We have recently proposed a neuroanatomical processing model as a heuristic guide to understand how interoceptive processing may contribute to optimal performance. In this model, we propose that optimal performers generate more efficient body prediction errors, that is, the difference between the value of the anticipated/predicted and the value of the current interoceptive state, as a way of adapting to extreme environments. First, information from peripheral receptors ascends via two different pathways: (1) the A-beta-fiber discriminative pathway that conveys precise information about the "what" and "where" of the stimulus impinging on the body, and (2) the C-fiber pathway that conveys spatially and time-integrated affective information (Craig 2007). These afferents converge via several way stations to the sensory cortex and the posterior insular cortex to provide a sense of the current body state.

Second, centrally generated interoceptive states, for example, via contextual associations from memory, reach the insular cortex via temporal and parietal cortex to generate body states based on conditioned associations (Gray and Critchley 2007). Third, within the insular cortex, there is a dorsal-posterior to inferior-anterior organization from granular to agranular, which provides an increasingly "contextualized" representation of the interoceptive states (Shipp 2005), irrespective of whether it is generated internally or via the periphery. These interoceptive states are made available to the orbitofrontal cortex for context-dependent valuation (Kringelbach 2005) and to the ACC for error processing (Carter et al. 1999) and action valuation (Rushworth and Behrens 2008). Fourth, bidirectional connections to the basolateral amygdala (Reynolds and Zahm 2005) and the striatum (Chikama et al. 1997), particularly ventral striatum (Fudge et al. 2005), provide the circuitry to calculate a body prediction error (similar to reward prediction error [Schultz and Dickinson 2000]) and provide a neural signal for salience and learning.

However, body prediction error differences may occur on several levels. For example, optimal performers may receive different afferent information via the C-fiber pathway that conveys spatially and time-integrated affective information (Craig 2007). Alternatively, optimal performers may generate centrally different interoceptive states (e.g., via contextual associations from memory), which are processed in the insular cortex via connections to temporal and parietal cortex to generate body states based on conditioned associations (Gray and Critchley 2007). Consistent with this idea, Williamson and colleagues (2002, 2006) suggest that the neural circuitry underlying central regulation of performance includes the insular and anterior cingulate cortex that interact with thalamic and brainstem structures which are important for cardiovascular integration (Williamson et al. 2006) as well as for the central modulation of cardiovascular responses (Williamson et al. 2002). Optimal performers may also differentially integrate interoceptive states within the insular cortex (which shows a clear gradient from the dorsal-posterior to ventral-anterior part) to provide an increasingly "contextualized" representation of the interoceptive state (Shipp 2005). This integration may occur irrespective of whether it is generated internally or via the periphery. The relative increase in activation in the mid-insula in adventure racers prior to experiencing a breathing load and the relatively attenuated activation after the load experience support the notion that the aversive interoceptive experience is less disruptive to these elite athletes compared to control subjects and may lead to relatively fewer changes in the subjective response to this stressor. Finally, optimal performers may generate different context-dependent valuation of the interoceptive states within the orbitofrontal cortex (Rolls 2004) leading to altered error processing in the anterior cingulate (Carter et al. 1998) and selection of different actions (Rushworth and Behrens 2008). The findings that both the ventral anterior cingulate and left anterior insula responses are important for the subjective effects of the breathing load support the notion that optimal performers may show different integration of aversive interoceptive stimuli. These results are consistent with those of Hilty and colleagues (2010) who reported that individuals who perform a handgrip exercise prior to task failure show increased activation in both the mid/anterior insular cortex and the thalamus. Thus greater activation and possibly a larger body prediction error might predict suboptimal performance.

The prevalent model of performance in sports physiology is the central governor model, which centers on perceived exertion (Borg and Dahlstrom 1962) (the subjective perception of exercise intensity) and has been used to explain performance differences in athletes (St. Clair and Noakes 2004). Recently, this model has been extended by Tucker and colleagues (2009) based on prior formulations by Hampson et al. (2001). Specifically, a system of simultaneous efferent feedforward and afferent feedback signals is thought to optimize performance by overcoming fatigue through permitting continuous compensation for unexpected peripheral events (Hampson et al. 2001). Afferent information from various physiological systems and external or environmental cues at the onset of exercise can be used to forecast the duration of exercise within homeostatic regulatory limits. This enables individuals to terminate the exercise when the maximum tolerable perceived exertion is attained. In this model, the brain creates a dynamic representation of an expected exertion against which the experienced exertion can be continuously compared (Tucker 2009) to prevent exertion from exceeding acceptable levels. The notion of a differential between expected and experienced exertion parallels our model of the body prediction error (Paulus et al. 2009). However, the degree to which peripheral input is necessary is still under debate. For example, Marcora and colleagues (2009) have developed a psychobiological model which proposes that perceived exertion is generated by a top-down or feedforward signal (Marcora 2008), that is, the brain—not the body—generates the sense of exertion. These investigators have argued that a centrally generated corollary discharge of the brain is critical for optimal effort (Marcora 2010) and that mental fatigue affects performance via altered perception of effort rather than afferent and body originating cardiorespiratory and musculoenergetic mechanisms (Marcora et al. 2009). Nevertheless, whether it is a purely central process, as suggested by Marcora (2008), or an interaction between afferent peripheral feedback and efferent central feedforward systems, the differential between the expected and observed, that is, the body prediction error, is the critical variable that moderates performance. The implementation of this process in the brain and its modulation by nature or nurture will be central to understand optimal performance.

REFRAMING RESILIENCE

Resilient individuals are able to generate positive emotions to help them cope with extreme situations (Tugade et al. 2004). According to Tugade and Fredrickson's (2004) broaden-and-build theory, positive emotions facilitate enduring personal resources and broaden one's momentary thought of action repertoire. That is, positive emotions broaden one's awareness and encourage novel, varied, and exploratory thoughts and actions, which, in turn, build skills and resources. For example, experiencing a pleasant interaction with a person you asked for directions turns, over time, into a supportive friendship. Furthermore, positive emotions help resilient individuals to achieve effective coping, serving to moderate stress reactivity and mediate stress recovery (Southwick et al. 2005). We suggest individuals that score high on self-reported resilience may be more likely to engage the insular cortex when processing salient information and are able to generate a body prediction error that enables them to adjust more quickly to different external demand characteristics.

In turn, a more adaptive adjustment is thought to result in a more positive view of the world and that this capacity helps maintain their homeostasis. This positive bias during emotion perception may provide the rose-colored glasses that resilient individuals use to interpret the world and achieve effective ways to bounce back from adversity and maintain wellness.

In the model of optimal performance in extreme environments, we started with the notion that these environments exert profound interoceptive effects, which are processed via the interoceptive system described above. The interoceptive system provides this information to systems that (1) monitor value and salience (orbitofrontal cortex and amygdala); (2) are important for evaluating reward (ventral striatum/ extended amygdala); and (3) are critical for cognitive control processes (anterior cingulate). Moreover, the more anterior the representation of the interoceptive state within the insular cortex, the more "textured," multimodal, and complex is the information that is being processed due to the diverse cortical afferents to the mid- and anterior insula. We hypothesized that the anterior insula, possibly in conjunction with the anterior cingulate, not only receives interoceptive information but also is able to generate a predictive model, which we have termed the body prediction error, which provides the individual with a signal of how the body will feel, similar to the *as if* loop in the Damasio (1994) somatic marker model. In this formulation, Damasio's theory extends the James Lang theory of emotion (Lang 1994) because the insula can instantiate body sensation without necessarily receiving peripheral inputs. The interoceptive information is thus "contextualized," that is, brought in relation to other ongoing affective, cognitive, or experiential processes, in relation to the homeostatic state of the individual, and is used to initiate new or modify ongoing actions aimed at maintaining the individual's homeostatic state. In this fashion interoceptive stimuli can generate an urge to act.

The neuroscience approach to understanding optimal performance in extreme environments has several advantages over traditional descriptive approaches. First, once the role of specific neural substrates is identified, they can be targeted for interventions. Second, studies of specific neural substrates involved in performance in extreme environments can be used to determine what cognitive and affective processes are important for modulating optimal performance. Third, quantitative assessment of the contribution of different neural systems to performance in extreme environments could be used as indicators of training status or preparedness. The observation that levels of resilience modulate the insular cortex and amygdala is a first step in bringing neuroscience approaches to a better understanding of what makes individuals perform differently when exposed to extreme environments and how to build resilience.

REFERENCES

Aglioti, S.M., P. Cesari, M. Romani, and C. Urgesi. 2008. "Action anticipation and motor resonance in elite basketball players." *Nature Neuroscience* 11(9):1109–1116.

Allman, J.M., A. Hakeem, J.M. Erwin, E. Nimchinsky, and P. Hof. 2001. "The anterior cingulate cortex. The evolution of an interface between emotion and cognition." *Annals of the New York Academy of Sciences* 935:107–117.

Augustine, J.R. 1996. "Circuitry and functional aspects of the insular lobe in primates including humans." *Brain Research: Brain Research Reviews* 22(3):229–244.

Bellec, P., V. Perlbarg, S. Jbabdi, M. Pelegrini-Issac, J.L. Anton, J. Doyon, and H. Benali. 2006. "Identification of large-scale networks in the brain using fMRI." *NeuroImage* 29(4):1231–1243.

Borg, G. and H. Dahlstrom. 1962. "A case study of perceived exertion during a work test." *Acta Societatis Medicorum Upsaliensis* 67:91–93.

Butti, C., M. Santos, N. Uppal, and P.R. Hof. 2013. "Von economo neurons: Clinical and evolutionary perspectives." *Cortex* 49(1):312–326. doi:10.1016/j.cortex.2011.10.004.

Campbell-Sills, L., S.L. Cohan, and M.B. Stein. 2006. "Relationship of resilience to personality, coping, and psychiatric symptoms in young adults." *Behaviour Research and Therapy* 44(4):585–599.

Carter, C.S., M.M. Botvinick, and J.D. Cohen. 1999. "The contribution of the anterior cingulate cortex to executive processes in cognition." *Nature Reviews Neuroscience* 10(1):49–57.

Carter, C.S., T.S. Braver, D.M. Barch, M.M. Botvinick, D. Noll, and J.D. Cohen. 1998. "Anterior cingulate cortex, error detection, and the online monitoring of performance." *Science* 280(5364):747–749.

Cauda, F., T. Costa, D.M. Torta, K. Sacco, F. D'Agata, S. Duca et al. 2012. "Meta-analytic clustering of the insular cortex: characterizing the meta-analytic connectivity of the insula when involved in active tasks." *NeuroImage* 62(1):343–355. doi:10.1016/j.neuroimage.2012.04.012.

Chang, L.J., T. Yarkoni, M.W. Khaw, and A.G. Sanfey. 2012. "Decoding the role of the insula in human cognition: functional parcellation and large-scale reverse inference." *Cerebral Cortex* 23(3):739–749. doi:10.1093/cercor/bhs065.

Chikama, M., N.R. McFarland, D.G. Amaral, and S.N. Haber. 1997. "Insular cortical projections to functional regions of the striatum correlate with cortical cytoarchitectonic organization in the primate." *Journal of Neuroscience* 17(24):9686–9705.

Cloutman, L.L., R.J. Binney, M. Drakesmith, G.J. Parker, and M.A. Lambon Ralph. 2012. "The variation of function across the human insula mirrors its patterns of structural connectivity: Evidence from in vivo probabilistic tractography." *NeuroImage* 59(4): 3514–3521. doi:10.1016/j.neuroimage.2011.11.016.

Craig, A.D. 2002. "How do you feel? Interoception: The sense of the physiological condition of the body." *Nature Reviews Neuroscience* 3(8):655–666.

Craig, A.D. 2003. "Interoception: The sense of the physiological condition of the body." *Current Opinion in Neurobiology* 13(4):500–505.

Craig, A.D. 2007. "Interoception and emotion: a neuroanatomical perspective." In *Handbook of Emotions*, Vol. 3, eds. M. Lewis, J.M. Haviland-Jones, and L. F. Barrett. New York: Guilford Press, pp. 272–290.

Craig, A.D. 2009. "How do you feel—Now? The anterior insula and human awareness." *Nature Reviews Neuroscience* 10(1):59–70.

Craig, A.D. 2010. "The sentient self." *Brain Structure & Function* 214(5/6):563–577. doi:10.1007/s00429-010-0248-y.

Damasio, A.R. 1994. "Descartes' error and the future of human life." *Scientific American* 271:144.

Davidson, R.J. 2000. "Affective style, psychopathology, and resilience: Brain mechanisms and plasticity." *American Psychologist* 55(11):1196–1214.

Deen, B., N.B. Pitskel, and K.A. Pelphrey. 2011. "Three systems of insular functional connectivity identified with cluster analysis." *Cerebral Cortex* 21(7):1498–1506. doi:10.1093/cercor/bhq186.

Etkin, A., T. Egner, and R. Kalisch. 2011. "Emotional processing in anterior cingulate and medial prefrontal cortex." *Trends in Cognitive Science* 15(2):85–93. doi:10.1016/j.tics.2010.11.004.

Friedl, K.E., S.J. Grate, S.P. Proctor, J.W. Ness, B.J. Lukey, and R.L. Kane. 2007. "Army research needs for automated neuropsychological tests: Monitoring soldier health and performance status." *Archives of Clinical Neuropsychology* 22(Suppl 1):S7–S14. doi:10.1016/j.acn.2006.10.002.

Fudge, J.L., M.A. Breitbart, M. Danish, and V. Pannoni. 2005. "Insular and gustatory inputs to the caudal ventral striatum in primates." *Journal of Comparitive Neurology* 490(2):101–118.

Gilbertson, M.W., L.A. Paulus, S.K. Williston, T.V. Gurvits, N.B. Lasko, R.K. Pitman, and S.P. Orr. 2006. "Neurocognitive function in monozygotic twins discordant for combat exposure: Relationship to posttraumatic stress disorder." *Journal of Abnormal Psychology* 115(3):484–495.

Gray, M.A. and H.D. Critchley. 2007. "Interoceptive basis to craving." *Neuron* 54(2):183–186.

Gu, X., P.R. Hof, K.J. Friston, and J. Fan. 2013. "Anterior insular cortex and emotional awareness." *Journal of Comparative Neurology*. doi:10.1002/cne.23368.

Haller, S., R. Kopel, P. Jhooti, T. Haas, F. Scharnowski, K.O. Lovblad et al. 2013. "Dynamic reconfiguration of human brain functional networks through neurofeedback." *NeuroImage* 81:243–252. doi:10.1016/j.neuroimage.2013.05.019.

Hampson, D.B., A. St Clair Gibson, M.I. Lambert, and T.D. Noakes. 2001. "The influence of sensory cues on the perception of exertion during exercise and central regulation of exercise performance." *Sports Medicine* 31(13):935–952.

Hariri, A.R., V.S. Mattay, A. Tessitore, B. Kolachana, F. Fera, D. Goldman et al. 2002. "Serotonin transporter genetic variation and the response of the human amygdala." *Science* 297:400–403.

Hendler, T., P. Rotshtein, Y. Yeshurun, T. Weizmann, I. Kahn, D. Ben Bashat et al. 2003. "Sensing the invisible: Differential sensitivity of visual cortex and amygdala to traumatic context." *NeuroImage* 19(3):587–600.

Hilty, L., L. Jancke, R. Luechinger, U. Boutellier, and K. Lutz. 2010. "Limitation of physical performance in a muscle fatiguing handgrip exercise is mediated by thalamo-insular activity." *Human Brain Mapping*. doi:10.1002/hbm.21177.

Hoge, C.W., J.L. Auchterlonie, and C.S. Milliken. 2006. "Mental health problems, use of mental health services, and attrition from military service after returning from deployment to Iraq or Afghanistan." *JAMA* 295(9):1023–1032.

Holroyd, C.B. and N. Yeung. 2012. "Motivation of extended behaviors by anterior cingulate cortex." *Trends in Cognitive Science* 16(2):122–128. doi:10.1016/j.tics.2011.12.008.

Hölzl, R., L.P. Erasmus, and A. Möltner. 1996. "Detection, discrimination and sensation of visceral stimuli." *Biological Psychology* 42(1/2):199–214.

Kringelbach, M.L. 2005. "The human orbitofrontal cortex: Linking reward to hedonic experience." *Nature Reviews Neuroscience* 6(9):691–702.

Kurth, F., K. Zilles, P.T. Fox, A.R. Laird, and S.B. Eickhoff. 2010. "A link between the systems: Functional differentiation and integration within the human insula revealed by meta-analysis." *Brain Structure & Function* 214(5/6):519–534. doi:10.1007/s00429-010-0255-z.

Lang, P.J. 1994. "The varieties of emotional experience: A meditation on james-lange theory." *Psychological Reviews* 101(2):211–221.

Larson, G.E., R.M. Highfill-McRoy, and S. Booth-Kewley. 2008. "Psychiatric diagnoses in historic and contemporary military cohorts: Combat deployment and the healthy warrior effect." *American Journal of Epidemiology* 167(11):1269–1276.

Limongi, R., S.C. Sutherland, J. Zhu, M.E. Young, and R. Habib. 2013. "Temporal prediction errors modulate cingulate-insular coupling." *NeuroImage* 71:147–157. doi:10.1016/j.neuroimage.2012.12.078.

Marcora, S.M. 2008. "Do we really need a central governor to explain brain regulation of exercise performance?" *European Journal of Applied Physiology* 1045:929–931.

Marcora, S.M. 2010. "Counterpoint: Afferent feedback from fatigued locomotor muscles is not an important determinant of endurance exercise performance." *Journal of Applied Physiology* 108(2):454–456.

Marcora, S.M., W. Staiano, and V. Manning. 2009. "Mental fatigue impairs physical performance in humans." *Journal of Applied Physiology* 106(3):857–864.

Marx, B.P., S. Doron-Lamarca, S.P. Proctor, and J.J. Vasterling. 2009. "The influence of predeployment neurocognitive functioning on post-deployment PTSD symptom outcomes among Iraq-deployed army soldiers." *Journal of the International Neuropsychological Society* 15(6):840–852.

Mohanty, A., A.S. Engels, J.D. Herrington, W. Heller, M.H. Ho, M.T. Banich et al. 2007. "Differential engagement of anterior cingulate cortex subdivisions for cognitive and emotional function." *Psychophysiology* 44(3):343–351. doi:10.1111/j.1469-8986.2007.00515.x.

Morey, R.A., C.M. Petty, D.A. Cooper, K.S. LaBar, and G. McCarthy. 2008. "Neural systems for executive and emotional processing are modulated by symptoms of posttraumatic stress disorder in Iraq war veterans." *Psychiatry Research* 162(1):59–72.

Ongur, D. and J.L. Price. 2000. "The organization of networks within the orbital and medial prefrontal cortex of rats, monkeys and humans." *Cerebral Cortex* 10(3):206–219.

Paulus, M.P., J.S. Feinstein, G. Castillo, A.N. Simmons, and M.B. Stein. 2005. "Dosedependent decrease of activation in bilateral amygdala and insula by lorazepam during emotion processing." *Archives of General Psychiatry* 62(3):282–288.

Paulus, M.P., E.G. Potterat, M.K. Taylor, K.F. Van Orden, J. Bauman, N. Momen et al. 2009. "A neuroscience approach to optimizing brain resources for human performance in extreme environments." *Neuroscience & Biobehavioral Reviews* 33(7):1080–1088.

Philippe, R.A. and R. Seiler. 2005. "Sex differences on use of associative and dissociative cognitive strategies among male and female athletes." *Perceptual & Motor Skills* 101(2):440–444.

Pollatos, O., W. Kirsch, and R. Schandry. 2005. "On the relationship between interoceptive awareness, emotional experience, and brain processes." *Brain Research: Cognitive Brain Research* 25(3):948–962.

Reynolds, S.M. and D.S. Zahm. 2005. "Specificity in the projections of prefrontal and insular cortex to ventral striatopallidum and the extended amygdala." *Journal of Neuroscience* 25(50):11757–11767.

Rolls, E.T. 2004. "Convergence of sensory systems in the orbitofrontal cortex in primates and brain design for emotion." *The Anatomical Record. Part A, Discoveries in Molecular, Cellular, and Evolutionary Biology* 281(1):1212–1225.

Rushworth, M.F. and T.E. Behrens. 2008. "Choice, uncertainty and value in prefrontal and cingulate cortex." *Nature Neuroscience* 11(4):389–397.

Schultz, W. and A. Dickinson. 2000. "Neuronal coding of prediction errors." *Annual Review of Neuroscience* 23:473–500.

Seal, K.H., G. Cohen, A.Waldrop, B.E. Cohen, S. Maguen, and L. Ren. 2011. "Substance use disorders in Iraq and Afghanistan veterans in VA healthcare, 2001–2010: Implications for screening, diagnosis and treatment." *Drug and Alcohol Dependence.* doi:10.1016/j.drugalcdep.2010.11.027.

Shackman, A.J., T.V. Salomons, H.A. Slagter, A.S. Fox, J.J. Winter, and R.J. Davidson. 2011. "The integration of negative affect, pain and cognitive control in the cingulate cortex." *Nature Reviews Neuroscience* 12(3):154–167. doi:10.1038/nrn2994.

Shipp, S. 2005. "The importance of being agranular: A comparative account of visual and motor cortex." *Philosophical Transactions of the Royal Society B: Biological Sciences* 360(1456):797–814.

Smith, T.C., M.A. Ryan, D.L. Wingard, D.J. Slymen, J.F. Sallis, and D. Kritz-Silverstein. 2008. "New onset and persistent symptoms of post-traumatic stress disorder self reported after deployment and combat exposures: Prospective population based US military cohort study." *British Medical Journal* 336(7640):366–371.

Southwick, S.M., M. Vythilingam, and D.S. Charney. 2005. "The psychobiology of depression and resilience to stress: Implications for prevention and treatment." *Annual Review of Clinical Psychology* 1:255–291. doi:10.1146/annurev.clinpsy.1.102803.143948.

Spencer, R.J., L.L. Drag, S.J. Walker, and L.A. Bieliauskas. 2010 "Self-reported cognitive symptoms following mild traumatic brain injury are poorly associated with neuropsychological performance in OIF/OEF veterans." *Journal of Rehabilitation Research and Development* 47(6):521–530.

Sponheim, S.R., K.A. McGuire, S.S. Kang, N.D. Davenport, S. Aviyente, E.M. Bernat, and K.O. Lim. 2011. "Evidence of disrupted functional connectivity in the brain after combat-related blast injury." *NeuroImage* 54(Suppl 1):S21–S29. doi:10.1016/j.neuroimage.2010.09.007.

Spreng, R.N., J. Sepulcre, G.R. Turner, W.D. Stevens, and D.L. Schacter. 2013. "Intrinsic architecture underlying the relations among the default, dorsal attention, and frontoparietal control networks of the human brain." *Journal of Cognitive Neuroscience* 25(1): 74–86. doi:10.1162/jocn_a_00281.

Sridharan, D., D.J. Levitin, and V. Menon. 2008. "A critical role for the right fronto-insular cortex in switching between central-executive and default-mode networks." *Proceedings of the National Academy of Sciences of the United States of America* 105(34):12569–12574. doi:10.1073/pnas.0800005105.

St. Clair Gibson, A. and T.D. Noakes. 2004. "Evidence for complex system integration and dynamic neural regulation of skeletal muscle recruitment during exercise in humans." *British Journal of Sports Medicine* 38(6):797–806.

Stein, M.B., A.N. Simmons, J.S. Feinstein, and M.P. Paulus. 2007. "Increased amygdala and insula activation during emotion processing in anxiety-prone subjects." *American Journal of Psychiatry* 164(2):318–327.

Sullivan, E.V., E. Muller-Oehring, A.L. Pitel, S. Chanraud, A. Shankaranarayanan, D.C. Alsop et al. 2013. "A selective insular perfusion deficit contributes to compromised salience network connectivity in recovering alcoholic men." *Biological Psychiatry* doi:10.1016/j.biopsych.2013.02.026.

Sutherland, M.T., M.J. McHugh, V. Pariyadath, and E.A. Stein. 2012. "Resting state functional connectivity in addiction: Lessons learned and a road ahead." *NeuroImage* 62(4): 2281–2295. doi:10.1016/j.neuroimage.2012.01.117.

Tang, Y.Y., M.K. Rothbart, and M.I. Posner. 2012. "Neural correlates of establishing, maintaining, and switching brain states." *Trends in Cognitive Sciences* 16(6):330–337. doi:10.1016/j.tics.2012.05.001.

Taylor, K.S., D.A. Seminowicz, and K.D. Davis. 2009. "Two systems of resting state connectivity between the insula and cingulate cortex." *Human Brain Mapping* 30(9): 2731–2745. doi:10.1002/hbm.20705.

Thomas, J.L., J.E. Wilk, L.A. Riviere, D. McGurk, C.A. Castro, and C.W. Hoge. 2010. "Prevalence of mental health problems and functional impairment among active component and national guard soldiers 3 and 12 months following combat in Iraq." *Archives of General Psychiatry* 67(6):614–623. doi:10.1001/archgenpsychiatry.2010.54.

Tucker, R. 2009. "The anticipatory regulation of performance: The physiological basis for pacing strategies and the development of a perception-based model for exercise performance." *British Journal of Sports Medicine* 43(6):392–400. doi:10.1136/bjsm.2008.050799.

Tugade, M.M. and B.L. Fredrickson. 2004. "Resilient individuals use positive emotions to bounce back from negative emotional experiences." *Journal of Personality and Social Psychology* 86(2):320–333. doi:10.1037/0022-3514.86.2.320.

Tugade, M.M., B.L. Fredrickson, and L.F. Barrett. 2004. "Psychological resilience and positive emotional granularity: Examining the benefits of positive emotions on coping and health." *Journal of Personality* 72(6):1161–1190. doi:10.1111/j.1467-6494.2004.00294.x.

Vaitl, D. 1996. "Interoception." *Biological Psychology* 42(1/2):1–27.

Van Wingen, G.A., E. Geuze, E. Vermetten, and G. Fernandez. 2011. "Perceived threat predicts the neural sequelae of combat stress." *Molecular Psychiatry*. doi:10.1038/mp.2010.132.

Vasterling, J.J., S.P. Proctor, P. Amoroso, R. Kane, T. Heeren, and R.F. White. 2006. "Neuropsychological outcomes of army personnel following deployment to the Iraq war." *JAMA* 296(5):519–529. doi:10.1001/jama.296.5.519.

Vasterling, J.J., M. Verfaellie, and K.D. Sullivan. 2009. "Mild traumatic brain injury and post-traumatic stress disorder in returning veterans: Perspectives from cognitive neuroscience." *Clinical Psycholgy Review* 29(8):674–684.

Vogt, B.A. 2005. "Pain and emotion interactions in subregions of the cingulate gyrus." *Nature Reviews Neuroscience* 6(7):533–544. doi:10.1038/nrn1704.

Vogt, B.A., G.R. Berger, and S.W. Derbyshire. 2003. "Structural and functional dichotomy of human midcingulate cortex." *European Journal of Neuroscience* 18(11):3134–3144.

Wald, I., G. Lubin, Y. Holoshitz, D. Muller, E. Fruchter, D.S. Pine et al. 2010. "Battlefield-like stress following simulated combat and suppression of attention bias to threat." *Psychological Medicine* 1–9. doi:10.1017/s0033291710002308.

Weston, C.S. 2012. "Another major function of the anterior cingulate cortex: The representation of requirements." *Neuroscience & Biobehavioral Reviews* 36(1):90–110. doi:10.1016/j.neubiorev.2011.04.014.

Williamson, J.W., P.J. Fadel, and J.H. Mitchell. 2006. "New insights into central cardiovascular control during exercise in humans: A central command update." *Experimental Physiology* 91(1):51–58.

Williamson, J.W., R. McColl, D. Mathews, J.H. Mitchell, P.B. Raven, and W.P. Morgan. 2002. "Brain activation by central command during actual and imagined handgrip under fypnosis." *Journal of Applied Physiology* 92(3):1317–1324.

Zhang, K., B. Johnson, D. Pennell, W. Ray, W. Sebastianelli, and S. Slobounov. 2010. "Are functional deficits in concussed individuals consistent with white matter structural alterations: Combined FMRI and DTI study." *Experimental Brain Research* 204(1):57–70. doi:10.1007/s00221-010-2294-3.

6 Neurotechnology and Operational Medicine*

Carey D. Balaban

CONTENTS

OVERVIEW

Operational medicine is the projection of societal (civilian and military), medical, and public health resources into the realms of homeland security medical operations, disaster relief, and humanitarian assistance. The effective management of these assets requires a system with the resilience and flexibility to respond to a changing threat landscape, particularly within the confines of an event. An event scenario evolves as a function of actions (and reactions) of response personnel, victims, and bystanders, each acting from different frames of reference that determine situation awareness. This communication presents elements of a neurotechnology approach to formalizing situation awareness for different roles in operational medicine contexts.

Operational medicine in the defense and intelligence domains is governed by the same principles as any military, intelligence, or counter-terrorist activity because it is essential to achieve situation awareness in the face of unpredictable natural events and/or deliberate actions of foes. A first principle is to know your foes. In operational medicine, one's foe is, first and foremost, the developing scenario, including its potential direct consequences and its potential collateral consequences. Awareness is achieved by engaging procedures that identify the evolving natural and/or man-made scenario, yet work prudently and safely to minimize its impact. These procedures are embedded in a process of situation assessment that yields a sufficiently clear

---

* This chapter is adapted with permission from Balaban, C.D. 2011. Neurotechnology and operational medicine. *Synesis: A Journal of Science, Technology, Ethics, and Policy* 2(1):T:45–T:54.

hypothesis for current operational safety and efficacy. A second principle is to know ourselves and our allies. This principle is implemented by developing a facile working knowledge of our resources, capabilities, and vulnerabilities, by understanding how they can be projected onto evolving scenarios, and by having an intuitive ability to detect when unexpected or unusual behavioral conditions are emerging. A third principle is to know what our foes expect to achieve. In the case of terrorism, one primary goal is to demonstrate the ineffectiveness of the operational medicine infrastructure, which contributes to creating a sense of anxiety, panic, and hopelessness in the population. A corollary of this principle is to know what our foes expect us to do. The assumption that foes will try to exploit our actions to reach their goals implies that our self-knowledge has high security value.

NEUROTECHNOLOGY AND FORMAL REPRESENTATIONS OF SITUATION AWARENESS

Situation awareness is a concept that is invoked often without explicit definition in operational contexts. This contribution explores applications of neurotechnology to the issue of adaptively establishing and maintaining situation awareness in different frames of reference, ranging from the relative macrocosm of operational command and control to the relative microcosm of the perception of personal health. This issue is viewed as analogous to an interaction of sensorimotor, interoceptive, and cognitive neural networks in the expression of comorbid features (including emotion and affect) of balance disorders, migraine, and anxiety disorders in conditions that include mild traumatic brain injury (Balaban and Thayer 2001; Balaban et al. 2011; Furman et al. 2013; Staab et al. 2013). This analogy can be rendered operational by implementing hybrid agent-based and discrete simulation tools that can be parameterized for each frame of reference and used for real-time decision support (Balaban et al. 2012; Wu et al. 2012).

The applications of this approach also will raise significant issues for security and intelligence communities. For example, this approach can generate families of trajectories for an individual patient's behavior in terms of latent (underlying) neural/neurochemical mechanisms, which can be compared to the current status to guide further treatment by improving the situational awareness of both the health providers and the patient. The same argument holds for a decision-maker in an application of the approach to a command and control center. In either case, the individual models become a metadata representation of the patient that can constitute a form of protected personal information and, for key personnel, a matter of potential high security and intelligence value.

NEUROTECHNOLOGY AND FORMAL OPERATIONAL REPRESENTATIONS OF SITUATION AWARENESS

The March 1995 issue of the journal *Human Factors* was truly a watershed event in the formalization of the concept of situation awareness as a cognitive construct. The definitions of situation awareness included "adaptive, externally directed consciousness" toward an environmental [external] goal (Smith and Hancock 1995), "up to the minute cognizance required to operate or maintain a system" (Adams et al. 1995), and "... just a label for a variety of cognitive processing activities that are critical to dynamic, event-driven, and multitask fields of practice..." (Sarter and Woods 1995). The general conceptual consensus was that situation awareness is the product of processes that map the knowledge, capacities, beliefs, and extrinsically directed goals (and criteria) of an agent onto the dynamic behavior of the environment.

Endsley (1995) provided a comprehensive framework for viewing situation awareness from a cognitive approach. The term situation awareness was defined as a state of knowledge that is the result of an adaptive, dynamic neurocognitive process that has been termed *situation assessment.* Situation assessment was represented as the product of purely cognitive processes, envisioned as rational agents (Figure 6.1a). A decision is produced by interactions among three different levels of rational processes (or agents), progressing from perception of the elements of a current situation (level 1) to comprehension of the elements in a context (level 2) to the projection or prediction of the future status (level 3) of a complex system, resulting in a decision. The decision then affects the instantaneous state of the evolving environment, which,

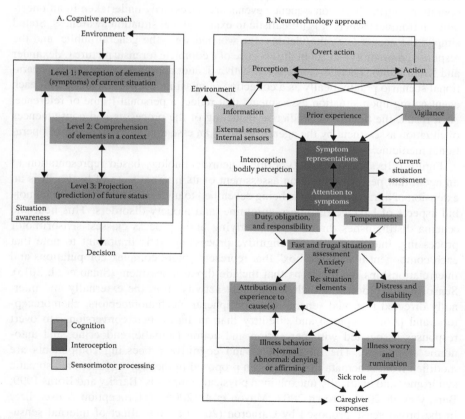

FIGURE 6.1 Approaches to achieving situation awareness. See text for detailed discussion.

in turn, influences the continuing situation assessment process. The holistic state of knowledge represented by the three levels of agents is termed situation awareness.

This modular approach is attractive because it extends inductively to the situation awareness of a group of interactive decision makers. For example, Endsley (1995) also defined Team Situation Awareness as the degree to which each member of a team possesses the situation awareness that is necessary for their responsibilities. Since the situation awareness of each team member can be envisioned as collection of three simple, smaller autonomous processes for situation assessment, the net construct parallels Marvin Minsky's (1986) metaphorical description of an artificial intelligence "society of mind." The quality of team coordination can be assessed by the quality of the situation awareness of team members with shared responsibilities. Conversely, potential vulnerabilities or weak links are produced by a team member lacking situation awareness for one element of a responsibility area. In this sense, the cognitive architecture in Figure 6.1a is one of the rational agents that serves as a basic building block of individual and group situation awareness.

Much thought and effort has been directed at developing operational situation assessment and intelligent data fusion processes that can approach an ideal of omniscience for command and control applications However, public health and medical operations during high consequence events are necessarily undertaken in an uncertain environment where it is unrealistic to expect global situation awareness. Stated simply, the agents (e.g., the victims, the worried well, the general public, and the response community) all act in the absence of a complete forensic picture (Alexander and Klein 2006). Therefore, one is left with a default approach: we regard an operational scenario pragmatically as a collection of quasi-independent agents, with each agent acting upon situation assessments that reflect a personal frame of reference. As a result, the fidelity of predictive modeling of the processes and consequences of situation assessment by the agents becomes an essential factor in effective operational medicine.

Figure 6.1b shows a prototype for a neurotechnology-based representation of an agent that performs situation assessment of its internal state of health. It is an extension of a framework that is being developed to understand the bases for comorbid aspects of balance disorders, migraine, and anxiety disorders. This heuristic schema distinguishes three basic underlying brain process classes: sensorimotor processing, interoception, and cognitive processing. It is important to note that each component is a "black box" that represents more complex computations and interrelationships for functions that include threat assessment (Staab et al. 2013). Sensorimotor processing includes afferent activity from the externally and internally directed neuronal sensors (e.g., peripheral mechanoreceptors, chemoreceptors, and photoreceptors) and circuitry that mediates their conversion into overt responses associated with perception and action (somatic, endocrine, and autonomic responses). The interoception and cognitive processing components are modified from schemata that have been proposed in the area of functional somatic syndromes and medically unexplained physical symptoms (Barsky and Boras 1999; Barsky et al. 2002; Brown 2004; Mayou et al. 2005). Interoception is used here in the broad sense proposed by Cameron (2002) as any effect of internal sensations on molar organic activity, even in the absence of awareness. In the case of disease perception, interoception can be envisioned as a process that maps sensor information onto symptom representations and modulates attention to those symptoms. In a more general decision-making sense, it can be envisioned as forming a "gut feeling" of subjective state along a continuum from well-being to discomfort to danger to panic. Craig (2009) has recently reviewed evidence for an important role of the insular cortex and amygdala in interoception. It is also modulated by prior experience (both conscious memory and classical conditioning) and is updated on the basis of results of upstream cognitive processes.

The selection of higher order cognitive processing components for the schema recognizes that human decision-making displays bounded rationality (Simon 1955), which is a necessary consequence of limits imposed by the structure of the environment, the structure of the perceived solution space (mental models), and human cognitive capabilities (Simon 1956). The schema itself is a form of "satisficing" (Simon 1955) that concludes a search of limited options as soon as a "good enough" criterion is reached. The instantaneous interoceptive status is subjected to initial situation assessment estimate of context, which is influenced by the individual's temperament and, in many contexts, sense of duty, obligation, and responsibility. In cognitive terms, this initial estimate is based upon a fast and frugal heuristic (Gigerenzer and Todd 1999; Goldstein and Gigerenzer 1999) or "rule of thumb" (Kahneman and Tversky 1979) assessment of the current situation within the context of their knowledge base and roles. Further analysis of this rapid estimate is used to attribute the status to a cause (e.g., "It must have been something I ate.") and to generate distress/disability, illness worry/rumination, and illness behaviors (Pilowsky 1993). The outcomes of these cognitive processes can influence interoception directly; alternatively, effects on emotional arousal can affect both interception and sensorimotor activity. From a neuroscientific perspective, these cognitive processes likely engage both (1) mechanisms for setting affective state that are associated with the ventrolateral prefrontal, orbitofrontal, and ventral anterior cingulate cortex and (2) mechanisms for regulation of the affective state that involve at least the dorsal (lateral and medial) prefrontal and dorsal anterior cingulate cortex (Balaban et al. 2011; Craig 2009; Phillips et al. 2003; Staab et al. 2013). The result of this process is an updating of anxiety-misery and fear-avoidance dimension (Sylvers et al. 2011) representations of the elements and context of the situation. The anxiety-misery dimension reflects future, sustained, and diffuse threats that one approaches (and probes) to improve awareness. The fear-avoidance dimension reflects specific, clear, and present threats that are avoided. Social interactions with caregivers and others (team members and mutual aid partners in a working environment), which can be influenced by self-labeling adoption of a *sick role* (Barsky and Boras 1999), can intervene in the environment in parallel to influences of one's own actions.

There are two approaches to scale this architecture to include local or global aspects of scenarios. First, the model of an individual can be generalized to encompass any response personnel, victim, or bystander in an operational medicine scenario by simply changing the term "symptom" to an analogous concept, "noteworthy aspects of the current environment or unfolding scenario." The task of the agent, then, is to develop the analog of a patient's interoceptive situation assessment relative to the current responsibility domain of concern, which is analogous to a patient's "body." For example, let us consider a basic dyad of first responder who is treating a victim. The responder (1) functions as a caregiver to the victim and (2) includes information about the patient with their own physiologic information. Conversely, the caregiver actions influence the behavior of the patient and the environment.

The second scaling approach proceeds from the microcosm of the body of the individual to the macrocosm of larger frames of reference within the scenario, which parallels classical Platonic microcosm–macrocosm relationships (Dampier 1949). Hence, a neurotechnology approach provides a scalable common framework for viewing situation assessment on two different scales: (1) the domains of individual health (vs disease) status, behavior, and community interactions and (2) in domains of mechanistic interconnections between sensorimotor, interoceptive, cognitive, and social (interpersonal interaction) subsystems. Each agent's *situation awareness* can then be defined operationally as the set of causes in the agent's experiential base that are hypothesized to be consistent with the current symptoms (or analogous conditions).

HYBRID MODELING AND SIMULATION APPROACHES TO ENHANCING SITUATION AWARENESS

Hybrid agent-based and continuous time simulation methods provide a flexible architecture for implementing *systems of systems* models of situation assessment for decision support. Time domain simulation methods, such as classical linear and non-linear systems approaches, are now used commonly in the neurosciences for modeling instantaneous neuronal signal processing and plasticity on the basis of underlying cellular phenomena (including enzyme and ion channel kinetics). Even the simplest linear systems approaches can provide heuristic insight into the dynamic aspects of judgments of the intensity of exteroceptive and interoceptive perceptual phenomena (McBurney and Balaban 2009). As such, they are appropriate for simulating the sensorimotor processing components for the neurotechnology approach (Figure 6.1b) as a specific systems model with parametric operations of change detectors (high-pass filters), level detectors (low-pass filters), and cumulative exposure detectors.

Agent-based models provide a far more flexible approach for simulating (and understanding) emergent properties of systems that can be characterized by complex rule-based behavior (Axelrod 1997; Bonabeau 2002). These methods have been employed widely in the social sciences (An et al. 2005; Batty et al. 2003; Deffaunt et al. 2005; Eguíluz et al. 2005; Epstein 2002) and have been applied to defense and intelligence medical issues such as pandemic and biowarfare scenarios (Carley et al. 2006; Ferguson et al. 2006; Halloran et al. 2008). Agent-based methods are appropriate for simulating the cognitive and interoceptive components for the neurotechnology approach (Figure 6.1b), as well as caregiver interactions.

Models that are hybrids of agent-based and continuous time components are becoming increasingly common for parsimonious simulation of complex and large problems. One of these hybrid systems is the *Dynamic Discrete Disaster Decision Simulation System* (D^4S^2). The D^4S^2 platform has been developed and patented (Balaban et al. 2012) by the Center for National Preparedness at the University of Pittsburgh for planning and decision support in one area of operational medicine, casualty clearance from an disaster scene (Wu et al. 2007a, 2007b, 2007c, 2008a, 2008b, 2012). This hybrid simulation architecture integrates an operations research simulation engine, a rules-based agent simulation, geographical information system (GIS) infrastructure data, graphical interfaces, and disaster information databases. In addition to standard simulations, the architecture has been used for evolutionary decision-making and frugal multicriterion optimization by constraining the heuristic search with a meta-modeling approach, nonlinear mixed integer programming (Wu et al. 2008b). A key feature of this architecture is the explicit definition of *situation awareness* as *the set of active hypotheses that is consistent with both the history of implemented rules (plans) and contextual information from the data base* (Balaban et al. 2012). This definition allows us to track the convergence of situation awareness to a small set of cause(s) and optimize the performance of agents to preserve options and improve resilience. These results can easily be extended to implement a neurotechnology approach to situation awareness in a hybrid simulation platform.

NEUROTECHNOLOGY PLATFORM FOR SITUATION AWARENESS: IMPLICIT HUMAN–MACHINE INTERFACES

Because simulation components are linked to findings and models related to brain activity, the neurotechnology approach to situation awareness is amenable to both experimental validation and to the future integration of validated real-time physiological measurements into a human–computer system. For example, the neuroscientific literature on interoception and emotional control (Craig 2009; Phillips et al. 2003) provides a theoretical basis for using methods such as near-infrared spectroscopy to identify vascular perfusion changes that are related to activity of the anterior insula (interoception), ventral prefrontal-orbitofrontal cortex (production of affective state) and dorsal prefrontal cortex (effortful regulation of affective state) during the situation assessment process. The sensorimotor components, on the other hand, are monitored by actions. The process of integrating these measures with simulation provides a roadmap for intermeshing physiological sensor, human response, and simulation features in the development of human–computer systems that achieve a high fidelity representation of an individual's situation assessment processes.

SOME OPERATIONAL APPLICATIONS FOR DEFENSE AND INTELLIGENCE

The previous sections have presented an academic background for using a neurotechnology-based hybrid simulation approach for modeling the instantaneous situation awareness of individuals and groups of individuals in an operational medicine response scenario. This section will discuss several examples of its applications to specific problem domains. It is not meant to be exhaustive; rather, it is designed to encourage inductive thinking about the broad scope of potential applications for the defense and intelligence communities.

Predicting help-seeking behavior. The help-seeking behavior of victims, the public at large, and the response community is an important consideration during an unfolding medical response scenario (Alexander and Klein 2006). Help-seeking behavior is determined by the current situation assessment, which maps onto the process termed interoception in the neurotechnology approach (Figure 6.1b). Inappropriate help-seeking manifests as the *worried well* phenomenon; it is an interoceptive hypervigilance-driven form of panic that creates bottlenecks in resource delivery by demanding assessment and treatment. Conversely, others may delay reporting symptoms, with negative implications for outcomes. Literature from both public health and social science domains provides ample evidence for predictable effects of factors such as personality, age, ethnicity, race, career role, socioeconomic status, and gender on the perception of illness and the likelihood of reporting potential illness to a response facility (Ayalon and Young 2005; Barratt 2005; Engel et al. 2002; Feldman et al. 1999; Ferguson et al. 2004; Kolk et al. 2002; Lefler and Bondy 2004; Marshall et al. 1994). Specifically, the behavior of individual agents can reflect distinct clusters of "Big 5" personality characteristics that impact the threshold for emergence of help-seeking behaviors (Feldman et al. 1999; Ferguson et al. 2004; Kolk et al. 2002; Marshall et al. 1994), the style of self-presentation (Barratt 2005), and the likelihood of seeking help from different resources (e.g., self-help books, religious institutions [Ayalon and Young 2005], internet resources, nurse practitioner hotlines, pharmacist, clinical facilities, or emergency responders). A hybrid simulation, neurotechnology approach is thus envisioned as a common platform for generating predictions from a formalization of the relationship between individual situation assessments and mass civilian and military responses to perceived threats and significant events (Aguirre 2005; Mawson 2005; Perry and Lindell 2003; Stokes and Banderet 1997).

The ability to predict help-seeking behavior in context can be an asset of particularly high value for responses to pandemics, bioterrorism, biological warfare, unsuspected chemical toxin exposure, or unsuspected radiological exposure. These scenarios are examples of latent events, which are detected only as the victims develop symptoms and conclude that they need to seek help. A special case is an "announced attack" scenario, where information about an impending or developing attack is released by terrorists to elicit panic responses in the public. The detection phase, defined as the period encompassing release (or infection), appearance of symptoms, illness, and first deaths, is a period when public responses to perceived symptoms and the societal milieu (including information, misinformation and disinformation) can have a profound impact on both (1) the ability to detect a significant latent event and (2) the resulting demands for response assets. When they are neither understood nor predictable, these individual variations become a significant component of noise (or "fog of war") that can impede the process of detection and the initiation of an effective response. However, a situation assessment simulation approach can be used to help identify sentinel populations (or features of multiple populations) to improve the speed and accuracy of detection of a latent event.

Decision support for diagnosis and treatment. The neurotechnology-based approach is designed to provide a mechanistically based, integrated overview of the progression of neurological and psychological signs of symptoms from the perspectives of both the patient's self-report and the clinical objective and subjective observations by medical and paramedical staff. This statement is hardly surprising because the approach is generalized from research directed at elucidating scientific bases for the comorbidity of balance disorders, anxiety disorders, and migraine and the clinical responses of the signs and symptoms to different therapeutic regimens (Balaban and Thayer 2001; Balaban and Yates 2004; Furman et al. 2003, 2005; Jacob et al. 1993, 1996; Marcus et al. 2005). The agent-based representation of the patient is intended to be a mechanistic, neurological and psychological hybrid model that explains the history and current clinical status. It can also be projected into the future to generate prognostic trajectories for the signs and symptoms on the basis of different sets of assumptions. On the one hand, these prognostic hypotheses can assist the physician in outcome-oriented case management by providing templates for the clinical course that indicate a good outcome or the need to correct treatment to account for other likely underlying factors. On the other hand, this approach may prove particularly useful as a research tool for untangling the interplay between neurological and psychological factors in the development of comorbid aspects of mild traumatic brain injury and posttraumatic stress disorder during acute, subacute, and chronic presentations (Hoffer et al. 2010).

Impact of competing of duties, obligations, and responsibilities. It is well known that a sense of duty, obligation, and responsibility can override considerations of personal well-being in difficult situations, sometimes to the benefit and sometimes to the detriment of the outcomes. The Milgram experiments (Milgram 1963) and the Stanford Prison Experiment (Zimbardo 2007) are prominent examples in the social psychology literature of sinister effects that can emerge in experimental social settings. They are counterbalanced by myriad cases of altruism and heroism in operational settings, including willful suppression of help-seeking behavior for others deemed more deserving. Serious consideration of these effects has been restricted to the anecdotal and forensic domains, with the purpose of generating ethical lessons and building a culture of esprit-de-corps.

More complex dilemmas arise in other operational medicine scenarios. The expected absenteeism of personnel in CBRN mass casualty scenarios likely reflects resolution of conflicting demands from multiple sets of duties, obligations, and responsibilities (immediate family vs community concerns). Because agent-based simulation methods can test the effects of rules for resolving these competing interests, the approach has the potential to build and validate structures for predicting implications of these competing interests in individual cases. These models can then be used to investigate the impact of difficult decisions by individuals on the outcome of an operational medicine response. Finally, the model predictions can be used prospectively to detect the effects of potentially deleterious individual decisions during a response so that prompt corrective actions can be taken. In particular, it allows us to improve situational awareness by identifying situations when acts of altruism/heroism can be beneficial or deleterious to a scenario outcome.

Individual agents and network simulation agents as private information and intelligence assets. High fidelity hybrid simulation representations of both individual patients and operational medicine response systems have the potential to be (1) a form of protected medical personal information and (2) a highly valued intelligence target for others. For example, the diagnostic criteria and features in the DSM 5 (APA 2013) provide useful templates and storylines for modeling individual behavioral agents. Let us assume that a neurotechnology-based hybrid simulation with individualized parameters (or data) is sufficient to predict the help-seeking behavior and the clinical courses of individual patients for a class of medical care scenarios. Within the context of the model, the parameters are a predictive meta-data representation of the behavior, and hence, a parsimonious representation of the patient's medical status in electronic personal records. This raises many questions. Does model representation (either parameters alone or parameters plus model) qualify as personal data that are subject to the privacy rule under the Health Insurance Portability and Accountability Act of 1996 (HIPAA)? Does a set of parameters (or parameters plus model), linking treatments with individual outcomes of many patients with a particular condition, qualify as a Patient Safety Work Product under Patient Safety and Quality Improvement Act of 2005? Could the models and data be used for biometric identification or profiling? Do the parameters for key decision makers have intelligence value, either for actions directed against the individuals, for sabotaging response capabilities or for revealing vulnerabilities to further a foe's operational goals? It is likely that the answers to these (and related) questions will depend upon the results obtained with neurotechnology-based simulation systems.

Building resilience into operational medicine. Neurotechnology-based computational hybrid models have the capability to facilitate the design of psychologically resilient operational networks. This goal can be realized by a direct application of research from the area termed "psychotraumatology," which examines factors that enhance psychological resilience in the face of traumatic experiences. *Growth through adversity* is a term that describes the positive adaptations and adjustments that can emerge in the process of living through traumatic and threatening situations (Joseph and Linley 2005). Linley (2003) has made the interesting assertion that three dimensions of wisdom contribute to these positive adaptations to adverse events: (1) the recognition of and ability to operate under conditions of uncertainty, (2) the development of a sense of connected detachment ("integration of affect and cognition"), and (3) the recognition and acceptance of human limitations. It is significant to note that these aspects of wisdom can emerge from the operations of the schema in Figure 6.1b. The first two dimensions represent interactions between interoceptive and cognitive components. The third dimension is equivalent to the cognizance of bounded rationality and the recognition that all human decisions are merely satisficing. The challenge is to design interfaces to convey this view to operational medicine responders and managers such that psychological resilience is embedded in the daily practices of operational medicine. If executed correctly, such a simulation platform has the potential to serve as an inductive teaching tool to inculcate the wisdom that lies at the heart of resilience in the face of adversity.

REFERENCES

Adams, M.J., Y.J. Tenney, and R.W. Pew. 1995. "Situation awareness and the cognitive management of systems." *Human Factors* 37:85–105.

Aguirre, B.E. 2005. "Emergency evacuations, panic, and social psychology." *Psychiatry* 68(2):121–128.

Alexander, D.A. and S. Klein. 2006. "The challenge of preparation for a chemical, biological, radiological or nuclear terrorist attack." *Journal of Postgraduate Medicine* 52:126–131.

An, L., M. Linderman, J. Qi, A. Shortridge, and J. Liu. 2005. "Exploring complexity in a human-environment system: An agent-based spatial model for multidisciplinary and multiscale integration." *Annals of the Association of American Geographers* 95(1):54–79.

APA. 2013. *Diagnostic and Statistical Manual of Mental Disorders: DSM-5*. Arlington, VA: American Psychiatric Association.

Axelrod, R. 1997. *The Complexity of Cooperation: Agent-Based Models of Competition and Collaboration*. Princeton, NJ: Princeton University Press.

Ayalon, L. and M.A. Young. 2005. "Racial group differences in help-seeking behaviors." *The Journal of Social Psychology* 145(4):391–403.

Balaban, C.D., B. Bidanda, M.H. Kelley, L.J. Shuman, K.M. Sochats, and S. Wu. 2012. United States Patent No. U.S. 8,204,836 B2. USPTO.

Balaban, C.D., R.G. Jacob, and J.M. Furman. 2011. "Neurologic bases for comorbidity of balance disorders, anxiety disorders and migraine: Neurotherapeutic implications." *Expert Reviews in Neurological Therapy* 11(3):379–394.

Balaban, C.D. and J.F. Thayer. 2001. "Neurological bases for balance-anxiety links." *Journal of Anxiety Disorders* 15:53–79.

Balaban, C.D. and B.J. Yates. 2004. "Vestibulo-autonomic interactions: A teleologic perspective." In *Springer Handbook of Auditory Research: The Vestibular System,* eds. S.N. Highstein, R.R. Fay and A.N. Popper. New York: Springer-Verlag, pp. 286–342.

Barratt, J. 2005. "A case study of styles of patient self-presentation in the nurse practitioner primary health care consultation." *Primary Health Care Research & Development* 6:329–340.

Barsky, A.J. and J.F. Boras. 1999. "Functional somatic disorders." *Annals of Internal Medicine* 130:910–921.

Barsky, A.J., R. Saintfort, and M.P. Rogers. 2002. "Nonspecific medication side effects and the nocebo phenomenon." *Journal of the American Medical Association* 287:622–627.

Batty, M., J. Desyllas, and E. Duxbury. 2003. "The discrete dynamics of small-scale spatial events: Agent-based models of mobility in carnivals and street parades." *International Journal of Geographical Information Science* 177:673–697.

Bonabeau, E. 2002. "Agent-based modeling: Methods and techniques for simulating human systems." *Proceedings of the National Academy of Sciences of the United States of America* 99:7280–7287.

Brown, R.J. 2004. "Psychological mechanisms of medically unexplained symptoms: An integrative conceptual approach." *Psychological Bulletin* 130:793–812.

Cameron, O.G. 2002. *Visceral Sensory Neuroscience*. New York: Oxford University Press.

Carley, K.M., D.B. Fridsma, E. Casman, A. Yahja, N. Altman, L.-C. Chen, B. Kaminsky, D. Nave. 2006. "BioWar: A scalable agent-based model of bioattacks." *IEEE Transactions on Systems, Man and Cybernetics—Part A: Systems and Humans* 36(2):252–265.

Craig, A.D. 2009. "How do you feel–now? The anterior insula and human awareness." *Nature Reviews Neuroscience* 10:59–70.

Dampier, W.C. 1949. *A History of Science and its Relations with Philosophy and Religion* (4th ed). Cambridge: Cambridge University Press.

Deffaunt, G., S. Huet, and F. Amblard. 2005. "An individual-based model of innovation diffusion mixing social value and individual benefit." *American Journal of Sociology* 110(4):1041–1069.

Eguíluz, V.M., M.G. Zimmermann, C.J. Cela-Conde, and M. San Miguel. 2005. "Cooperation and the emergence of role differentiation in the dynamics of social networks." *American Journal of Sociology* 110(4):977–1008.

Endsley, M.R. 1995. "Toward a theory of situation awareness in dynamic systems." *Human Factors* 37(1):32–64.

Engel, C.C.J., J.A. Adkins, and D.N. Cowan. 2002. "Caring for medically unexplained physical symptoms after toxic environmental exposures: Effects of contested causation." *Environmental Health Perspectives* 110(Suppl 4):641–647.

Epstein, J.M. 2002. "Modeling civil violence: An agent-based computational approach." *Proceedings of the National Academy of Sciences of the United States of America* 99(Suppl 3):7243–7250.

Feldman, P.J., S. Cohen, W.J. Doyle, D.P. Skoner, and J.M.J. Gwaltney. 1999. "The impact of personality on the reporting of unfounded symptoms and illness." *Journal of Personality and Social Psychology* 77(2):370–378.

Ferguson, E., J.E. Cassaday, and G. Delahaye 2004. "Individual differences in the temporal variability of medically unexplained symptom reporting." *British Journal of Health Psychology* 9:219–240.

Ferguson, N.M., D.A. Cummings, C. Fraser, J.C. Cajka, P.C. Cooley, and D.S. Burke. 2006. "Strategies for mitigating an influenza pandemic." *Nature* 442:448–452.

Furman, J.M., C.D. Balaban, R.G. Jacob, and D.A. Marcus. 2005. "Migraine-anxiety associated dizziness (Mard): A new disorder?" *Journal of Neurology, Neurosurgery & Psychiatry* 76:1–8.

Furman, J.M., D.A. Marcus, and C.D. Balaban. 2003. "Migrainous vertigo: Development of a pathogenetic model and structured diagnostic interview." *Current Opinions in Neurology* 16:5–13.

Furman, J.M., D.A. Marcus, and C.D. Balaban. 2013. "Vestibular migraine: Clinical aspects and pathophysiology." *Lancet Neurology* 12:706–715.

Gigerenzer, G. and P.M. Todd. 1999. "Fast and erugal heuristics: The adaptive toolbox." In *Simple Heuristics That Make Us Smart*, eds. G. Gigerenzer, P.M. Todd, and The ABC Research Group. New York: Oxford University Press, pp. 3–34.

Goldstein, D.G. and G. Gigerenzer. 1999. "The recognition heuristic: How ignorance makes us smart." In *Simple Heuristics That Make Us Smart*, eds. G. Gigerenzer, P.M. Todd, and The ABC Research Group. New York: Oxford University Press, pp. 37–58.

Halloran, M.E., N.M. Ferguson, S. Eubank, I.M. Longini, D.A. Cummings, B. Lewis, S. Xu et al. 2008. "Modeling targeted layered containment of an influenza pandemic in the United States." *Proceedings of the Natational Academy of Sciences of the United States of America* 105(12):4639–4644.

Hoffer, M.E., C.D. Balaban, K.R. Gottshall, B.R. Balough, M.R. Maddox, and J.R. Penta. 2010. "Blast exposure: Vestibular consequences and associated characteristics." *Otology & Neurotology* 31(2):232–236.

Jacob, R.G., J.M. Furman, and C.D. Balaban. 1996. "Psychiatric aspects of vestibular disorders." In *Handbook of Neurotology/Vestibular System*, eds. R.W. Baloh and G.M. Halmagyi. Oxford: Oxford University Press, pp. 509–528.

Jacob, R.G., J.M.R. Furman, D.B. Clark, J.D. Durrant, and C.D. Balaban. 1993. "Psychogenic dizziness." In *The Vestibulo-Ocular Reflex and Vertigo*, eds. J.A. Sharpe and H.O. Barber. New York: Raven Press, pp. 305–317.

Joseph, S. and P.A. Linley. 2005. "Positive adjustment to threatening events: An organismal valuing theory of growth through adversity." *Review of General Psychology* 9:262–280.

Kahneman, D. and A. Tversky. 1979. "Prospect theory: An analysis of decision under risk." *Econometrica* 47:263–291.

Kolk, A.M.M., G.J.F.P. Hanewald, S. Schagen, and C.M.T.G. van Wijk. 2002. "Predicting medically unexplained physical symptoms and health care utilization. A symptom-perception approach." *Journal of Psychosomatic Research* 52:35–44.

Lefler, L.L. and K.N. Bondy. 2004. "Women's delay in seeking treatment with myocardial infarction: A meta-analysis." *Journal of Cardiovascular Nursing* 19(4):251–268.

Linley, P.A. 2003. "Positive adaptation to trauma: Wisdom as both process and outcome." *Journal of Traumatic Stress* 16:601–610.

Marcus, D.A., J.M. Furman, and C.D. Balaban. 2005. "Motion sickness in migraine sufferers." *Expert Opinion in Pharmacotherapy* 6(15):2691–2697.

Marshall, G.N., C.B. Wortman, R.R.J. Vickers, J.W. Kusulas, and L. Hervig. 1994. "The five-factor model of personality as a framework for personality-health research." *Journal of Personality and Social Psychology* 67(2):278–286.

Mawson, A.R. 2005. "Understanding mass panic and other collective responses to threat and disaster." *Psychiatry* 68(2):95–113.

Mayou, R., L.J. Kirmayer, G. Simon, K. Kroenke, and M. Sharpe. 2005. "Somatoform disorders: Time for a new approach in DSM-V." *American Journal of Psychiatry* 162:847–855.

McBurney, D.H. and C.D. Balaban. 2009. "A heuristic model of sensory adaptation." *Attention, Perception, & Psychophysics* 71(8):1941–1961.

Milgram, S. 1963. "Behavioral study of obedience." *Journal of Abnormal and Social Psychology* 67:371–378.

Minsky, M. 1986. "Society of mind." *Whole Earth Review* (Summer) 51:4–10. http://www.wholeearth.com/issue-electronic-edition.php?

Perry, R.W. and M.K. Lindell. 2003. "Understanding citizen response to disasters with implications for terrorism." *Journal of Contingencies and Crisis Management* 11(2):49–60.

Phillips, M.L., C.W. Drevets, S.L. Rauch, and R. Lane. 2003. "Neurobiology of emotion perception I: The neural basis of normal emotion perception." *Biological Psychiatry* 54:504–514.

Pilowsky, I. 1993. "Dimensions of illness behaviour as measured by the illness behaviour questionnaire: A replication study." *Journal of Psychosomatic Research* 37:53–62.

Sarter, N.B. and D.D. Woods. 1995. "How in the world did we ever get into that mode? Mode error and awareness in supervisory control." *Human Factors* 37:5–19.

Simon, H.A. 1955. "A behavioral model of rational choice." *The Quarterly Journal of Economics* 69(1):99–118.

Simon, H.A. 1956. "Rational choice and the structure of the environment." *Psychological Reviews* 63(2):129–138.

Smith, K. and P.A. Hancock. 1995. "Situation awareness is adaptive, externally directed consciousness." *Human Factors* 37:137–148.

Staab, J.P., C.D. Balaban, and J.M. Furman. 2013. "Threat assessment and locomotion: Clinical applications of an integrated model of anxiety and postural control." *Seminars in Neurology* 33:297–306.

Stokes, J.W. and L.E. Banderet. 1997. "Psychological aspects of chemical defense and warfare." *Military Psychology* 9(4):395–415.

Sylvers, P., S.O. Lilienfeld, and J.L. LaPrairie. 2011. "Differences between trait fear and trait anxiety: Implications for psychopathology." *Clinical Psychology Review* 31:122–137.

Wu, S., L. Shuman, B. Bidanda, M. Kelley, K. Sochats, and C.D. Balaban. 2007a. "Disaster policy optimization: A simulation based approach." *Proceedings of the Industrial Engineering Research Conference*, Norcross, GA: Institute of Industrial Engineers, pp. 872–877.

Wu, S., L. Shuman, B. Bidanda, M. Kelley, K. Sochats, and C.D. Balaban. 2007b. "Embedding GIS in disaster simulation." *Proceedings of the 27th Annual ESRI International User Conference*, Paper UC1847, Redlands, CA:ESRI.

Wu, S., L. Shuman, B. Bidanda, M. Kelley, K. Sochats, and C.D. Balaban. 2007c. "System implementation issues of dynamic discrete disaster decision simulation system (D4S2)—Phase I." *Proceedings of the 2007 Winter Simulation Conference*, eds. S.G. Henderson, B. Biller, M.-H. Hsieh, J. Shortle, J.D. Tew and R.R. Barton, pp. 1127–1134. http://ieeexplore.ieee.org/stamp/stamp.jsp?tp=&arnumber=4419712

Wu, S., L. Shuman, B. Bidanda, M. Kelley, K. Sochats, and C.D. Balaban. 2008a. "Agent-based discrete event simulation modeling for disaster responses." *Proceedings of the 2008 Industrial Engineering Research Conference*, Norcross, GA: Institute of Industrial Engineers, pp. 1908–1913.

Wu, S., L. Shuman, B. Bidanda, O. Prokopyev, M. Kelley, K. Sochats, and C.D. Balaban. 2008b. "Simulation-based decision support system for real-time disaster response management." *Proceedings of the 2008 Industrial Engineering Research Conference*, Norcross, GA: Institute of Industrial Engineers. pp. 58–63.

Wu, S., L.J. Shuman, B. Bidanda, C.D. Balaban, and K. Sochats. 2012. "A hybrid modelling framework to simulate disaster response decisions." *International Journal of Advanced Intelligence Paradigms* 4(1):83–102. doi:10.1504/IJAIP.2012.046968.

Zimbardo, P. 2007. "From heavens to hells to heroes." *Inquisitive Mind*, No. 4. http://www.in-mind.org/article/from-heavens-to-hells-to-heroes.

7 "NEURINT" and Neuroweapons

Neurotechnologies in National Intelligence and Defense*

Rachel Wurzman and James Giordano

CONTENTS

INTRODUCTION

Advances in neuroscience have progressed rapidly over the past two decades. The field has become increasingly interdisciplinary and has been a nexus for the development and use of a wide range of technological innovations (i.e., neurotechnology). While usually considered in medical contexts, many neurotechnologies may also be viably engaged as weapons. Such "neuroweapons" are obviously of great interest in and to national security, intelligence and defense (NSID) endeavors, given both the substantial threat that these

---

* This chapter is an expanded and updated version of Giordano and Wurzman (2011), adapted with permission from *Synesis: A Journal of Science, Technology, Ethics, and Policy* 2(1):T55–T71.

technologies pose to the defense integrity of the United States and its allies and the via-bility of these approaches in the U.S. NSID armamentarium. The landmark 2008 report entitled *Emerging Cognitive Neuroscience and Related Technologies* by the *ad hoc* Committee on Military and Intelligence Methodology for Emergent Neurophysiological and Cognitive/Neural Science Research in the Next Two Decades (National Research Council of the National Academy of Sciences 2008 [hereafter referred to as the 2008 NAS *ad-hoc* committee report]) summarized the state of neuroscience as relevant to the (1) potential utility for defense and intelligence applications; (2) pace of progress; (3) pres-ent limitations; and (4) threat value of such science. In characterizing the challenges to advancing defense-oriented neurotechnologies—as well as maintaining the United States' international competitive edge—the committee noted that a significant problem was the "... amount of pseudoscientific information and journalistic oversimplifica-tion related to cognitive neuroscience" (NAS 2008). More recently, a series of Strategic Multilayer Assessment (SMA) conferences considered the potential impact of neurosci-entific understanding of aggression, decision-making, and social behavior on policy and strategy pertaining to NSID deterrence and influence campaigns (Canna and Popp 2011; Sapolsky et al. 2013). These reports highlight (1) how neuroscientific insights to individ-ual, collective, and intergroup social behavior might be used to finesse an understanding of threat environments in an ever-increasingly interdependent and changing environ-ments; (2) the utility of neuroscience and neurotechnologies (i.e., "neuro S/T") for NSID analysis and operations in the context of conflicts with state and nonstate actors; and (3) how neuroscientific understanding of aggression may influence strategies for deterrence.

Given the relative nascence of neuroscience and much of neurotechnology, the development and use of neuroweapons discussed in this chapter are incipient, and in some cases, the potential utility of these approaches is speculative. Yet the pace of progress in neuro S/T continues to increase as the creation of new tools and theories continue to build upon one another. It is notable that since the publication of an earlier iteration of this chapter (Giordano and Wurzman 2011), many ideas described are presently under development. Accordingly, any such speculation must acknowledge that neurotechnological progress in these areas is real, and therefore, consideration of the potential trajectories that neurotechnologies-as-weapons might assume is both important and necessary. As well, such discussion must entail a pragmatic view of the capabilities and limitations of these devices and techniques and the potential pitfalls of—and caveats to—their use. Herein, we address (1) the possible ways that neuro-technologies can be utilized as weapons; (2) the NSID aims that might be advanced by neuroweapons; (3) the specific import of neuroscience to intelligence collection and analysis as "NEURINT"; and (4) some of the consequences and/or implications of developing neurotechnologies toward these ends.

WHAT IS A NEUROWEAPON?

A weapon is formally defined as "a means of contending against an other" and "... something used to injure, defeat, or destroy" (Merriam-Webster Dictionary 2008). Both definitions apply to neurotechnologies used as weapons in intelligence and/or defense sce-narios. Neurotechnology can support intelligence activities by targeting information and technology infrastructures, to either enhance or deter accurate intelligence assessment, the ability to efficiently handle amassed, complex data, and human tactical or strategic efforts. The objectives for neuroweapons in a traditional defense context (e.g., combat) may be achieved by altering (i.e., either augmenting or degrading) functions of the nervous system, so as to affect cognitive, emotional and/or motor activity, and capability (e.g., perception, judgment, morale, pain tolerance, or physical abilities and stamina). Many technologies (e.g., neurotropic drugs; interventional neurostimulatory devices) can be employed to produce these effects. As both "nonkinetic" (i.e., providing means of contending) and "kinetic" weapons (i.e., providing means to injure, [physically] defeat, or destroy), there is particularly great utility for neuroweapons in irregular warfare, where threat environments are "asymmetric, amorphous, complex, rapidly changing, and uncertain" and require greater "speed and flexibility in US intelligence gathering and decision-making" (Canna and Popp 2011, 1).

As implements that target, measure, interact with, or simulate nervous system structure, function and processes, the use of neurotechnologies as weapons are by no means a new innovation, *per se*. Historically, such weapons have included nerve gas and various drugs. Weaponized gas has taken several forms: lachrymatory agents (aka. tear gases), toxic irritants (e.g., phosgene, chlorine), vesicants (blistering agents; e.g., mustard gas), and paralytics (e.g., sarin). The escalation of the 2013 conflict in Syria involving the use of weaponized gas demonstrated the ongoing relevance of nervous system targets. Yet these may seem crude when compared to the capabilities of the more sophisticated approaches that can be used today—or in the near future, as novel targets and more powerful delivery mechanisms are discovered (Romano et al. 2007). Pharmacologic stimulants (e.g., amphetamines) and various ergogenics (e.g., anabolic steroids) have been used to augment combatant vigilance, and sedatives (e.g., barbiturates) have been employed to alter cognitive inhibition and facilitate cooperation during interrogation (Goldstein 1996; Abu-Qare et al. 2002; Moreno 2006; McCoy 2006; Romano et al. 2007). Sensory stimuli have been applied as neuroweapons: some to directly transmit excessively intense amounts of energy to be transduced by a sensory modality (e.g., sonic weaponry to incapacitate the enemy), while others cause harm by exceeding the thresholds and limits of tolerable experience by acting at the level of conscious perception (e.g., prolonged flashing lights, irritating music, and sleep deprivation to decrease resistance to interrogation) (McCoy 2006). Even the distribution of emotionally provocative propaganda as psychological warfare could be considered to be an indirect form of neuroweapon (Black 2009).

While such an expansive consideration may be important to evaluate the historicity, operational utility, and practical (and ethical) implications of neurotechnology-as-weapons, in this chapter we primarily seek to provide a concise overview of neuroweapons and therefore restrict discussion to applications of emergent technologies of cognitive neuroscience, computational neuroscience, neuropharmacology, neurotoxicology, and neuromicrobiology. The former approaches (e.g., cognitive and computational neuroscience;) could be used for more indirect (yet still neurofocal) applications, including the enablement and/or enhancing of human efforts by simulating brain functions, and the classification and detection of human cognitive, emotional, and motivational states to augment intelligence, counterintelligence, or human terrain deployment strategies. The latter methods (e.g., neuropharmacology, neurotoxicology, and neuromicrobiology) have potential utility in more combat-related or special operations' scenarios.

This chapter will also consider neuro S/T contributions in a novel intelligence framework, which we call "NEURINT" (i.e., neural intelligence). The human dimension presents unprecedented challenges in current and projected conflicts, which generates more critical needs for improved strategic intelligence. Neuroscientific contributions to strategic intelligence are potentially large, are categorically distinct, and may (more) directly address challenges of the human dimension (e.g., sociocultural dynamics that affect human behavior). We posit that these possibilities require discussion beyond neurotechnological applications for traditional intelligence, as they instantiate a productive conflation of intelligence analysis and operations, with particular utility for asymmetric conflicts and irregular warfare.

CONTENDING AGAINST POTENTIAL ENEMIES: NEUROTECHNOLOGY WITHIN INFORMATION INFRASTRUCTURES AND INTELLIGENCE STRATEGY

Those neurotechnologies that enhance the capabilities of the intelligence community may be considered weapons in that they provide "... a means of contending against an other" (Merriam-Webster Dictionary 2008). Certain neurotechnologies may be particularly well suited to affect performance in and of the intelligence community. While communication technologies generate valuable sources of intelligence to provide strategic insight into human and social domains of conflicts, the volume and complexity of such information also generate steep challenges for analysts and their assistive technologies. As the volume of available information swells, the tasks of both human analysts and the technologies they use are becoming evermore reciprocal and interdependent. Without technology to preprocess and sort large quantities of complicated information, human analysts cannot obtain a cohesive picture from which to draw necessary inferences about the capabilities and intentions of (friendly, neutral, or hostile) intelligence targets. The widespread and inexpensive use of sophisticated communication technology and difficulty of allocating resources to gather intelligence-focal "signals" over evermore increasing, nonrelevant "noise" have made more coherent collection and interpretation of intelligence information a priority (NAS 2008; Pringle and Random 2009). Yet, information technology presently requires human programming and implementation of human-conceived (and biased) models to parse the volume and types of information collected. Moreover, some information remains problematic to collect (e.g., attitudes and intentions of human subjects). Neurotechnologies that would facilitate and enhance collection and interpretation capabilities might decrease the fallibility of "human weak links" in the intelligence chain. Neurally yoked, advanced computational strategies (i.e., brain–machine and machine–brain interfaces; BMI/MBI, respectively) can be applied to the management and integration of massed data. Similarly, neurotechnologies can be developed to manage the increasingly significant problem of the sheer volume of cyber-based communications that has threatened intelligence systems with inundation.

The principal neurotechnologies that can be used in these tasks are distributed human–machine systems that are either employed singularly or linked to networked hierarchies of sophisticated BMIs, to mediate access to, and manipulation of signal detection, processing, and/or integration. Neurotechnologic innovations that are capable of processing high volume, complex data sets are based upon computational biology as physiomimetic computing hardware (NAS 2008). Such hardware leverages analog, rather than digital components, with "a continuous set of values and a complex set of connections," based on an understanding of neural networks as more than mere binary switches. An analog circuit approach would address current "modeling and simulation challenges," be smaller and "... easy for the US— and its adversaries—to construct" (NAS 2008). As well, given the analog nature of the magnetic fields used for real-time computing, a small, portable, physiomimetic computer of this type might be uniquely valuable for applications of high-density information processing (Watson 1997; Konar 1999; von Neumann 2000; Giles 2001; Arbib 2003; Siegelmann 2003; Schemmel et al. 2004; Trautteur and Tamburrini 2007). The Systems of Neuromorphic and Adaptable Plastic Scalable Electronics (SyNAPSE) program at the Defense Advanced Research Projects Agency (DARPA) has explored these possibilities (Pearn 1999). Table 7.1 provides an overview of several other such NSID research programs with a neuroscientific focus. (Information used in the discussion of DARPA/IARPA programs was derived from the respective program websites, which are listed in the "Online sources" section of the references.)

Information systems could conceivably be conjoined so that neural mechanisms for assigning and/or detecting salience (i.e., processes involving cortical and limbic networks) may be either augmented or modeled into neurotechnologic devices for rapid and accurate detection of valid (i.e., signal vs. noise) information within visual (e.g., field sensor, satellite and unmanned aerial vehicle [UAV]-obtained images) and/or auditory aspects (e.g., narratives, codes) of human intelligence (HUMINT) or signal intelligence (SIGINT). Formulating and testing credible hypotheses while monitoring large amounts of information could be accomplished by computational cognitive frameworks that are capable of both self-instruction (e.g., using the Internet as a "training environment") and learning from experience (e.g., via direct access to the operational environment). This articulates a form of artificial intelligence (AI) that functions to mimic human neural systems in cognition. The 2008 NAS ad hoc committee identified such technology as a potential threat, but one that remains largely theoretical—at least at present (NAS 2008).

Nevertheless, efforts are already underway with these aims (Table 7.1). For example, the Mind's Eye Program (MEP) at DARPA is already attempting to build a "smart camera with machine-based visual intelligence," capable of learning "generally applicable and generative representations of action between objects in a scene". For text-based environments, DARPA's Machine Reading Program seeks to replace "expert and associated knowledge engineers with unsupervised or self-supervised learning systems that can "read" natural text and insert it into AI knowledge bases (i.e., data stores especially encoded to support subsequent machine reasoning)." The Intelligence Advanced Research Project Activity's (IARPA) program titled Knowledge Representation by Neural Systems (KRNS) is aimed at "understanding how the brain represents conceptual knowledge to lead to building new analysis tools that acquire, organize, and yield information with unprecedented proficiency". Another IARPA program, Integrated Cognitive Neuroscience Architecture for Understanding Sensemaking (ICArUS), seeks to understand how humans conduct sense-making under various conditions and how bias affects computational models

TABLE 7.1

Overview of NSID Research Programs with a Neuroscientific Focus

Category	Program	Principal Development	Goals and Objectives
Physiomimetic computing	SYNAPSE[a] (Systems of Neuromorphic and Adaptable Plastic Scalable Electronics)	Nanoscale neuromorphic electronic systems	Develop electronic neuromorphic machine technology that scales to biological levels of microcircuitry and information processing capacity Process extremely high volume, complex datasets to assist problem solving and intelligence analysis Development of architecture and design tools, fabrication process, and large-scale computer simulations of neuromorphic electronic systems to inform and validate design prior to fabrication (DARPA/SYNAPSE website)
Systems with automated learning capabilities	Mind's Eye Program[b]	Smart camera with machine-based visual intelligence	Automate "the ability to learn generally applicable and generative representations of action between objects in a scene directly from visual inputs, and then reason over those learned representations " (DARPA/MEP website)
	Machine Reading Program (this program is now closed)[b]	Automated text-analysis system capable of self-instruction and learning from experience	Develop "unsupervised or self-supervised learning systems that can "read" natural text and insert it into (...) data stores that are encoded to support subsequent machine reasoning" (DARPA/MRP website)
Understanding and modeling cognition to create new analysis tools	KRNS[c] (Knowledge Representation by Neural Systems)	Systems capable of (both predictive and interpretive) association of patterns of neural activity and particular concepts	Develop novel theories "that explain how the human brain represents diverse types of conceptual knowledge within spatial and temporal (dynamic) patterns of neural activity" Understand "how the brain's representation of an individual concept varies as a function of semantic context" and "how combinations of multiple individual concepts are represented in the brain" (IARPA/KRNS website)

ICArUS[c] (Integrated Cognitive Neuroscience Architecture for Understanding Sensemaking)	Computational models to operate in a geospatial task environment with functional architecture that conforms closely to the human brain	Understand how human sense-making operates in optimal conditions and how cognitive biases related to attention, memory, and decision-making contribute to optimal analysis performance. Develop a model of "sense-making" that "borrows" human capabilities on a idealized conceptual (as opposed to mechanical) basis. (IARPA/ICArUS website)
Cybernetic systems and components[f]		
RE-NET[d] (Reliable Neural Interface Technology)	Implanted interfaces with the central and peripheral nervous systems with improved reliability	Understand the "interactions between biotic and abiotic systems" and determine "what mechanisms lead to interface failure," to improve the long-term (e.g., decades) reliability of implanted neural interfaces. Build complete systems that interface with the central and peripheral nervous systems, to "develop a deeper understanding of how motor-control information is conveyed from neural tissue through implanted interfaces and electronics to efficient and robust decoding algorithms" as well as enhance the speed and resolution of sensing/stimulating capabilities (DARPA/RE-NET website)
REMIND[d] (Restorative Encoding Memory Integration Neural Device)	Biomimetic models of the hippocampus to serve as a neural prosthesis for memory impairment	"Determine quantitative methods to describe the means and processes by which short-term memory is encoded" "Demonstrate the ability to restore performance on a short-term memory task in animal models" (DARPA/REMIND website)
RAM[d] (Restoring Active Memory)	"Implantable neural device for human clinical use to restore specific types or attributes of memory"	Mitigate the long-term consequences of TBI on memory Develop models of how neurons code for conscious recall of knowledge such as events, times, and places "Explore new methods for analysis and decoding of neural signals in humans to understand how neural stimulation could be applied to facilitate recovery of memory encoding following brain injury" (DARPA/RAM website)

(Continued)

TABLE 7.1

(Continued) Overview of NSID Research Programs with a Neuroscientific Focus

Category	Program	Principal Development	Goals and Objectives
	SUBNETS[d] (Systems-Based Neurotechnology for Emerging Therapies)	Closed-loop diagnostic and therapeutic neural interfaces for evaluating and treating complex systems-based brain disorders	Develop new neurotechnology with the capability to detect and model how brain systems function in depression, compulsion, debilitating impulse control, and chronic pain and then adapt these models to closed-loop neural interfaces for simultaneous recording, analysis, and stimulation in multiple brain regions "Create one of the most comprehensive datasets of systems-based brain activity ever recorded" (DARPA/SUBNETS website)
	Revolutionizing Prosthetics Program[d]	Neurally interfaced, anthropomorphic, advanced modular prosthetic arm systems with near-natural range of motion, dexterity, and control	Further integrate the prosthetic arm(s) thus far developed with peripheral and central nervous systems to achieve brain control of the artificial limb Prevent limb loss through refined application of dextrous hand capabilities to small robotic systems used in manipulating unexploded ordnance (DARPA/Revolutionizing Prosthetics website)
Fundamental neural systems modeling	REPAIR[d] (Reorganization and Plasticity to Accelerate Injury Recovery)	Decompartmentalized models of neural computation subserving brain function that are able to integrate from neuronal to interregional brain activity with high specificity	"Uncover the mechanisms underlying neural computation and reorganization to improve modeling of the brain and our ability to interface with it" "Determine the sequencing of neural signaling from initial cues to task completion and correlate these neuron-level signals to changes in brain activity" (DARPA/REPAIR website)

	Neuro-FAST[b] (Neuro Function, Activity, Structure, and Technology)	Novel optical methods for unprecedented visualization, recording, and decoding of brain activity	"Overcome the dual challenges of achieving single-neuron resolution while simultaneously being able to analyze activity from large numbers of neurons to acquire detailed modeling of the dynamic wiring of neural circuits that cause behavior"
			Combine advances in genetics, optical recording, brain–computer interfaces, and tissue processing (i.e., CLARITY) to permit simultaneous individual identification of specific cell types, registration of connections between organizations of neurons, and tracking of firing activity in awake, behaving subjects
			(DARPA/Neuro-FAST website)
Cultural norms recognition and insight	SCIL[c] (Sociocultural Content in Language)	Novel designs, algorithms, methods, techniques, and technologies to automatically identify social actions and characteristics of groups by examining the language used by group members	Automated correlation of social goals of a group (identified using existing social science theories) with language used by the group's members
			"Gain insight into and evidence for the nature and status of the group and the roles and relationships of its members"
			"Attempt to generalize the behaviors across cultures in order to highlight the contrasts and similarities in accomplishing social goals"
			(IARPA/SCIL website)
	Metaphor Program[c]	Tools to help recognize norms across cultures and reveal contrasting stances	"Develop automated tools and techniques for recognizing, defining, and categorizing linguistic metaphors associated with target concepts and found in large amounts of native-language text and validate the resulting metaphors through social science methods"
			Apply developed methodologies to "characterize differing cultural perspectives associated with case studies of the types of interest to the Intelligence Community"
			(IARPA/Metaphor website)

(Continued)

TABLE 7.1

(Continued) Overview of NSID Research Programs with a Neuroscientific Focus

Category	Program	Principal Development	Goals and Objectives
Neural causes and biopsychosocial effects of narratives and social identities	Narrative Networks Program[d]	Understanding how narratives influence human cognition and behavior for applications in international relations and security contexts	"Develop quantitative analytic tools to study narratives and their effects on human behavior in security contexts" "Analyze the neurobiological impact of narratives on hormones and neurotransmitters, reward processing, and emotion–cognition interaction" "Develop models and simulations of narrative influence in social and environmental contexts, develop sensors to determine their impact on individuals and groups, and suggest doctrinal modifications" (DARPA/Narrative Networks website)
	SMISC[b] (Social Media In Strategic Communication)	Tools to analyze patterns and cultural narratives expressed in social media, to augment capabilities to "counter misinformation or deception campaigns with truthful information"	Research "linguistic cues, patterns of information flow and detection of sentiment or opinion in information generated and spread through social media [and crowd sourcing] to "track ideas and concepts" as well as "model emergent communities and analyze narratives and their participants" Create a closed and controlled environment with large amounts of data for collection (such as a closed social media network or role playing game) in which analysis tools can be developed and tested by performing experiments (DARPA/SMISC website)
	SSIM[d] (Strategic Social Interaction Module)	Training techniques to impart "basic human dynamics' skills and proficiencies needed to enter into social encounters, regardless of the cultural, linguistic, or other contextual parameters" to warfighters	"Identify and codify the constitutive elements of successful social interaction skills" "Develop a training simulator with virtual social interaction space operating in support of human-based training" "Develop techniques to measure the effectiveness of training and wider longitudinal outcomes" (DARPA/SSIM website)

| Behavioral and semantic cues processing | DCAPS[b] (Detection and Computational Analysis of Physiological Signals) | A "general metric of psychological health" capable of detecting "subtle changes associated with posttraumatic stress disorder, depression and suicidal ideation." | Develop novel algorithms to extract "distress cues" from patterns within "data such as text communication, daily patterns of sleeping, eating, social interaction and online behaviors and nonverbal cues such as facial expression, posture, and body movement"

Correlate distress markers derived from neurological sensors with distress markers derived from algorithmic analysis of "sensory data inherent in daily social interaction" (DARPA/DCAPS website) |
| | TRUST[e] (Tools for Recognizing Useful Signs of Trustworthiness) | A subjective perceptual process to "assess whom can be trusted under certain conditions and in contexts relevant to the intelligence community, potentially even in the presence of stress and/or deception" | Conduct "research to bring together sensing and validated protocols to develop tools for assessing trustworthiness by using one's own ('Self') signals to assess an other's ('Other') trustworthiness" (IARPA/TRUST website) |

DARPA: Defense Advanced Research Projects Agency; IARPA: Intelligence Advanced Research Project Activity.

a DARPA—Defense Sciences Office

b DARPA—Information Innovation Office

c IARPA—Office of Incisive Analysis

d DARPA—Biological Technology Office

e IARPA—Office of Smart Collection

f These programs are for strictly clinical aims; however, such accomplishments in the clinical sphere may lead to enabling technologies such as BMIs with dual-use applications.

whose functional architecture "closely conforms to that of the human brain". Such computational cognitive frameworks may "borrow" human capabilities, not by mimicking processes in the brain (which may not be sufficiently well understood to begin with), but by involving conceptual components of idealized neurally modeled systems that are linked in ways that enable performance of similar—if not more rapid and advanced—neurocognitive functions.

Moreover, neurally coupled hybrid systems could be developed that yoke computational interfaces to human neuronal activity, so as to optimize Bayesian-like predispositions to certain types of stimuli (Sato 2011). This would limit input data sets to more critical features and thereby allow more efficient (i.e., rapid and accurate) detection, observation, orientation (and decisions) by the human user. This conjoinment and reciprocity could be used to enhance the feature detection and intelligence capacities of both machine and human systems. A prototypical example of such neurotechnology, the Revolutionary Advanced Processing Image Detection (RAPID) system, is discussed by Stanney and colleagues (elsewhere in this volume) as a training adjunct to support image analysis that uses neural signatures to "eliminate the need for assessing behavioral responses" and "improve the speed of image throughput." RAPID deploys eye tracking technology and electroencephalography event-related potentials (EEG/ERPs) in an individual viewing an image, to "track the imagery analysis process and automatically identify specific [areas of interest] within reviewed images."

Enhancement of neural and cognitive capabilities may be achieved through some form of cybernetics, broadly considered as a feedback and feedforward system that obtains iterative reassessment and modification capacities through ongoing interactions between an agent and its environments (Wiener 1948). According to the classification scheme of Clynes and Kline (1961), the use of human–machine interfaces can be regarded as a level 2 or level 3 cybernetic organism (i.e., a cyborg) in that it entails both natural and artificial systems that are functional, portable, and/or biologically integrated. Cybernetic and cyborg systems can be seen as sophisticated distributed human–machine networks, such as integrated software or robotic augmentations to human-controlled activity, that would fuse and coordinate the distinct cognitive advantages of humans and computers. As stated in the NAS *ad hoc* committee report, these systems could assist "... advanced sensory grids, and could control unmanned autonomous systems, advanced command posts and intelligence analyst workbenches, coordination of joint or coalition operations, logistics, and information assurance" with consequences that "enhance the cognitive or physical performance of war fighters and decision-makers or allow them to coordinate the actions of autonomous systems with much-improved effectiveness" (NAS 2008). These systems would be of evident utility to multiple forms of intelligence acquisition and processing at both tactical and strategic levels. One DARPA program, Reliable Neural Interface Technology (RE-NET), while not conceived for such application, might nevertheless lay groundwork to develop stable physical components for BMI that might be used in this way. Two other programs, Restorative Encoding Memory Integration Neural Device (REMIND) and Restoring Active Memory (RAM), strictly address the clinical problem of memory loss and attempt to restore memory function with the creation of cybernetic neural prostheses. If successful, the creation of neurobiomimetic cybernetic technology will confirm the feasibility of engineering human–machine systems that could be of value in several of the aforementioned NSID tasks and challenges.

NEUROTECHNOLOGICAL APPLICATIONS FOR STRATEGIC INTELLIGENCE

Strategic intelligence is defined as gathering and analyzing information regarding the capacities and intentions of foreign countries; it may also encompass political intelligence, given that "… [political intelligence] is at once the most sought-after and the least reliable of the various types of intelligence. Because no one can predict with absolute certainty the effects of the political forces in a foreign country, analysts are reduced to making forecasts of alternatives based on what is known about political trends and patterns" (Pringle 2009). The complex dynamics of political forces that contribute to such predictive difficulty are due, in part, to the numerous and varied agents involved, all of whose actions are individually determined. Thus, understanding the biopsychosocial factors that influence individual and group dynamics and being able to detect these variables with high ecological validity (i.e., "in the field," under real-world conditions) are important to both descriptive/analytic and predictive intelligence approaches (*vide infra*).

A combination of (1) advanced sociocultural neuroscientific models of individual–group dynamics based upon theories of complexity adapted for use(s) in anthropology, (2) sufficient computing and BMI frameworks (perhaps as speculated above), and (3) certain forms of neuroimaging to accurately detect the mental states and decision-biases of key or representative individuals might enable dramatically improved forecasting of behavior patterns that are influential to sociopolitical change. These forecasts could include the description of: mental states of specific agents/actors, the propagation dynamics of an idea or cultural construct, and/or node–edge interactions of individuals and cohorts—any and all of which might be viable to subsequently identify specific targets for manipulation (via other neuroweapons).

However, *intentions*, as opposed to corresponding cognitive and/or emotional states and their associated neuronal signatures, are difficult to detect using existing neurotechnologies. This affects and alters the modeling approaches that could—and should—be used to describe or predict individual or group activities. As well, it is important to consider the potential of technological interventions to alter events. Here, lessons may be garnered from experience with psychological warfare (Goldstein 1996). Sometimes, techniques and tactics will induce unintended, if not frankly contrary effects and results. Given the overarching applications of neurologically and psychologically viable approaches, there is interest in neurotechnology to augment the role, capability, and effect(s) of psychological operations (PSYOPs) as a "force multiplier" in both political and military tactics. This trend began with the 1985 Department of Defense (DoD) PSYOP's master plan and has been accelerated by the challenges posed by insurgencies in the present conflicts in Iraq, Afghanistan, Lybia, and Syria (Paddock 1999).

Such challenges emphasize the problems of cultural intelligence and how these generate psychosocial obstacles to achieving tactical ends. Tactical deficits may be related to the military approach to psychological–political warfare as being centered upon a "conflict of ideas, ideologies, and opinions" while not adequately emphasizing notions such as "cultural and political symbols, perceptions and emotions, behavior of individuals and groups under stress, and cohesion of organizations and alliances" (Lord 1996). Even if aware of such variables, we might still be flummoxed in influencing "the minds and hearts" of enemy combatants, because of the failure to correctly define and predict which factors may influence aspects of psychological warfare (such as the severance or formation of alliances and collectives' reactions to the threat of integrity).

Thus, an appeal of neurotechnology is its (theoretical) potential for use in (1) defining substrates and mechanisms operative in culturally relevant cognitions and behaviors and (2) directly affecting perceptions, emotions, behaviors, and affiliative tendencies. The most obvious possibility is the use of neurotechnology to assess and affect cognitive capability, emotions, and/or motivation. Various forms of neuroimaging have been considered, as have the concomitant use of neurogenetics and neuroproteomic approaches in this regard. However, cognitive and emotional effects in individuals and across a population are complicated and can often be unpredictable. Hence, a main criticism of neuroimaging is that although relatively valid and reliable in assessing individual mechanisms and substrates of cognition and emotion under controlled (i.e., experimental) situations, the ecological validity of such protocols is questionable, and thus neuroimaging may be of limited value in depicting more subtle cognitive-emotional and motivational states, such as deception in "real-world" scenarios (Uttal 2001; Illes and Racine 2005). Adding to this is that neuroimaging is not a subtle technique, and protocols for assessing cognitive-emotional variables would need to be explicitly concerned with the ways that the testing environment affects individuals being evaluated. Neurogenetics and neuroproteomics could enable assessment of predispositional variables and even phenotypic characteristics that influence cognitions, emotions, and behaviors, but these approaches are of only limited predictive value, given the nonlinear relationship of genetics to both phenotype(s) and the ultimate expression of cognitive states and behavioral actions (Ridley 2003; Wurzman and Giordano 2012).

Arguably, a more culturally invariant framework for conceptualizing cultural norms is required before we can understand how they interact with neural substrates to influence behavior. A significant problem is that cultural norms are often opaque, and thus, it becomes difficult to recognize such influences on behavior. Programs such as Sociocultural Content in Language (SCIL) and the Metaphor program at IARPA are geared toward obtaining insight to how to better recognize norms across cultures. The Narrative Networks program at DARPA employs a specifically neuroscientific approach to understanding and modeling the influence of narratives in social and environmental contexts and seeks to "develop sensors to determine their impact on individuals and groups". Ultimately, it is hoped (and quite likely) that a better understanding of the neural causes and effects of narratives will contribute significant insight to the reciprocal (neuro)biological, psychological, and sociocultural effects on brain development, function, and behavior, in a way that can be leveraged for operational or analysis purposes.

NEURINT: NEURAL INTELLIGENCE AS A NOVEL COLLECTION DISCIPLINE AND ANALYSIS TOOL

Consensus expressed at the recent SMA conferences (on influence strategies of state and nonstate actors as well as the impact of social and neurobiological sciences on key aspects of national security) endorsed the need for better frameworks to analyze and influence human factors that contribute to heightened complexity and uncertainty in threat environments. In particular, there is an increased awareness of the need to explicitly comprehend the often concealed influence of social and cultural norms on perception, cognition, decision-making, and behavior in individuals and collectives. In this sense, there is a specific need for intelligence that speaks to identity—not in the *idem-identity* sense (i.e., about sameness, or *what* constitutes someone; as in having constant ownership of a body), but rather in the *ipse-identity* sense (i.e., about selfhood and *who* someone is, as in their "biographical, embodied, temporal, and narrative dimensions") (Ajana 2010). Such intelligence would not seek to reduce information into a type or a category (e.g., a "trace" of a "what" the subject is), but instead engages iterative reflection of biographical stories (e.g., an "echo" of "who" the subject is).

The Social Media in Strategic Communication (SMISC), Strategic Social Interaction Module (SSIM), and Narrative Networks programs at DARPA reflect a growing awareness of the critical roles that social identities, oft-hidden cultural norms, and narratives occupy in providing necessary context during the collection, analysis, and utilization of strategic intelligence at the levels of individuals and groups. Furthermore, there is implicit recognition of the neural basis of such effects, operating both upon the subject and the analyst or decision-maker. Other DARPA and IARPA programs seize the opportunity for neuroscience and neurotechnology to augment understanding of neurophysiological and neuropsychological processing of behavioral and semantic cues, given and perceived (Table 7.1). For example, DARPA's Detection and Computational Analysis of Psychological Signals (DCAPSs) program, although intended for a clinical as opposed to an intelligence context, seeks to better understand distress cues by correlating distress markers derived from neurological sensors with those from algorithms that extract "distress cues" out of patterns within data "such as text and voice communications, daily patterns of sleeping, eating, social interactions and online behaviors, and nonverbal cues such as facial expression, posture and body movement." As such, the program aims to "[extract and analyze] 'honest signals' from a wide variety of sensory data inherent in daily social interactions". Similar methods could be applied to other emotional, cognitive, and perceptual experiences (beyond trauma response) to generate a better understanding of motivations, intentions, beliefs, values, experiences, and actions. Such improvements could be powerfully leveraged for strategic advantage during kinetic operations or intelligence analysis.

Importantly, this source of intelligence is inextricable from influences afforded by the social, cultural, and psychological milieu of the individual subject(s). Thus, the aim is ontologically distinct from mere biometric applications to intelligence in that it does not seek to read body signals to categorize a behavior or expression in terms of "what" biopsychological state is presented. On the one hand, it has been pointed out that although "biometric technology recognizes the fact that bodies are indeed biographies, it hardly offers an outlet for *listening* to those biographies" (Ajana 2010); on the other hand, NEURINT accesses the interaction between the "story" and the "attribute" (or the "who" and the "what") represented by an individual's narrative and biometric data. An important factor here is the assumption that relationships between biometric patterns and neural activity are individualistic; thus, the utility in understanding these variables is not to identify the "what" of a person (e.g., typing or categorizing, or otherwise reducing according to patterns digital data), but instead, is to be found in recognizing their contingency (e.g., between the brain, body, and biography). In other words, it invokes the comment of Mordini and Ottolini (2007) that "[b]ody requires mind, not in the trivial sense that you need a neurological system to animate the body, but in the profound sense that the very structure of our body is communicational [...] We do not just need words. We are words made flesh."

By first cross-correlating putative neural mechanisms subserving both experiences and an individual's biometric patterns, NEURINT collection shifts the process from one of "reading" (off) the body to one of "listening" (in)to the body. Biometric analyses alone are often used to verify identification and thus "reduce singularity and uniqueness to sameness" (Ajana 2010). A complementary understanding of the relationships of biometrics (as well as the embodied experiences they reflect) to neurological signals prevents the inadvertent "reduction of the story to its attributes." This requires that any biometric or behavioral indicators that are collected and analyzed (with an aim to draw inferences about subjective phenomena in target populations) must first be studied using rigorous research methods to establish a neural framework for understanding such phenomena.

On the other hand, the analysis of NEURINT is also inextricable from influences afforded by the social, cultural, and psychological milieu of the individual *analyst(s)* (as well as the target subject[s]). Therefore, as an analysis tool, NEURINT does not yield products with predictive validity that can be considered independently. However, its outcomes do dynamically enhance analysis and utility of HUMINT and SIGINT/COMINT (of which NEURINT may be considered to be essentially comprised.) This is due to the fact that the analyst's own cognitive filters are subject to the neurobiological effects of cultural norms and narratives. By its contingent nature, NEURINT engages the analyst in a hermeneutic, interpretive process that neither seeks nor attains a stable meaning for the data, but instead maintains an open process of reinterpretation and expandability. At first glance, this seems to negate its utility as a source of actionable intelligence. Yet it is this process by which NEURINT remains irreducible to neural or biometric phenomena or (conversely) to any purely subjective construct. Thus, it has a unique essence that renders it a distinct form of intelligence. Furthermore, given that the complex human domain into which it seeks to provide operationally relevant insight entails similarly unstable, open processes. NEURINT is revealed as both agile and flexible. This is due to its constructive nature, which confers the advantage of novel ways to (1) mitigate the analyst's (sociocultural) biases and/or (2) compensate for ways that the analyst's own narratives and identity influence the perception of threats, or how meaning is assigned to observations. The process by which an analyst interprets NEURINT *explicitly* engages principles of how the analyst's and subject's personal narratives influence their mutual projection and perception of the others' narrative (i.e., that which is experienced by the subject and that which is assigned/attributed to the subject by the analyst). While the loop remains open, the process itself permits insight specifically into *how the two cultural narratives interact to influence one another's sense of identity in an individualistic context.*

Provided that the basic inferences are rooted in an empirically derived and neurotechnologically enabled dynamic understanding of how neural signals and biometric indicators correspond with subjective phenomena experienced by the subject (in a correlated fashion), the process of NEURINT analysis may be used to provide insight about identity and active narratives in target populations. In turn, these may suggest tools, strategies, or direct interventions for improving identification, communication, and rapport, which thereby enhance collection and nuance the analyses of HUMINT and SIGINT/COMINT. One example of a research program essentially aligned with the principle strategy of NEURINT is IARPA's Tools for Recognizing Useful Signs of Trustworthiness (TRUST) program. TRUST leverages intersubject variability and dynamic interaction between a sensor and its target to validate a subjective perceptual process for assessing a behavioral trait or tendency in a target.

Beyond direct measurements of neurological activity, NEURINT may be collected as narratives from electronic sources or as human biometric observations during social interaction or surveillance. Fundamentally, NEURINT provides an additional layer of context to HUMINT and SIGINT by suggesting which neural systems and processes are engaged at the time of the observed behavior. In this way, it might be used to train personnel to recognize key biometric indicators enabling the strategic engagement with, or manipulation of others' psychological state to best advantage in strategic kinetic and nonkinetic deployments. Another possibility is that NEURINT might provide for real-time identification of sacred narratives being invoked during an interview, which might then specifically guide later interpretation, filtering, and analysis of information. High-level analysis of narrative patterns in a NEURINT context could also assist with deception detection—including self-denial—which is much less detectable through biometrics. NEURINT may be of value for optimizing communication with individuals or groups by catering to cognitive styles or perceptual sensitivities. Finally, an additional tier of insight may be afforded by systematically relating evidence-supported inferences about the analyst's cognition and perceptions (i.e., based on biometric signals or possible proxy linguistic indicators) to those inferred from observations of the subject.

At present, specific NEURINT methodologies have yet to be developed. However, their potential is tantalizing. While NEURINT research and its enabling technologies require sophisticated equipment, the collection and analysis of NEURINT may not need to assume a highly technical form for operational deployment, which might overcome obvious obstacles such as equipment size and the lack of ecological validity. Of course, today's limitations often represent the challenges and opportunities for tomorrow's technology, and ongoing work is dedicated to use of a more convergent scientific and technological paradigm to compensate for extant constraints and limitations, so as to create technologies that are effective and easily employed/deployed in operational settings (Giordano 2011).

NEUROWEAPONS IN COMBAT SCENARIOS

A considerably more imposing possibility is to "change minds and hearts" by altering the will or capacity to fight through the use of neuropharmacologic, neuromicrobiological, and/or neurotoxic agents that (1) mitigate aggression and foster cognitions and emotions of affiliation or passivity; (2) incur morbidity, disability, or suffering and in this way "neutralize" potential opponents; or (3) induce mortality. James Hughes (2007) has identified six domains of neurocognitive function that can currently be pharmacologically manipulated; these are (1) memory, learning, and cognitive speed; (2) alertness and impulse control; (3) mood, anxiety, and self-perception; (4) creativity; (5) trust, empathy, and decision-making; and (6) waking and sleeping. As well, movement and performance measures (e.g., speed, strength, stamina, motor learning) could also be enhanced or degraded (Pringle and Random 2009).

NEUROTROPIC DRUGS

As mentioned previously, the use of neuropharmacological agents to affect cognitive, emotional, and behavioral abilities is certainly not novel (*vide supra*). However, an expanded fund of neuroscientific knowledge, namely, the increased understanding of molecular and systems-based, structure–function relationships of the brain, has fortified depiction of substrates and mechanisms that are viable pharmacologic targets. Such knowledge, when coupled to advancements in pharmaceutical technology, has allowed discovery and development of neurotropic agents with greater specificity, potency, and bioavailability.

In general, drugs that have utility in combat and/or special operational settings include (1) cognitive and motoric stimulants such as the chain-substituted amphetamine, methylphenidate (Hoag 2003), and the novel dopaminergic reuptake blocker and histamine and orexin potentiating agent modafinil (Buguet et al. 2003); (2) somnolent agents such as the barbiturates, benzodiazepines, and certain opiates (Albucher and Liberzon 2002); (3) mood-altering agents such as the azaspirone anxiolytics (e.g., buspirone; Albucher and Liberzon 2002), beta-adrenergic antagonists (e.g., propranolol, which has been considered for its effects in decreasing agitation and anxiety associated with traumatic events; Albucher and Liberzon 2002), as well as dopamine and serotonin agonists (which at higher doses have been shown to induce fear and psychotic symptoms including paranoia; Davis et al. 1997); (4) "affiliative" agents such as the neurohormone oxytocin (Gimpl and Fahrenholz 2001) and the substituted amphetamines (e.g., methylenedioxy methamphetamine, MDMA—"ecstasy"; Murphy et al. 2006); and (5) epileptogenics, such as acetylcholine (ACh) and gamma aminobutyric acid (GABA) receptor antagonists

(Rubaj 2003). The actions and effects of these categories of drugs are provided in Table 7.2.

While some of these agents can be used to enhance the neurocognitive and motor performance of (one's own) troops (e.g., low dose of stimulants and mood-altering drugs), others have apparent utility against hostile forces (e.g., somnolent, psychotogenic, affiliative, and epileptogenic agents). Moreover, while a "weapon" is characteristically considered to be a tool used to incur injury, agents such as oxytocin and/or MDMA may actually reduce or prevent harm inflicted on an opponent by decreasing their desire to fight or making them more amenable to affiliation. These effects are wholly consistent with the more formal definition of a weapon, as "... a means of contending against an other" (Merriam-Webster Dictionary 2008). To paraphrase Kautilya: the person who becomes a friend is no longer an enemy (Kautilya 1929). Yet, this too can be viewed as potentially harmful in that drug-induced effects upon cognition and emotion may alter the identity, autonomy, and free will of others, and in doing so, are exercises of "biopower" (Foucault 2007; Rippon and Senior 2010; see also Chapter 12). Nevertheless, we opine that when attempting to balance benefits, risks, and harms within contexts of *jus ad bello* and *jus in bellum*, such outcomes—while poweful—may need to be considered as less injurious than either more profound forms of neuropsychiatric insult or those produced by more "traditional" weaponry.

To be sure, the use of drugs to affect cognitive, emotional, or behavioral function carries many risks of abuse (e.g., excessive doses or too-frequent use), misuse, unintended interactions with other drugs, foods and situations, and alterations in social behavior (Hughes 2007). Additionally, the effects of any drug depend on an individuals' particular combination of genes, environment, phenotype, and the presence or absence of both physiological and psychopathological traits, and these can vary widely within a given population. Therefore, despite the relatively small size of the military, considerable diversity still exists in the aforementioned characteristics, and this would need to be accounted for, as would any variations in those populations in which the use of neurotropic agents is being considered.

Thus, it is probable that any neurotropic agent would produce variable responses and effect(s) in a population reflective of individual geno- and phenotypes, as well as biological variations in given individuals over time. This could incur an increased likelihood of unanticipated effects. Therefore, it is important that pharmaceutical research, development, testing, and evaluation (RDTE) of such agents engage the time and resources required to maximize desirable drug actions and effects based upon individual and group geno- and phenotypes and assess potentially adverse and/or unwanted side effects. Of course, adverse effects could also be exploited for use on an enemy population. In light of this, drug design would require resources necessary for evaluation and measurement of geno- and phenotypic characteristics that could be optimized to selectively employ particular drugs within a population. By targeting these characteristics, it would be possible to mass deliver agents and still achieve some significant measure of individualized effect(s). Current efforts in "personalized medicine" may afford steps toward

TABLE 7.2

Neuropharmacologic Agents

Category	Type/Drugs	Principal Actions/Effects	Side Effects/Very High Dose Effects
Cognitive/ motoric stimulants	CNS stimulants (e.g., amphetamines[a], methylphenidate[a], pemoline[b])	Increase DA/NE turnover/release, increase arousal, increase attention, elevate mood, induce rebound depression and anxiety	Loss of appetite, insomnia, dizziness, agitation, increased heart rate, dry mouth, high-frequency tremor, or restlessness
	Eugeroics (e.g., modafinil, adrafinil)	Increase DA turnover, decrease DA reuptake, elevate hypothalamic histamine levels, potentiate action of orexin, wakefulness and decreased fatigue, increase arousal, increase attention, few autonomic side effects	Excitation or agitation, insomnia, anxiety, nervousness, aggressiveness, confusion, tremor, palpitations, sleep disturbances, nausea
	Non-stimulant cognitive enhancers (Ampakines[f], e.g., ampalex, farampator, phenotropil)	Potentiate AMPA receptor-mediated neurotransmission, increase attention, increase alertness, no PNS stimulation, enhance learning and memory, increase tolerance to cold and stress	Possible headache, somnolence, nausea, or impaired episodic memory (farampator)
	Other nootropics (Racetams[f], e.g., piracetam, oxiracetam, aniracetam)	Potentiate muscarinic ACh receptor activity, activate NMDA/glutamate receptors co-localized with ACh receptors, nonspecific increase in neuronal excitability (via AMPA and NMDA-receptor activation), enhance learning and memory (nootropic), anti-emetics, anti-spasmodics	CNS stimulation, excitation, depression, dizziness, sleep disturbances
	Monoamine reuptake inhibitors[a,b] (DA, e.g., buproprion[b], atomoxetine, reboxetine; 5-HT/NE, e.g., venlafaxine, mirtazapine)	DA/NE reuptake inhibitors, antidepressant effects, decrease anxiety, increase concentration	Dry mouth, blurred vision, dizziness, buproprion causes seizures at high doses
Somnolent and tranquilizing agents	Benzodiazepines[c] (e.g., lorazepam, prazepam, clonazepam, oxazepam, diazepam, midazolam, alprazolam)	Increase GABA binding, increase neural inhibition, increase somnolence, decrease arousal, decrease reaction time/coordination, motoric lethargy, anterograde amnesic effects	Blurred vision, headache, confusion, depression, impaired coordination, trembling, weakness, memory loss

Azaspirones[b] (e.g., buspirone, gepirone)	$5HT_{1A}$ receptor agonists, decrease NE activity, decrease arousal, decrease agitation, decrease anxiety	Dizziness, nausea, headache, nervousness, insomnia, lightheadedness
Adrenergics[c] (e.g., clonidine, guanfacine)	Stimulate α-adrenergic receptors, NE autoreceptor agonist, sedative effects, reduce heart rate, relax blood vessels	Dry mouth, dizziness, constipation
Barbiturates (e.g., phenobarbitol, mephobarbitol, thiopental, amobarbital, secobarbital, pentobarbital)	Bind to α subunit of $GABA_A$ receptors, Inhibit AMPA receptor-mediated glutamate activity, high doses may inhibit neurotransmitter release, block ligand-gated cation channel receptors (nicotinic, glycine, $5-HT_3$ receptors), decreased arousal, decreased concentration, impaired coordination, slurred speech, decreased anxiety	Potential for lethal overdose via respiratory depression
Muscle relaxants (e.g., carisoprodol, cyclobenzaprine, metaxalone)	CNS-acting muscle relaxers, anti-cholinergic, potentiation of NE, $5HT_2$ antagonist, increase descending 5-HT activity, analgesic effects, increase somnolence	Potential anti-cholinergic effects
Imidazopiridyne hypnotics (e.g., zolpidem, zopiclone)	Potentiate $GABA_A$ receptors, increase in slow-wave sleep, facilitate sleep initiation (not maintenance), anterograde amnesia	Hallucinations, sleep walking
Antipsychotics/neuroleptics (dopamine antagonists, e.g., chlorpromazine, haloperidol, fluphenazine, thioridazine, loxapine, thiothixene, pimozide)	Block D_2 family of DA receptors/autoreceptors, decrease DA release, decrease agitation, increase sedation, hypothalamic-mediated effects on metabolism, body temperature, alertness, muscle tone, and hormone production	Sedation, blood pressure or heart rate changes, dizziness, cognitive dulling
Atypical antipsychotics[d] (e.g., clozapine, risperidone, olanzapine)	Block D_2 family receptor types expressed in limbic-system; decrease DA release; fewer extrapyramidal effects; some are anti-cholinergic (clozapine, perphenazine); may block $5-HT_{2A}$, α1 adrenergic, and H_1 histamine receptors	Tardive dyskinesia (uncontrollable writing movements) Neuroleptic malignant syndrome

(*Continued*)

TABLE 7.2
(Continued) Neuropharmacologic Agents

Category	Type/Drugs	Principal Actions/Effects	Side Effects/Very High Dose Effects
	Anticonvulsants[c] (e.g., gabapentin, pregabalin)	Anticonvulsant, anxiolytic, analgesic in neuralgia, CNS depressant	Somnolence, hypertonicity, abnormal gait, coordination, or movements, vision problems, confusion, euphoria, and/or vivid dreams
Mood-altering agents	Other monoamine antagonists[e] (e.g., reserpine, tetrabenazine)	Prevent DA, NE, and 5-HT transport into synaptic release vesicles, deplete DA in presynaptic terminals, depress mood and motor activity, akasthesia, restlessness	Nausea and vomiting, severe depression, drowsiness, dizziness, and nightmares
	Dissociatives (e.g., PCP, ketamine)	NMDA antagonists, alters monoamine release/reuptake, inhibit DA release in frontal cortex through presynaptic NMDA receptors (PCP), increase DA release/inhibits DA reuptake in limbic areas (PCP), dissociative anesthesia (marked by catalepsy, amnesia, and analgesia), induce/exacerbate psychosis (with euphoria/dysphoria, paranoia, delirium, multisensory hallucinations), ketamine induces anesthesia via σ- and μ-opioid receptors	Muscle weakness, ataxia, loss of consciousness (ketamine), self-destructive or self-injurious behavior, impaired judgment
	Beta-blockers[e] (e.g., propranolol)	β-adrenergic antagonist, decrease autonomic stress response, decrease performance anxiety, reduce posttraumatic agitation and anxiety	Dizziness/light-headedness, drowsiness/fatigue insomnia
	Psychedelics (Tryptamine alkaloids, e.g., ibogaine, yohimbine, psilocybin, LSD)	5-HT$_{2A}$ agonists, decrease 5-HT reuptake, loss of coordination, psychedelic/hallucinogenic effects, σ$_2$ receptor, nicotinic receptor, NMDA receptor antagonism	Dry mouth, nausea, vomiting
	Dopamine agonists (e.g., L-DOPA, tyramid, pargyline, benztropine, apomorphine* ropinerole*, bromocriptine*)	Inhibit DA reuptake/breakdown or increase DA release or act as DA receptor agonists*, may induce psychosis	Induce nausea and vomiting, cause abnormal movements, increase compulsive behavior, enhance paranoia, and/or fear response

Affiliative agents	Amphetamine derivatives (e.g., 3,4-methylenedioxy-methamphetamine, aka., MDMA)	Increase net release of monoamine neurotransmitters (5-HT and NE > DA), decrease 5-HT reuptake, increased wakefulness, endurance, postponement of fatigue and sleepiness, sense of euphoria and heightened well-being, sharpened sensory perception, greater sociability and, extroversion, increased tolerance and desire for closeness to other people	Thermal dysregulation, muscle tension, headache, dry mouth, over-arousal (flight of ideas, insomnia, agitation, hyperactivity), panic attacks, delirium, depersonalization mild hallucinations, or brief psychosis is possible. Neurotoxic with chronic exposure
	Oxytocin[f]	Acts at oxytocin receptors, intracellular Ca^{2+} release, evokes feelings of contentment, calmness, and security, reduces anxiety, increases trust and human bonding, decreases fear, stimulates uterine contractions in pregnancy	Pro-social feelings enhanced may sometimes include ethnocentrism, territoriality, and xenophobia
Epileptogenics	Inverse $GABA_A$ receptor agonists (non-volatile, e.g., DMCM, sarmazenil; volatile e.g., flurothyl[f])	Act at benzodiazepine-binding site of $GABA_A$ receptors, decrease $GABA$-binding affinity of $GABA_A$ receptors, decrease frequency of chloride channel opening (to decrease inhibition), convulsant, stimulant, and anxiogenic effects, inhalation of flurothyl elicits seizures, flurothyl mechanism not well-understood	Anxiety (low doses), seizure
	Competitive $GABA_A$ antagonists (e.g., gabazine, bicuculline)	Bind competitively to $GABA_A$ receptor at GABA-binding site, reduce synaptic inhibition of neurons by reducing chloride conductance through receptor channel, convulsant effects	
	Non-competitive $GABA_A$ antagonists (e.g., picrotoxin bergamide, pentetrazol a.k.a., PTZ)	PTZ increases neuronal excitability by affecting Na^+, K^+, and Ca^{2+} currents via NMDA, $5-HT_{1A}$, $5-HT_3$, glycine receptors, and L-type Ca^{2+} channels; anxiogenic; circulatory and respiratory stimulants; high doses cause convulsions	
	Glycine antagonists (e.g., strychnine, tutin)	ACh receptor antagonists; initial effects of nervousness, restlessness, muscle twitching, neck stiffness give way to pupil dilation and convulsions of increasing severity; death by asphyxiation	Highly toxic

(Continued)

TABLE 7.2
(Continued) Neuropharmacologic Agents

Category	Type/Drugs	Principal Actions/Effects	Side Effects/Very High Dose Effects
	Mixed GABA antagonists/ glutamate agonists (e.g., cyclothiazide)	Positive allosteric modulators of AMPA, inhibit the desensitization of AMPA receptors, potentiate glutamate currents, negative allosteric modulator of $GABA_A$, convulsant without neurotoxic effects	Seizures at higher doses
	Ionotropic glutamate receptor agonists (e.g., kainic acid)	Enhance glutamate effects through kainate receptors, CNS stimulant, excitotoxic convulsant	Neuronal damage
	Muscarinic agonists (e.g., pilocarpine)	Non-selective muscarinic ACh receptor agonist, systemic injection leads to chronic seizures	Excessive sweating and salivation, bronchospasm and increased bronchial mucus secretion, vasodilation, bradycardia
	Nonspecific cholinergic agonists (e.g., physostigmine)	Acetylcholinesterase inhibitor, increases synaptic ACh levels, indirectly stimulates nicotinic and muscarinic ACh receptors	Convulsant at high doses
	Local anesthetics (e.g., lidocaine, prilocaine)	Block fast voltage gated sodium (Na^+) channels; inhibit neuronal firing; moderate systemic exposure causes CNS excitation, symptoms of nervousness, dizziness, blurred vision, tinnitus and tremor; seizures followed by depression	With increasingly heavier exposure: Drowsiness, loss of consciousness, respiratory depression, and apnea

ACh, acetylcholine; AMPA, alpha amino-3-hydroxy-5-methyl isoxazoleproprionic acid; DA, dopamine; GABA, gamma amino butyric acid; 5-HT, 5-hydroxytryptamine; (serotonin); NE, norepinephrine; NMDA, n-methyl d-aspartate.

a Also elevates mood or antidepressant effects; potential as mood-altering agent(s)

b Also lowers seizure thresholds; potential epileptogenic effects, especially at high doses

c Also anxiolytic; potential mood-altering agent(s)

d Also decreases agitation/psychosis and/or antidepressant effects; potential as mood-altering agent(s)

e Significant sedation side effects; potential as somnolent agent(s)

f Mechanism not well understood, or effects inconclusively established at this time realizing these possibilities (Calabrese and Baldwin 2001; Nebert et al. 2003). As well, certain physiochemical obstacles to delivery have been overcome by the use of nano-neurotechnologies that have allowed molecular conjugation or encapsulation of ligands, ligand precursors, and biosynthetic enzymes capable of crossing the blood–brain and blood–cerebrospinal fluid barriers and thus permitting greater access to, and higher bioavailability in the brain (NAS 2008).

NEUROMICROBIOLOGICAL AGENTS

A number of microbiological agents directly target or indirectly affect the central nervous system (CNS), and these are certainly employable as neuroweapons (Watson 1997; NAS 2008). Of particular note are (1) the viral encephalitides, such as the *Alphavirus* genus of the Togaviridae family that cause Venezuelan, Eastern, and Western equine encephalitis (Smith et al. 1997); (2) the anaerobic bacterium *Clostridium botulinum*, the seven strains of which produce specific neurotoxins (Nagase 2001); and (3) the sporulating bacillus, *B. anthracis* that causes anthrax (Nagase 2001) (Table 7.3).

Here too, use of nanotechnology to preserve stable forms of pathogenic agents could be important to more durable aerosolized neurobioweapons (McGovern and Christopher 1999). While capable of inducing large-scale infections in a given population, such mass casualty effects may not be required or desired. Instead, using these agents in more punctate approaches might be advantageous. Such techniques include (1) inducing a small number of representative cases (with high morbidity and/or mortality) that could incur mass public reaction (e.g., panic and/or paranoia) and impact upon public health resources—inclusive of a strained public–governmental fiduciary; (2) targeting specific combatants to incur health-related effects upon operational infrastructures; or (3) "in close" scenarios in which particular individuals are targeted for effect to incur more broadly based manifestations and consequences (e.g., diplomats and heads of state during negotiation sessions to alter cognitive, emotional, and behavioral functions and activities).

NEUROTOXINS

Of the aforementioned scenarios in which neuroweapons could be leveraged, the latter two are prime situations for the use of organic neurotoxins. These agents are extracts or derivatives of peptides found in mollusks (i.e., conotoxins), puffer fish and newts (i.e., tetrodotoxin), dinoflagellate algae (i.e., saxitoxin), blue-ringed octopus (i.e., maculotoxin), and certain species of cobra (i.e., naja toxins). As depicted in Table 7.4, all are potent paralytics, acting through mechanisms of ionic blockade (e.g., acetylcholine receptor antagonism; or direct inhibition of sodium, calcium, or potassium channels) in the peripheral nervous system and/or at the neuromuscular junction (Cannon 1996), to induce flaccid paralysis and cardiorespiratory failure. Being peptides, the stability of these agents vary, but can be enhanced through chemical modifications such as structural cyclization, disulfide bridge modification, and substitution of various residues (Craik and Adams 2007), thereby increasing their utility.

TABLE 7.3
Neuromicrobiologic Agents

Category	Agent	Action	Effects
Viral encephalitides	Togaviridae—*Alphavirus* (e.g., Venezuelan, Eastern, and Western equine encephalitis viruses) Bunyaviridae—*Orthobunyavirus* (e.g., La Crosse virus [LCV] and James Canyon virus [JCV] variants of California encephalitis virus) Flaviviridae—*Flavivirus* (e.g., Powassan virus)	Arthropod vectors (mosquitoes, ticks) Single-stranded RNA genomes Invade the brain via vascular endothelial cells Replicates in neurons of cerebral cortex Causes neuronal necrosis	Encephalitis: headache, nausea, vomiting, lethargy, seizures, coma, and possible paralysis or focal signs of neuropathy LCV: primarily children; <1% mortality; but most have persistent neurological sequelae (e.g., recurrent seizures, partial to full hemiparalysis; cognitive and neurobehavioral abnormalities)
Microbial encephalitides	Amoebic parasite (e.g., trophozoites of *Nuegleria fowleri*)	Amoeba penetrate nasal mucosa after insufflation or ingestion of infected water Invades CNS via olfactory nerve Progressively necrotizes brain tissue	Primary amoebic meningoencephalitis (PAME) Progressive meningitis resulting in encephalitis Death from respiratory failure consequential to inflammation and/or necrotization of the brainstem Mortality in 2–4 weeks
	Protozoan parasite (e.g., *Toxoplasma gondii*)	Feline hosts Zoonotic transmission via human contact or insufflation of fecal matter Ingested; becomes active in immunocompromised host Encodes the DA synthetic enzyme tyrosine hydroxylase Increases CNS DA levels Induces polyfocal neuroinflammation	Behavioral changes resembling those of DA reuptake inhibitor type antidepressants and stimulants Loss of impulse control; agitation, confusion Encephalitis

Bacterial toxigenics	Agent	Mechanism	Effects
	Bacillus anthracis (i.e., anthrax toxin)	Bacterial spores are inhaled and uptaken by macrophages; Spores become active bacilli and rupture macrophages, releasing bacteria into bloodstream, where the anthrax toxin is released; Anthrax toxin enables bacteria to evade the immune system, proliferate, and ultimately cause polyfocal effects	Anthracic meningitis; Inhibition of protective immune responses; Cell lysis and destruction; Bleeding and death
	Clostridium botulinum (i.e., Botulinum toxin) (also, C. butyricum, C. baratii, C. argentinense)	Decreased ACh release; Decreased neuromuscular transmission	Induces flaccid paralysis and cardio-respiratory failure
	Several genera of Cyanobacteria (e.g., Anabaena, Aphanizomenon, Oscillatoria, Microcystis, Planktothrix, Raphidiopsis, Arthrospira, Cylindrospermum, Phormidium) (i.e., anatoxin-α)	Stimulates nicotinic ACh receptors; Mimics ACh causing irreversible stimulation of NMJ	Permanent contraction of muscles; Loss of coordination, twitching, convulsions, and rapid death by respiratory paralysis
	Gambierdiscus toxicus (i.e., ciguatoxin)	Lowers the threshold for activating voltage-gated sodium channels	Causes headache, muscle aches, weakness, dizziness, itching, nightmares, and/or hallucinations. Causes paresthesias (e.g., sensation of burning or "pins-and-needles," reversal of hot and cold temperature sensation, or unusual tastes) Very low mortality; recovery within 1 month.

ACh, acetylcholine; CNS, central nervous system; DA, dopamine; NMJ, neuromuscular junction.

TABLE 7.4
Organic Neurotoxic Agents

Origin	Agent	Mechanism
Marine cone snails (genus *Conus*, e.g., *Conus geographus*)	σ-Conotoxin	Inhibits the inactivation of voltage-dependent sodium channels; prolonged opening
	κ-Conotoxin	Inhibits the inactivation of voltage-dependent sodium channels; prolonged opening
	μ-Conotoxin	Inhibits voltage-dependent sodium channels in muscles
	ω-Conotoxin	Inhibits N-type voltage-dependent calcium channels
Symbiotic bacteria found in rough-skinned newts, pufferfish, and procupinefish (e.g., *Pseudoalteromonas tetraodonis*, certain species of *Pseudomonas* and *Vibrio*)	Tetrodotoxin	Prevents action potential firing in neurons Binds near the pore and blocks voltage-gated fast sodium channels on presynaptic terminals
Symbiotic bacteria found in blue-ringed octopus (*Hapalochlaena maculosa*)	Maculotoxin (a venomous form of tetrodotoxin)	Presynaptic sodium channel pore blockade Inhibition of action potential
Shellfish/mollusks (e.g., *Saxidomus giganteus* contaminated by algal blooms ["red tides"], e.g., *Alexandrium catenella*)	Saxitoxins (e.g., saxitoxin, gonyautoxin, neosaxitoxin)	Selective pore blocker of neuronal voltage-gated sodium channels Water soluble and dispersible by aerosols Inhalation causes death within minutes
Cobra snake (Genus *Naja*)	Naja toxins (e.g., α-cobratoxin)	Block nicotinic ACh receptors at NMJ Also selective antagonist to α7-nicotinic ACh receptors in the brain
Krait snakes (*Bungarus multicinctus*)	Bungarotoxins (e.g., α-bungarotoxin)	Irreversible and competitive binding to NMJ nicotinic ACh receptors Also selective antagonist to α7-nicotinic ACh receptors in the brain
Mamba snakes (*Dendroaspis*)	Dendrotoxins (e.g., α-dendrotoxin, σ-dendrotoxin)	Block voltage-gated (A-type) potassium channels at nodes of Ranvier in motor neurons Prolong duration action potentials Increase ACh release at NMJ
Australian taipan snakes (*Oxyuranus scutellatus*)	Taipoxin	Induced increasing blockade of presynaptic ACh release from motor neurons

ACh, acetylcholine; NMJ, neuromuscular junction.

In all cases, paralysis occurs rapidly after introduction of a small dose of the agent, and the victim remains conscious until overcome by shock and/or respiratory arrest. As well, except for the naja toxins (for which there are a number of species-specific antivenins, each of which differ in effectiveness), antidotes are not available; rapid triage for cardiorespiratory support is required to prevent mortality (although the effects of tetrodotoxin can be mitigated with edrophonium) (Masaro 2002).

PRACTICAL CONSIDERATIONS, LIMITATIONS, AND PREPARATIONS

The use of such neuroweapons, especially if apparent, is unlikely to result in lasting peace. Yet, the distribution of a neurotropic drug or neuropathological agent throughout a population could create a societal burden that significantly impacts the means, economic resources, and/or motivations to fight. But there is also the risk of a number of "spillover effects." First, given the environment(s) in which most current warfare is conducted, it would be nearly impossible to completely protect a civilian population from the effects of neuroweapons. If the agent has a known antidote or may be inoculated against, effects might be relatively easy to counter—but this would depend upon the integrity of the public health infrastructure of the town or country in which the agents are employed (and/or the capability of military forces to provide medical assistance to those civilians that are affected). Second, if an antidote is not available, then there is risk of both serious collateral injury to the civilian population and to one's own troops should they be exposed to the agent. Third, there is risk to much broader populations if stocks of the agent were purloined from secure storage or should a neuromicrobiological agent mutate while being employed (NAS 2008). Evidently, these considerations prompt ethical and legal concerns that must be addressed—and resolved—through the formulation of guidelines and policies, as discussed elsewhere in this volume.

While the use of neurotechnology in NSID applications may be relatively new, the concept of using psychological science to develop weaponry is not. In some ways, it may be as difficult to distinguish between neurotechnical and psychological warfare as it is to discriminate structure from function as relates to brain and mind. "Changing minds and hearts" may not be a task that is best addressed by neurotechnologies as weapons. Instead, cultural sensitivity and effective communication might be a more desirable approach (Masaro 2002; Freakley 2005; NAS 2008). Still, neurotechnology will be evermore viable in translating the nuances of social cognition and behaviors and thus gaining a deeper understanding as to why certain principles or violations are more or less likely to induce violence and/or strong opposition. Neurotechnology could—and likely will—play an increasingly larger role in exploring the relationships between culture and neuropsychological dynamics in and between populations.

But here we pose a caveat: Ignoring unresolved ambiguities surrounding issues of the "brain–mind" and "reductionist/antireductionist" debate (i.e., as connoted by the "neuro" prefix) when employing scientific evidence as rationale for employing neurotechnologies as weapons may lead to erroneous conclusions that may profoundly affect the intelligence and defense community—and the public. An example is the current interest in using functional neuroimaging (fMRI) for detecting deception (NAS 2008). The validity of this approach depends on the accuracy with which such technologies can detect psychological states relevant to deception (such as anxiety vs. something more abstract, e.g., cognitive dissonance). How science portrays the relationship between patterns that may be detected by neuroimaging and what those patterns actually represent depends upon (neuroscientific) interpretation of the validity of the technology (to actually do what it is intended) and, in light of this, the *meaning* of data and information, as constituting viable knowledge (Uttal 2001; Farah 2005; McFate 2005; Illes 2006; Giordano 2012).

An illustration of how (mis)conceptions of causal relationships of the brain and cognition can constitute a rationale for employing technologies or tactics is reflected by the following quote, from the May 4, 2009, issue of *Newsweek*. In this context, the speaker is asking the contractor who will replace him about a given approach to interrogation.

> ... I asked [the contractor] if he'd ever interrogated anyone, and he said 'no, but that didn't matter', the contractor shot back, 'Science is science. This is a behavioral issue.' He told me he's '... a psychologist and ... knows how the human mind works' (Isikoff 2009).

This is relevant to the use of neurotechnology as it reflects a social tendency to concretize contingent neuroscientific understanding as "truth." To be sure, neuroscience is an iterative enterprise. Thus, applications and use(s) of neurotechnology remain works-in-progress. Still, neurotechnology can be used to create weapons that may have an unprecedented capacity to alter the cognitions, emotions, beliefs, and behaviors of individuals, and groups—if not societies. Thus, the potential "power" of neurotechnology as weaponry lies in the ability to assess, access, and change aspects of a definable "self." As with any weapons, they pose threats to autonomy and free will and can do so to an extent that psychological weapons alone could not.

It is foolhardy to think that the technological trend that compels the use of neurotechnology as weapons will be impeded merely by considerations of (1) the burdens and risks that might arise as science advances ever deeper into the frontiers of the unknown; (2) the potential harms that such advances could intentionally and/or unintentionally incur; and (3) the ethico-legal and social issues instantiated by both the positive and negative effects and implications of these advances. This is because a strong driving (or "pushing") force of both science and technology is the human desire(s) for knowledge and control. At the same time, environmental events, market values, and sociopolitical agendas create a "pulling force" for technological progress and can dictate direction(s) for its use. Both former and latter issues are important to national security and defense. In the first case, the use of contingent knowledge could evoke unforeseen consequences that impact public safety, and the power conferred by scientific and technological capability could be used to leverage great effect(s). In the second case, the intentional use of these technologies by individual agents or groups in ways that are hostile could incur profound public threats.

Thus, a simple precautionary principle in which risk–benefit ratios determine the trajectory and pace of technological advancement is not tenable on an international level, as there is the real possibility—if not probability—that insurgent nations and/or groups could fund and covertly conduct RDTE of neuroweapons, beyond the auspices and influence of extant international guidelines and policies. Instead, we argue for a process that entails some measure of precaution, together with significant preparedness. As detailed in this book and elsewhere, such preparedness requires knowledge of (1) what technological accomplishments can be achieved (given the incentives and resources afforded); (2) whether such work is being prepared and/or undertaken; (3) groups involved in such work; (4) overt and/or covert intention(s) and purposes; (5) what possible scenarios, effects, and consequences could arise from various levels of technological progress; and (6) what measures can and should be taken to counter threats imposed by such progress and its effects (see Chapters 1, 8, and 15; Isikoff 2009). For this approach to work, surveillance (i.e., intelligence) is necessary, and thus, the development and use of many of the aforementioned neurotechnological developments becomes increasingly important. Although international governance of neurotechnological RDTE may be difficult, what can be governed and regulated are the ways in which neuroscientific and neurotechnological RDTE efforts are conducted and employed by the U.S. agencies. In this regard, ethical questions need to be prudently addressed and balanced with the interests of public (i.e., national) security and protection.

In an ideal world, science and technology would never be employed for harmful ends; but we should not be naïve and succumb to the dichotomy of ought versus is. Neuroscience can—and will—be engaged to effect outcomes relevant to NSID operations by countries and nonstate entities. As history has shown, a dismissive posture that fails to acknowledge the reality of threat increases the probability of being susceptible to its harms. In an open society, it is the responsibility of the government to protect the polis. Hence, there is a duty to establish proactive, defensive knowledge of these scientific and technological capabilities and the vulnerabilities that they exploit, inorder to recognize how neuroscience and neurotechnology could be used to wage hostile acts, and to develop potential countermeasures to respond appropriately (Giordano et al. 2010). But a meaningful stance of preparedness also mandates rigorous analyses and address of the ethico-legal and social issues that such use—and/or misuse—of neuroscience and neurotechnology generate, and guidelines and policies must be formulated to effectively direct and govern the scope and conduct of research and its applications in this area.

ACKNOWLEDGMENTS

The authors are grateful to Sherry Loveless for assistance on the final preparation of this manuscript.

REFERENCES

Abu-Qare, A.W. and M.B. Abou-Donia. 2002. "Sarin: Health effects, metabolism, and methods of analysis." *Food and Chemical Toxicology* 40(19):1327–1333.

Ajana, B. 2010. "Recombinant identities: Biometrics and narrative bioethics." *Journal of Bioethical Inquiry* 7:237–258.

Albucher, R.C. and I. Liberzon. 2002. "Psychopharmacological treatment in PTSD: A critical review?" *Journal of Psychiatric Research* 36(6):355–367.

Arbib, M. 2003. *The Handbook of Brain Theory and Neural Networks*. 2nd edition. Cambridge, MA: MIT Press.

Black, J. 2009. "The ethics of propaganda and the propaganda of ethics." In *The Handbook of Mass Media Ethics*, eds., L. Wilkins and C.G. Christians. New York: Routledge, pp. 130–148.

Buguet, A., D.E. Moroz, and M.W. Radomski. 2003. "Modafinil—Medical considerations for use in sustained operations." *Aviation, Space, and Environmental Medicine* 74(6):659–663.

Calabrese, E.J. and L.A. Baldwin. 2001. "Hormesis: U-shaped dose responses and their centrality in toxicology." *Trends in Pharmacological Sciences* 226:285–291.

Canna, S. and G. Pop. 2011. "Strategic Multilayer Assessment (SMA)." *Paper presented at the 5th Annual SMA Conference at the National Institutes of Health*, Bethesda, MD, November.

Cannon, S.C. 1996. "Ion-channel defects and aberrant excitability in myotonia and periodic paralysis" *Trends in Neurosciences* 19(1):3–10.

Clynes, N. and M. Kline. 1961. *Drugs, Space, and Cybernetics: Evolution to Cyborg*. New York: Columbia University Press.

Craik, D.J. and D.J. Adams. 2007. "Chemical modification of conotoxins to improve stability and activity." *ACS Chemical Biology* 2(7):457–468.

Davis, L.L., A. Suris, M.T. Lambert et al. 1997. "Post-traumatic stress disorder and serotonin: New directions for research and treatment." *Journal of Psychiatry & Neuroscience* 22(5):318–326.

Farah, M.J. 2005. "Neuroethics: The practical and the philosophical." *Trends in Cognitive Sciences* 9(1):34–40.

Freakley, B.C. 2005. "Cultural awareness and combat power." *Infantry Magazine* 94:1–2.

Foucault, M. 2007. *Security, Territory, Population: Lectures at the Collège de France 1977–1978*. New York: Palgrave Macmillan.

Giles, J. 2001. "Think like a bee." *Nature* 410(6828):510–512.

Gimpl, G. and F. Fahrenholz. 2001. "The oxytocin receptor system: Structure, function, and regulation." *Physiological Reviews* 81(2):629–683.

Giordano, J. 2011. "Integrative convergence in neuroscience: Trajectories, problems and the need for a progressive neurobioethics." In *Technological Innovation in Sensing and Detecting Chemical, Biological, Radiological, Nuclear Threats and Ecological Terrorism*. eds., Vaseashta, E. Brahman and P. Sussman. New York: Springer.

Giordano J. 2012. "Neuroimaging in psychiatry: Approaching the puzzle as a piece of the bigger picture(s)." *American Journal of Bioethics Neuroscience* 3(4):54–56.

Giordano J., C. Forsythe, and J. Olds. 2010. "Neuroscience, neurotechnology and national security: The need for preparedness and an ethics of responsible action." *American Journal of Bioethics Neuroscience* 1(2):10–13.

Giordano, J. and R. Wurzman. 2011. "Neurotechnologies as weapons in national intelligence and defense—An overview." *Synesis: A Journal of Science, Technology, Ethics, and Policy* 2(1):T55–T71.

Goldstein, F.L. and B.F. Findley. 1996. *Psychological Operations: Principles and Case Studies*. Montgomery, AL: Air University Press.

Hoag, H. 2003. "Neuroengineering: Remote control." *Nature* 423:796–798.

Hughes, J. 2007. "The struggle for a smarter world." *Futures* 29(8):942–954.

Illes, J. 2006. *Neuroethics: Defining the Issues in Theory, Practice, and Policy*. Oxford: Oxford University Press.

Illes, J. and E. Racine. 2005. "Imaging or imagining? A neuroethics challenge informed by genetics." *American Journal of Bioethics* 5(2):5–18.

Isikoff, M. 2009. "We could have done this the right way: How Ali Soufan, an FBI agent, got Abu Zubaydah to talk without torture." *Newsweek*, April 25. http://www.newsweek .com/id/195089.

Kautilya. 1929. *Arthashastra*. 3rd edition, Translated by Shamasastry R. Mysore, India: Weslyan Mission Press.

Konar, A. 1999. *Artificial Intelligence and Soft Computing: Behavioral and Computational Modeling of the Human Brain*. Boca Raton, FL: CRC Press.

Lord, C. 1996. "The psychological dimension in national strategy." In *Psychological Operations: Principles and Case Studies*, ed., C. Lord. Montgomery, AL: Air University Press, pp. 73–90.

Masaro, E.J. 2002. *Handbook of Neurotoxicology*. Totowa, NJ: Humana Press.

McCoy, A.W. 2006. *A Question of Torture: CIA Interrogation, From the Cold War to the War on Terror*. New York: Owl Books.

McFate, M. 2005. "The military utility of understanding adversary culture." *Joint Force Quarterly* 38:42–48.

McGovern, T.W. and G.W. Christopher. 1999. "Cutaneous manifestations of biological warfare and related threat agents." *Archives of Dermatology* 135:311–322.

Merriam-Webster Dictionary. 2008. "Weapon." http://www.merriam-webster.com/dictionary/ weapon.

Mordini, E. and C. Ottolini. 2007. "Body identification and medicine: Ethical and social considerations." *Annali dell'Istituto Superiore di Sanità* 43(1):5–11.

Moreno, J. 2006. *Mind Wars*. New York: Dana Press.

Murphy, P.N., W. Wareing, and J.E. Fisk. 2006. "Users' perceptions of the risks and effects of taking ecstasy (MDMA): A questionnaire study." *Journal of Psychopharmacology* 20(3):447–455.

Nagase, H. 2001. "Metalloproteases." *Current Protocols in Protein Science* 21(4):1–13.

National Research Council of the National Academy of Sciences (NAS). 2008. *Emerging Cognitive Neuroscience and Related Technologies*. Washington, DC: National Academies Press.

Nebert, D.W., L. Jorge-Nebert, and E.S. Vesell. 2003. "Pharmacogenomics and 'individualized drug therapy': High expectations and disappointing achievements." *American Journal of Pharmacogenomics* 3(6):361–370.

Paddock, A.H. 1999. "No more tactical information detachments: US military psychological operations in transition." In *Psychological Operations: Principles and Case Studies*, eds., F.L. Goldstein and B.F Findley. Montgomery, AL: Air University Press, pp. 25–50.

Pearn, J. "Artificial brains: The quest to build sentient machines." DARPA SyNAPSE Program. 1999. http://www.artificialbrains.com/darpa-synapse-program. Accessed October 17, 2013.

Pringle, R.W. and H.H. Random. 2009. s.v. "Intelligence." *Encyclopedia Britannica*. http://www.britannica.com/EBchecked/topic/289760/intelligence.

Ridley, M. 2003. *Nature via Nurture: Genes, Experience, and What Makes Us Human*. London: HarperCollins.

Rippon, G. and C. Senior. 2010. "Neuroscience has no place in national security." *American Journal of Bioethics Neuroscience* 1(2):37–38.

Romano, J.A., B.J. Lukey, and H. Salem. 2007. *Chemical Warfare Agents: Chemistry, Pharmacology, Toxicology, and Therapeutics*. 2nd edition. Boca Raton, FL: CRC Press.

Rubaj, A., W. Zgodziński, and M. Sieklucka-Dziuba. 2003. "The epileptogenic effect of seizures induced by hypoxia: The role of NMDA and AMPA/KA antagonists." *Pharmacology, Biochemistry, and Behavior* 74(2):303–311.

Sapolsky, R.M., J.A. Gunn, C.F. Gunn et al. 2013. "Topics in the neurobiology of aggression: Implications to deterrence." *Paper presented at the 6th Annual Strategic Multi-Layer (SMA) Conference at the National Institutes of Health*, Bethesda, MD, February.

Sato, Y. and K. Aihara. 2011. "A Bayesian model of sensory adaptation." *PLoS One* 64:e19377.

Schemmel, J., S. Hohnmann, K. Meier, and F. Schurmann. 2004. "A mixed-mode analog neural network using current-steering synapses." *Analog Integrated Circuits and Signal Processing* 38(2):233–244.

Siegelmann, H. 2003. "Neural and super-turing computing." *Minds and Machines* 13(1):103–114.

Smith, J.F., K. Davis, M.K. Hart et al. 1997. "Viral Encephalitides." In *Medical Aspects of Chemical and Biological Warfare*, eds., F.R. Sidell, E.T. Takafuji, and D.R. Franz. Washington, DC: Borden Institute, pp. 561–589.

Stanney, K.M., K.S. Hale, S. Fuchs et al. 2011. "Training: Neural systems and intelligence applications." *Synesis: A Journal of Science, Technology, Ethics, and Policy* 2:38–44.

Thomsen K. 2010. "A Foucauldian analysis of 'a neuroskeptic's guide to neuroethics and national security'." *American Journal of Biology* 1(2):29–30.

Trautteur, G. and G. Tamburrini. 2007. "A note on discreteness and virtuality in analog computing." *Theoretical Computer Science* 371(1/2):106–114.

Uttal, W. 2001. *The New Phrenology: The Limits of Localizing Cognitive Processes in the Brain*. Cambridge, MA: MIT Press.

Von Neumann, J. 2000. *The Computer and the Brain*. 2nd edition. New Haven, CT: Yale University Press.

Watson, A. 1997. "Neuromorphic engineering: Why can't a computer be more like a brain?" *Science* 227(5334):1934–1936.

Wiener, N. 1948. *Cybernetics: On Control and Communication in the Animal and the Machine*. Cambridge, MA: MIT Press.

Wurzman, R. and J. Giordano. 2012. "Differential susceptibility to plasticity: A 'Missing link' between gene-culture co-evolution and neuropsychiatric spectrum disorders?" *BMC Medicine*, 10:37.

ONLINE RESOURCES

Biological Technology Office—DARPA."Narrative networks program." http://www.darpa.mil/Our_Work/BTO/Programs/Narrative_Networks.aspx. Accessed Oct. 17, 2013.

Biological Technology Office—DARPA. "Neuro Function, Activity, Structure, and Technology (Neuro-FAST) program." http://www.darpa.mil/Our_Work/BTO/Programs/Neuro_Function_Activity_Structure_and_Technology_NeuroFAST.aspx. Accessed June 1, 2014.

Biological Technology Office—DARPA. "Reorganization and Plasticity to Accelerate Injury Recovery (REPAIR) program." http://www.darpa.mil/Our_Work/BTO/Program/Reorganization_and_Plasticity_to_Accelerate_Injury_Recovery_REPAIR.aspx. Accessed June 1, 2014.

Biological Technology Office—DARPA. "Restoring Active Memory (RAM) program." http://www.darpa.mil/Our_Work/BTO/Programs/Restoring_Active_Memory_RAM.aspx. Accessed June 1, 2014.

Biological Technology Office—DARPA. "Revolutionizing Prosthetics program." http://www .darpa.mil/Our_Work/BTO/Programs/Revolutionizing_Prosthetics.aspx. Accessed June 1, 2014.

Biological Technology Office—DARPA. "Strategic social interaction modules (SSIM)." http://www.darpa.mil/Our_Work/BTO/Programs/Strategic_Social_Interaction_Modules_SSIM.aspx. Accessed Oct. 17, 2013.

Biological Technology Office—DARPA. "Systems-Based Neurotechnology for Emerging Therapies (SUBNETS) program." http://www.darpa.mil/Our_Work/BTO/Programs/Systems-Based_Neurotechnology_for_Emerging_Therapies_SUBNETS.aspx. Accessed June 1, 2014.

Biological Technology Office—DARPA. "Systems of neuromorphic adaptive plastic scalable electronics (SyNAPSE) program." http://www.darpa.mil/Our_Work/DSO/Programs/ Systems_of_Neuromorphic_Adaptive_Plastic_Scalable_Electronics_(SYNAPSE) .aspx. Accessed Oct. 17, 2013.

Information Innovation Office—DARPA. "Detection and computational analysis of physiological signals (DCAPS) program." http://www.darpa.mil/Our_Work/I2O/Programs/ Detection_and_Computational_Analysis_of_Psychological_Signals_(DCAPS).aspx. Accessed Oct. 17, 2013.

Information Innovation Office—DARPA. "Mind's eye program." http://www.darpa.mil/Our_ Work/I2O/Programs/Minds_Eye.aspx. Accessed Oct. 17, 2013.

Information Innovation Office—DARPA. "Social media in strategic communication (SMISC) program." http://www.darpa.mil/Our_Work/I2O/Programs/Social_Media_in_Strategic_ Communication_(SMISC).aspx. Accessed Oct. 17, 2013.

Office of Incisive Analysis—IARPA. "Integrated Cognitive-Neuroscience architectures for understanding sensemaking (ICARUS) program." http://www.iarpa.gov/Programs/ia/ ICArUS/icarus.html. Accessed Oct. 17, 2013.

Office of Incisive Analysis—IARPA. "Knowledge representation in Neural systems (KRNS) program." http://www.iarpa.gov/index.php/research-programs/krns. Accessed Oct. 17, 2013.

Office of Incisive Analysis—IARPA. "Metaphor program." http://www.iarpa.gov/index.php/ research-programs/metaphor. Accessed Oct. 17, 2013.

Office of Incisive Analysis—IARPA. "Sociocultural content in language (SCIL) program." https://www.fbo.gov/?s=opportunity&mode=form&tab=core&id=72cc15a91610fc6b9 2acc73b6980279e&_cview=1. Accessed Oct. 17, 2013.

Office of Smart Collection—IARPA. "Tools for recognizing useful signals of trustworthiness (TRUST)." http://www.iarpa.gov/index.php/research-programs/trust. Accessed Oct. 17, 2013.

8 Brain Brinksmanship
Devising Neuroweapons Looking at Battlespace, Doctrine, and Strategy

Robert McCreight

CONTENTS

BACKGROUND AND BRINKSMANSHIP

As the twentieth century came to a close some years ago, the United States possessed an interesting array of military might that stood apart from all other nations because the once vaunted superpower rivalry with the Soviet Union had ended. Despite numerous theories about the reasons for that ending, there is little doubt that in economic and military terms, along with democracy's appeal, the United States emerged as the world's singular superpower. However, the term *superpower* is ultimately measured using a combination of economic, social, cultural, scientific, or other epochal ingredients mixed in with the sheer gravitas of military power. The cold war's end meant that the United States found itself in something of an atypical global leadership position afforded by its military capability finding itself dominant in a very uncertain and unstable world where military strength, by itself, was no guarantee of security.

The tragic events of the 9/11 attack verified that conventional defense and the idea of "fortress America" were woefully inadequate. Prior strategic thinking about security, thwarting attacks, and preparing for enemy behavior were eroded in the classic and conventional sense and instead redefined, and dramatically new frontiers of threat and deterrence had to be considered (National Commission on Terrorist Attacks 2004). Even the definition of security itself had become less clear and more multifaceted with the rise of cyber systems and ever-advancing scientific technologies.

Terrorism threats aside for a moment, many strategic analysts might argue that by dint of military prowess and advanced weapons systems, including especially nuclear, stealth, and unmanned aerial vehicles, the United States stood alone as a fearsome global force. Many experts still believe that the United States would remain the sole superpower as the twentieth century ended (Art et al. 2012; Associated Press 2012; Kagan 2012; U.S. Department of Defence 2012). Instead as the first decade of the twenty-first century elapsed, there are no illusions to that effect. It became clear that other nations were eagerly in pursuit of near equivalent superpower status. In turn, the United States continued to modernize and update its arsenal. That relentless desire remains unabated today.

The array of twenty-first century weaponry, which has emerged in different realms of the globe, reflects an expanding and modernized military among a variety of European, Asian, and Middle Eastern states. True enough, some of these nations exhibit the typical array of tanks, warplanes, bombers, artillery, howitzers, rockets, bombs, mines, submarines, and super carriers. Several countries, such as China, Russia, India, Israel, Iran, Malaysia, Thailand, and Brazil, have embarked on advanced weapons systems with the explicit goal of directly keeping pace with the United States. Some nations can even lay claim to chemical, biological, and nuclear weaponry that have been enhanced and modified (Alexander and Heal 2003; Dockery 2007; Auslin 2012).

More recently, we can see some nations acquiring newer weapons involving lasers, acoustic waves, directed energy, satellite-based, drone attack craft, cyberstrike systems, neutron radiological devices, stealth subs, and all weather jet fighters among many others (Arms Control Association 2012; Burgess 2012). Given tendencies toward technology sharing and proliferation, this array of weapons systems will soon likely be within the arsenals of several nations.

Human history and tradition have taught us that there can be a suitable defense found, and in many cases an effective deterrent developed, to most new weapons. The irony is that as each new weapon system is developed and deployed, competing advances in rival weaponry, better defense systems, and deterrent strategies escalate proportionally to match, neutralize, or overcome each new weapon that comes along. Nevertheless, the perennial search continues for a perfectly invincible weapon, that is, 100% reliable and effective: The quest involves a superweapon against which no adequate defense or deterrent can be found. The search for a perfect, invincible weapon seems embedded in human military history (O'Connell 1989; Herbst 2006), as such a weapon would be a strategic game changer and likely revolutionize the balance of global power.

Pursuit of a cutting-edge all but invincible weapon continues. What seems elusive and difficult to discern is whether a whole new weapons system may emerge during the period 2015–2025. A key concern is whether that decade will reveal a new type of weapon which could be a strategic geopolitical game-changer. Owing to breakthroughs in global science and technology, as well as the dynamic influence of convergent technologies, it is conceivable that new weapons would be devised and tested that seem to be border-line science fiction today.

Would this new decade usher in the dawn of neuroweapons? (Giordano and Wurzman 2011; see also Chapter 7). Indeed, such innovations would completely redefine and revolutionize the potential battlespace, if not, they would trigger a renewed appreciation for what weapons really are. Human brains, thoughts, and ideas would ostensibly become "targets" if one could devise and harness a genuine neurocognitive weapon. Thoughts, beliefs, perceptions, ideas, and behaviors could be made directly vulnerable to external threat and control for the first time in human history.

It is fair to claim the complete reordering of global power and the epochal reprioritizing of strategic arsenals would be transformed in unexpected ways, and perhaps to an extent, never seen before. Targeting with the mind to unravel its secrets is one thing, steering human thought, shaping decisions, influencing behavior, altering perceptions, and penetrating the subconscious is quite another. Worse, we must allow for the distinct possibility that external and undetected neuroweapons attacks of some sophistication could possibly originate in governments, organizations, cybernetic machines, criminal or terrorist groups, or even individuals.

Neuroweapons defy easy explanation and definition. An agreed global definition of this term does not exist and differences about core components, structure, design, and intent will no doubt continue as populations wrestle with the notion that such a category of weapons can be created at all. Until quite recently, *a facile and broadly applicable definition of neuroweaponry—and neuroweaponology—has remained somewhat obscure*. However, ongoing research has launched a reasonable, lucid, and well-established definition of these techniques and technologies along with the ways that neuroscience might be employed to "... contend against hostile others." (Giordano and Wurzman 2011; see also Chapter 7).

As time passes, we must come to grips with the reality of neuroweapons development, emerging as it does alongside advances in synthetic biology, and the merger of nanobiotechnology with robotic–cybernetic self-aware logic. Each discrete avenue of advanced science will relentlessly pursue its own form of progressive perfection and one day some of those endeavors will combine and integrate. Distinctions between laboratory curiosity and applied research will dissolve. Spillover and synergistic effects between areas of advanced research must also be considered (Giordano 2012; Vaseashta 2012).

Futurist James Canton foresees possible applications of neurotechnology including stem cell therapy for memory repair, brain–machine implants to overcome paralysis, silicon nano-retinas to provide sight, genomic neurotherapy to reprogram disease-causing genes, enabling neurons that control robotic arms and legs, and neural engineering to rewire brains and combat mental illness. These are beneficial breakthroughs to be welcomed, but there is a darker side embedded in dual-use neuroscience and weaponization research (Canton 2010; see also MIT Media Lab 2010).

Definitional issues are important, but accepting the reality of neuroweapons as part of our collective future is an equally important milestone. While diverse definitions may eventually emerge, consider one fairly neutral version that synergizes the existing lexicon and is broad enough to capture the essence of this concept. Here, the claim is made that neuroweaponry encompasses all forms of interlinked cybernetic, neurological, and advanced biotech systems, along with the use of synthetic biological formulations and merged physiobiological and chemical scientific arrangements, designed expressly for offensive use against human beings. Neuroweapons are intended to influence, direct, weaken, suppress, or neutralize human thought, brainwave functions, perception, interpretation, and behaviors to the extent that the target of such weaponry is either temporarily or permanently disabled, mentally compromised, or unable to function normally.

Those who would argue that neuroweapons verge on obscene science fiction would have to demonstrate that such devices and approaches are technically impossible to create, that cleverly designed collaborating systems could never produce neuroweapons, or that they are so far beyond the scope of human reason and intelligence as to stand at the threshold of being beyond human capability. Some may argue that neuroweapons could potentially negate human thought, redirect will, or weaken spirit and therefore should be outlawed before they are created. They may even expect society and government to enact systems and security measures to thwart their present development, before further research brings us closer to the realities of weaponization. (Rees and Rose 2004; Glannon 2006; see also Chapter 14).

The problem is that none of those statements about neuroweaponry are reliable, valid, or true. Neuroweaponry is, in fact, possible to create. It seems likely that sophisticated engineered systems' integration of biophysical, nanochemical, cyberdynamic, and related elements could—or more accurately will—be engineered to produce a neuroweapon. To claim that neuroweaponry is beyond human intelligence and logic also fails to address or acknowledge the current state of engineered integration and merging of diverse technologies. Such trends must be seen as possible, probable, and a strategic risk of epochal global impact that must prepared for, else we should expect instead to suffer dire consequences for collective ignorance of its import (Giordano et al. 2010).

If the central goal is to manipulate human thought, emotions, and behavior through a combination of psychopharmacological, biotechnical, and cybernetic activities and synergized systems to steer, influence, and shape thought and conduct—then we must be and remain alert to such potential goals and progress toward them to date. Bringing the perfect mixture of scientific, engineering, psychological, behavioral, and medical properties to influence, redirect, sublimate, subdue, and repress any thoughts or actions that would ordinarily be seen as aggressive, hostile, or murderous merits global interest. What nation would hesitate to develop and field a weapon that could control, shape, or redirect human thoughts and actions—given the power such a weapon would yield?

Nations that currently possess sophisticated or advanced weapons have built arsenals based upon knowledge that military forces on all sides pose both offensive threats and defensive capabilities, while simultaneously offering deterrent potential against attacks. Nations may be hesitant to be overtly provocative, engage in hostilities, or even consider preemptive attacks knowing that some form of retaliation is likely. Instead, they may engage in elaborate postures that foster rising tensions, but that may also inadvertently yield some unexpected advantages. For example, new offensive missiles unveiled by one nation may stimulate deployment of advanced missile defense systems in other states, while others may choose instead to deploy their own rival missile systems. Nevertheless, nations may—and often do—go to extraordinary lengths to avoid inadvertent warfare and seek to increase pressures without resorting to a military strike or some form of preemptive action. There is a parallel thrust and parry set of maneuvers each nation may persuade or invoke so as to influence or redirect another nation away from open warfare—or lead them to the brink. This is referred to in the global security and international relations literature as *brinksmanship.*

Former Secretary of State John Foster Dulles defined (his policy of) brinkmanship as "... the ability to get to the verge without getting into the war is the necessary art." (Sheply 1956). During the cold war, this was used by the United States to coerce the Soviet Union into military restraint. Since then it has been used numerous times to coerce a response from one nation, but it also contains an inherent risk that sheer coercive influence may not produce the desired effect or may instead trigger unwanted outcomes. Used by governments, this is not a game, but a political tactic. If there is a "winner" in brinkmanship, it is never the most reckless player, but the one who has achieved a specific goal. The point of brinkmanship, however, is to illustrate that a "win" by either side is impossible. Taking a conflict over the brink will result in unacceptable losses by both sides. Brinksmanship can confer concessions with little sacrifice or loss.

In an unequal contest of power, brinkmanship generally favors the side that has the least to lose. Brinksmanship should not be confused with bluffing because it deliberately exaggerates and mischaracterizes what the bluffer is trying to conceal. Bluffing in a card game is a nonhostile form of psychological leverage intended to elicit a response—to get others to fold and retreat even though the bluffer has nothing, in fact, to sustain the bluff. By contrast, brinksmanship's effectiveness derives from the implied genuine threat it represents rather than a display of an illusory or theatrical one. Can a state asserting possession of a neuroweapon exert strategic leverage over another? Arguably yes, but only if the implied threat is real and a genuine neuroweapon exists. Strategic coercive intimidation becomes a possibility whether or not a neuroweapon is actually used. It is the mere threat of use which triggers concessions.

Brinksmanship deserves to be explored as the reality of neuroweapons emerges over the next decade. Prodding an opponent in subtle, invisible, and unverifiable ways has great strategic value. Rational choice, objective analysis of options, and development of plans and strategies may all be at risk. Brinksmanship arises when one contemplates the degree to which danger, threat, and opportunity are potentially misperceived or misunderstood. Can leaders really grasp their strategic options? can they assess their opponents' reactions? and can they estimate what is effective versus merely a foolish gesture? Do they know if and where the red-lined boundaries of leveraging weaponized neuroscience really exist?

Geopolitical gambles in recent history provide evidence that brinksmanship is part of the global game and the assertion of geostrategic power (see Chapter 16). Is it the mere threat of weapons' use that sometimes elicits desired behavior from an opponent? Does building a nuclear reactor to enrich uranium compel neighboring states to assume that nuclear bomb-making is at hand? Does it alter the way in which the state electing to "go nuclear" is treated and regarded by the international community? Does it trigger as many gestures to favor and persuade as it does to invite criticism and sanctions? Persuading a state with nuclear energy to forsake nuclear weapons involves an exquisite blend of diplomatic, economic, and political leverage with no implied guarantee of success. As a state builds steadily toward a new weapons' capability of intimidating effect, the array of external pressures to stem hostile activity grows in proportion to the presumed new threat. This is a sophisticated game we understand very well.

Multiple sources provide evidence that brinksmanship in normal diplomatic analysis and discussion of regional security affairs typically entails strategic risk-taking (Chen 2011; Dareini and Jakes 2012; Farre 2012). In such a context, it is seen as the practice, especially in international relations, of taking a dispute to the verge of conflict in the hope of forcing the opposition to make concessions or to capitulate toward a favorable position. It is also sometimes seen as the technique or practice of maneuvering a dangerous situation to the limits of tolerance or safety in order to secure the greatest strategic advantage, especially by creating diplomatic crises. Such maneuvering can include genuine threats, phony threats or mixed messages that deliberately favor the sponsor in an effort to keep opponents off balance or unaware of the extent of concessions progressively made. Hidden agendas are paramount as brinksmanship plays out.

Herein what is offered is a perspective on brinksmanship and its relevance to neuroweaponry in order to equate it with the advantages, issues, and leverages that aspiring nuclear states currently can derive from adopting a posture where simply the potential for weapons development is at least as great as actually possessing them. We can examine the cases of Iran and North Korea, for example, to see how much leverage the apparent possession or near acquisition of nuclear weaponry provides. International efforts to divert, subdue, or deflect aspiring nuclear states from their apparent power trajectories always carry a proverbial package of carrots and sticks which, over time, grant the "aggressor state" some degree of geopolitical leverage and influence. Such states can derive greater concessions, despite their pariah status in the global community, among nations seeking to keep the number and variety of nuclear powers stable and in this way preserve the geopolitical status quo. The chief concern is that even today we are witnessing the gradual and nascent emergence of neuroweapons as an element of this power calculus. Those who can develop, acquire, and refine such weapons will have more than a strategic advantage.

On a political–psychological perspective, significant and substantive work focused on the research and development of neuroweaponry almost equates to the actual possession of neuroweapons. Threshold knowledge of neuroscientific manipulation of brain function and neurotechnologic influence of thought and behavior is well within the realm of reason and possibility. In so doing, dramatic leverage in regional and global security affairs is likely to be accrued by states embarking on this path to obtain an influential if not decisive neuroweapon. Whether they actually achieve such neurotechnical feats or not is irrelevant. Nations meaningfully engaged in serious and deliberate research and development of neurological technologies can be regarded with the same degree of strategic attention as those states embarking on a nuclear energy program. It is inherently a dual-use endeavor and deserves some measure of international review and scrutiny.

In some instances, inside certain states, the overall effort to develop and devise some form of neuroweaponry has been going on for decades, openly rooted in legitimate medical science research, and proving that neurological research steered toward military purposes will be much harder to demonstrate. It is the sheer capacity for dual use that ought to provoke wider global attention.

This crucial issue is being raised prior to a substantive discussion of neuroweapons in order to illustrate how the global mix of offensive and defensive measures and research related to the gradual evolution of neuroweapons could trigger an open-ended, while inadvertent, frenetic arms race. Finding powerful measures and inducements to steer aspiring states away from developing and engaging neuroweapon research may prove too daunting to place on the global agenda. Switching from hardware-based weapons to neural software-based weapons seems as much a paradigmatic shift as anything else.

Mere possession of a credible and viable neuroweapons research program can result in global reaction toward, and some accommodation with, the nation so engaged. This is because it conjures up images of a weapon that could provide a geostrategic edge to its possessor, making conventional defensive measures of comparatively limited value. What combination of restraints, self-imposed discipline, or global opprobrium could dissuade an aspiring state from embarking on neuroweapons' research (especially if a state saw this as the great equalizer)? Some states may even expect that acquiring this capacity elevates them to superpower status. Few, if any, could resist the lure of instant geopolitical power, and it remains to be seen if this technology can be stolen, mimicked, or reverse engineered to produce valid copies. If copying, theft, and proliferation are possible—what then? A new currency in the global balance of power arises (Giordano and Benedikter 2012).

The darker dimensions of neuroweaponry go beyond external manipulation of brain function and human performance by neurochemical agents, neurotoxins, or neuromicrobials to include remote influence of directed thought, external acoustic and brainwave interference, and subtle mood-altering stimulants from afar (Giordano and Wurzman 2011; see also Chapter 7). Like the speculative frontiers of psychological warfare itself, memory, perception, cognition, analysis, and thought can be influenced by coercion, pressure, intimidation, and dominance that are blended carefully and intended to render targets passive, confused, fearful, and hesitant. Whether this can affect thousands versus one person at a time remains to be seen. Is the target a population, an army, a bevy of generals, societal leaders, scientific experts, or a head of state? The political and social ramifications are truly epochal and severe.

As futurist James Canton (2012) has noted, we find ourselves locked in a perpetual dilemma. Neuroscience research must advance because human health, longevity, and scores of health-related issues may be promoted and sustained. The entire globe, all of humanity, stands to gain from neuroscientific breakthroughs. However, like the insidious nature of all dual-use science, neuroscience holds keys to doors that may open to nefarious avenues toward controlling thought, perception, emotion, and behavior to such an extent that will enable scientific fact to realize ideas that were heretofore merely science fictional.

EVOLVING NEUROWEAPONRY

The U.S. Department of Defense retains a keen interest in the emerging field of *neural network enhanced performance projects* where the modest and unimpeachable aim is to augment human sensory threat assessment, or radically improve warfighter performance and combat effectiveness. This work has been progressing for well over two decades

(Clancy 2006; Bulletin of Atomic Scientists 2008; Hearns 2008; Huang and Kosal 2008). Obviously, if Western nations within NATO (e.g., the United States and its allies) are committed to neuroscience research, it is reasonable that People's Republic of China, Russia, Iran, and North Korea would be similarly invested in such work. It should be expected that other nations will also develop new technologies that apply biotechnology to anticipate, find, fix, track, identify, and characterize human intent and physiological status. For example, research involving transcranial pulsed ultrasound technology that could be fitted to troops' battle helmets would allow soldiers to manipulate brain functions to boost alertness, relieve stress, or perhaps even reduce the effects of brain insult. Manipulating the brain to enhance warfighting capabilities and maintain mental acuity on the battlefield has long been a field of interest for the Defense Advanced Research Projects Agency (DARPA) and various military research labs (Hoag 2003; Adams 2005; Smith and Bigelow 2006; Boyle 2010), but this remains relatively limited in scope.

It is not unrealistic to expect that certain technologies, packaged in a warfighter's equipment array, could allow soldiers to stimulate different regions of the brain, helping to relieve battle stress or to enhance alertness during long periods without sleep. Soldiers might relieve pain from injuries or wounds without resorting to pharmaceuticals. Such ends could be seen as a benign and legitimate pursuit of advanced technology.

To a great extent, this research and development is largely defensive in nature, designed to enhance protective options for individual soldiers. Its undeniable benefits, however, deserve to be weighed against its long-term offensive potential. Turning to psychological warfare experiences and precepts may be somewhat helpful, albeit to a limited extent. Looking at the immediate and long-term effect of neuroweapons and their foundations in brain science, Giordano and Wurzman (Chapter 7) note:

> Given the relative nascence of neuroscience and much of neurotechnology development and use of neuroweapons are incipient, and in some cases, their utility is speculative. But speculation must acknowledge that neurotechnological progress is real, and therefore consideration of neurotechnologies-as-weapons is both important and necessary.

So, like most scientific research, there is an inherent dual-use devil embedded in all aspects of neuroscientific advancement. Opportunities for benefits, well-being, and tools to improve human life abound, but so does the darker side of escalating use, misuse, and abuse that arises from weaponization. Brain science progresses at a pace that may outstrip the time needed for ethical reflection, the formulation of groundrules and guidelines, and mechanisms for governance (Giordano and Benedikter 2011). Regrettably, this may allow malevolent diversion and misuse to flourish alongside benevolent achievement and application. One could speculate that deliberate manipulation of news items, editorials, media tautologies, and ongoing perceptual warfare campaigns—like a psychological operations (PSYOPs) program—could subtly steer a point of view driven by economic and/or geopolitical interests.

NEUROWEAPONRY AND THE GLOBAL BATTLESPACE

The Department of Defense defines a "battlespace" as "... the environment, factors, and conditions which must be understood to successfully apply combat power, protect the force, or complete the mission. This includes the air, land, sea, space, and the included enemy and friendly forces, facilities, weather, terrain, the electromagnetic spectrum, and information environment within the operational areas and areas of interest." (U.S. Department of Defense 2011). So where this definition is the allowance for neuroweaponry or similar kinds of mind-altering, influencing, or redirecting systems of modern warfare? Should we assume it is axiomatically included within the broad ambit of the definition provided?

Such a definition falls short even when dealing with the known spectrum of cyberwarfare. One aspect related to this categorical problem is defining the strategic frontiers for future national defense purposes. Other strategic frontiers beyond outer and cyberspace include frontiers in nanospace, genomic space, and "neurospace." Conceivably, these are legitimate domains for future conflict because they are as yet ungoverned spaces akin to their geographical counterparts. They reflect a no-man's land of unrestricted activity where hostile and benign activities are equally permissible. No formal doctrinal definitions of these new realms of security activity and frontiers of geostrategic thought have been devised thus far. If these are genuine strategic frontiers, a main issue is determining where neuroscience fits into the existing operational definition of redefined battlespace. Certainly, we have seen both outer space and cyberspace progressively redefined as part of the strategic battlespace.

In the physical world, battlespace is well known and its parameters defined. Similarly, an act of aggression or war in the physical sense is just as well defined. That is not the case when it comes to potentially novel and unexplored battlespaces. Federal officials, military leaders, policy scholars, and security experts are all looking at this issue and struggling to answer the question—what constitutes an act of war in these new battlespaces? Kevin Coleman (2008) argues:

> The contemporary definition of "battlespace" is, in my opinion, too confining. Battlespace is often defined as a three-dimensional area—width, depth, and airspace. Its fourth dimension of time and distance, tempo and synchronization, is also already considered, as is the radio frequency (RF) spectrum. The battlespace's fifth dimension is cyberspace, an area where battles will be fought anonymously but tenaciously. However, the overlooked but critical and dynamic factor of the twenty-first century battlespace is the human factor. This proposed "sixth dimension" of human factors includes leadership, motivation, ingenuity, and patience—factors that shape every aspect of the battlespace from the application of force through the effect of bandaging a child's hand.

Thus, it is fair to raise questions of whether neuroweapons constitute yet a seventh dimension beyond the cognitive and behavioral, or do weaponized neurotechnologies lie within the sixth dimension residing as it does in the broad array of human factors as described? Arguments that place neuroweaponry in the seventh dimension are rooted in psychology, perception, and interpretation of phenomenon that go beyond behavioral and autonomic responses and actions. These are thoughts, behaviors, and perceptions that are formulated on the basis of interpreted reality and external stimuli. According to Richardson, this dimension deals with complex thinking and situational assessment of novel conditions, options, and activities that engage emotion, motivation, ingenuity, and patience. It is the frontier of self-consciousness and rational mental operations (Posner and Russell 2005; Davidson and Begley 2012). This underscores one of the many issues embedded in neuroscientific research—the assessment of control networks within the brain. While there remains much to discover, the question is whether the pace of research will prompt global interest in advancing neuroscience to the extent that the likelihood of incurring risks of incipient neuroweapons has rendered neuroethical preparation and prevention to be little more than a race against time.

Following Richardson's theory, the seventh dimension tackles metathought and strategic analysis issues as its fundamental driving force: it is cumulative, derivative, and hierarchical in nature. Here, an argument can be made that there are no boundaries, geospatial limits, or shielding zones to perimetize defending the "mind." The extent to which the mind is unprotected, open, and exploitable by external elements is as the body was before the invention of armor, sword, and shield.

Neuroweaponry makes the global landscape a potential level playing field. It is no longer an inherent strategic advantage to possess killer satellites, long-range missiles, laser weapons, and advanced weapons of mass destruction (WMDs) if cognitions, emotions, and behaviors can be accessed and channeled toward outcomes independent of other extant weapons system. The global battlespace will be dramatically altered and in light of this, it will be necessary to design and implement systems to protect humans from neural interference if impending neurowarfare is to be regarded as realistic and eventual.

A paper by Tim Thomas (1998) of the U.S. Army's Command and General Staff school entitled "The mind has no firewall" delved into the degree to which Russian military scientists and a specific researcher, Chernisev, had focused upon what was termed "psychotropic weapons." Such weapons included the following:

- A "psychotropic generator," which produced a powerful electromagnetic probe capable of being sent through telephone lines, TV and radio networks, supply pipes, and incandescent lamps.
- An autonomous generator device that operated in the 10–150 Hz band, which at the 10–20 Hz band formed an infrasonic oscillation destructive to living creatures.
- A nervous system generator, designed to paralyze the central nervous systems of insects, which—upon further development—could exert similar effects on humans.
- Ultrasound pulses, which were supposedly capable of effecting bloodless internal manifestations without leaving external evidence.
- Noiseless cassettes that placed infra-low frequency voice patterns over music that could be subconsciously detected and which were claimed to be used as "bombardments" with computer programming to treat alcoholism or smoking.
- A "25th-frame effect," wherein each 25th frame of a movie reel or film footage contained a message that would incur subconscious effects on cognition, emotions, and/or behavior. This technique was advocated to curb smoking and alcoholism, but was also noted to possess wider applications if applied on a TV or movie audience or a computer operator.
- Psychotropic drugs as medical preparations used to induce trance, euphoria, or depression. Referred to as "slow-acting mines," such agents could

elicit symptoms of headache, auditory and visual hallucinations, dizziness, abdominal pain, cardiac arrhythmia, and cardiovascular arrest and could be administered to individuals, groups, a politician's food, or a population's drinking water supply (Thomas 1998).

While this research is over 15 years old, it is illustrative of international efforts to analyze human cognition so as to exert influence and control over thought, emotions, and actions. Thomas emphasized that, "The point to underscore is that individuals in … other countries … believe these means can be used to attack or steal from the data-processing unit of the human body."

Some argue that mind–machine interfacing is well within our grasp. They posit that extrapolations of information dominance and areas where artificial intelligence, cybernetics, nanotechnology, and advanced biotechnology can be usefully blended into something new and unique is at hand. This has received some degree of serious international attention, yet skeptics may still counter that the capacity for engaging in neuroweaponry is decades away (Morris 2009; Syd 2009; Yokum and Rossi 2009; Pickersgilla 2011). What degree of authenticity can be extended to these notions? DaVinci could describe and theorize flight centuries before it was ever attempted, does this render his ideas and sketches as valid scientific speculation or fiction? In the same context, we must ask what ideas about types of neuroweapons mean when contemplating special operations, PSYOPs, and intelligence gathering.

Have the concept and fundamental assumptions undergirding these endeavors changed through the advent of neuroweapons? If, in fact, an applied technology of neuroweaponry is realistic, does this not force a rethinking of those activities? Have we opened up a realm of inquiry and influence at the level of the synapse that may be extendable to the conduct of the self in society? Is this the new battlespace upon which to focus, or is this merely a frontier domain of science that is not well defined or understood? Should we risk waiting until the tangible first evidence of neuroweapons research has landed on the front page of our major newspapers and CNN? What measures should we be contemplating even now to anticipate and prepare for this eventuality?

In response, let me support the invocation provided by Giordano in this volume and elsewhere (Giordano et al. 2010) that the time to face the reality of operationalized neuroweapons is now, and not at some unforeseen point in the future, or as an "after the fact" reaction.

To wit, the new battlespace is the brain itself. It is the substance that gives rise to our perceptions and thoughts and seems completely vulnerable to manipulation. Is this a problem for which no rational defensive doctrine can be developed because the offensive applications far outstrip anything designed to insulate or protect? Are we defenseless against the presumptive actions of neuroweapons (and those who may employ them)? How would we discern and sense an attack against our own minds? Would we even recognize an attack if perceptions were somehow altered and viewpoints reduced to blithering neutrality?

If one posits a battlespace similar to the cyberwarfare domain, where does it originate and how can forensic science find, determine, and pinpoint the attacker? What type, form, and extent of new doctrine would be needed to reflect this new battlefield? What policies and agreed protocols would provide a governing framework for global engagement in neuroweapons activity? What are the overall implications for this new science in terms of classical geopolitical strategy?

Could it be argued that neuroscience research that focuses on control of thought, perception, behavior, and emotion necessitates regulation and governance by virtue of intent and potential application in war? Can we sort benign research from malevolent research and can we imagine new thresholds for action if and when we discover that neuroscience is being used to control human thought and behavior to an "unacceptable" degree? By whose standards? Under what conditions? (For further discussion, see Chapters 10 through 15.)

One problem is that a legitimate neuroscientific battlespace exists well before the development of the weapons themselves. This is not novel. In looking at history, did any nation have doctrine and strategy on the uses of airpower in 1912? In 1951, did any nation have developed military doctrine on the use of intercontinental ballistic missiles before those missiles demonstrated their capabilities? In both cases, doctrine was developed not upon recognition of a potential battlespace, but upon the emergence of the battlefield technology. Is this a reasonable guide to the future with regard to neuroweaponry, given the speed of expected research and development? In effect, doctrine suitable for weaponizable neuroscience and technology may not emerge until well after the new technology has displayed proof of concept.

TOWARD EMERGING DOCTRINAL STRATEGY—MITIGATING RISKS OF CONFLICT

Faced with these sobering possibilities how will the training and force doctrine of global militaries be changed? What constitutes asymmetric warfare in the age of neuroscience and neuroweaponry? Is this a new era that combines cybersystems and brain functions? Must we develop a defensive strategy based on risks that our forces could be victimized by neuroweapons used by an enemy state? Or, should we explicitly embark on a program of neuroweaponry strategy that completely revises the doctrine and principles of traditional conflict? How does the emergence of neuroweaponry alter our conventional understanding of warfare? Will it feature split autonomy and control by the military and civilian leaders? Does it require authentication and two-party key control? Who can or should participate; which agency and which officials decide?

Is this a human rights issue or something even more profound? External penetration of and influence over the minds of others does not readily lend itself to clarity of purpose or meaning in law, politics, society, or warfare. The power to influence or direct the thoughts and behaviors of others without them knowing crosses a threshold in human behavior and criminal conduct we have never seriously encountered or examined. Is using neuroweaponry to any degree, however, small or large in scale, whether domestic or transnational, automatically a formal act of war? Can we know whether civil insurrections, staged coups, urban riots, or border uprisings were naturally occurring or externally induced? Such questions are neither esoteric, inappropriate, or premature.

Neuroweaponry in the hands of an aggrandizing state leader or rogue nation is one thing. Having it available for terrorists and criminals, or excluded from the arsenals of less developed states is quite another. Today, there are no ironclad assurances against either scenario. Like nuclear weapons, neuroweapons convey a special status to those who possess them, especially if the technology cannot be readily shared, reverse engineered, or stolen through espionage or commercial theft. It places a relatively invincible weapon in the hands of a state or actor that may use this technology for domination or to simply extract submission and surrender from foes and enemies. Unlike nuclear weapons, sophisticated mechanical, bio-physical, and related scientific systems may not be necessary to develop or acquire neuroweapons.

When a new weapons system is devised, it is almost axiomatic that the sponsoring military must develop doctrine—a set of objectives and criteria—to define the most effective use and focus of this technology in a warfare setting. This compels raising additional questions that must be used to inform and formulate any doctrine, strategy, and use guidelines. When, how, and to what degree will neuroweaponry be used? Could it been seen as a WMD? Should it be seen that way? Could it reside only inside a closed and concealed command and control system not unlike the launch codes for nuclear missiles? Who could have access and who should be trained? How would this weapons system be deployed alongside conventional, or strategic weaponry? Is there an obligation to safeguard and protect its existence or must it be globally declared? If the mere existence of a neuroweapon is proven, how would it be revealed or demonstrated to a skeptical world and press?

The global arms control record is sobering. Despite bilateral agreements on nuclear arsenals and nuclear testing, proliferation of nuclear weapons technology still haunts us. Worse, we know from bitter experience that treaties dealing with chemical and biological weapons have been of limited coercive value, have not stemmed advanced research, and are unlikely to thwart actual use if a nation declared that employing such weapons was warranted. This brings us to the threshold question: when and under what circumstances would neuroweapons be warranted? Given the likelihood that some neuroweapons' research will continue in the future, do we know confidently that the actual emergence of such weapons can and will be somehow constrained? Recall that first use of atomic weapons preceded written doctrine about its use by at least a decade.

Assuming that several nations acquire this new weaponry within years of each other, does this mean that the era of "neural conflicts" would then be fully upon us? Would it stimulate serious international discussion of a treaty or similar mechanism to curb excessive and warlike uses of neuroscientific research? Would it instead signal that the first nation who holds such a weapon can neutralize, overcome, and vanquish any other armed power? How would neuroweapon strategies be adopted by an insurgent opponent or renegade power? Is it reasonable to expect a period where neuroweapons would be tested (upon both among witting and/or unwitting subjects)?

ISSUES INVOLVING CONTROL, PROLIFERATION, AND DETERRENCE

From this, we can safely conclude that like most other weapons systems in modern history, there will gradually but inevitably be an arms race to competitively adopt and deploy neuroweaponry. Risks of proliferation will alarm the community of nations possessing this technology. In turn, this will lead to global outcries for control, restrictions, and curbs on such weapons. This may trigger the usual underground behavior among some nations to proliferate and share technologies with the aim of rivaling extant neuroscience capability with an aggressive edge. Anything underground that has a criminal taint will, of course, find some appeal among transnational criminals and terrorists. Putting the proverbial genie back in the bottle will be difficult, if not futile.

It will become the kind of technology that contains many of the same inherent risks as nuclear energy. There are peaceful uses, and there are darker outcomes. Speculating for a moment on the universal appeal of neuroscience in its most benign forms, many nations would seek access to such technology for allegedly peaceful purposes. There will be the corresponding problem of curbing excursions into neuroweaponry and keeping activities utterly peaceful in a global system. Absent that, an unregulated environment would allow research at least on defensive measures against neuroweaponry. However, when some degree of international neuroweaponry equivalence and balance is established, systems must be devised to develop countermeasures, defensive doctrine, and deterrence systems.

Today, global research and development in neurotechnology or what may be regarded as cognitive science continues in a relatively unrestricted environment. It is to be expected that offshoots of this research will be applicable to affect and control individuals or even populations. How can—or should—such research be controlled so as to optimize its positive effects and minimize its dubious or frankly injurious outcomes? What constraints, rules, and approaches seem best to steer such research away from the darker side of neuroscience? One has to wonder whether any ethical imperatives and research boundaries will operate at all. Openly progressive research done in bioscience and synthetic biology with some restrictive measures imposed by consensus may be one approach to consider (e.g., see Chapter 14).

Clearly, it is important to ask what can be done to address this emerging technology. Some argue that programs of neuroscience education and training are needed. Some would insist that students learn about the potential societal impact of their work, its ethical and legal contours, and the specific ways it could be used by militaries, or terrorists, to create weapons. Others may favor revision of international treaties, particularly the chemical and biological weapons conventions, adjusted to account somehow for the new scientific realities. Brain chemistry is still a nascent field of inquiry, and there is some speculation that the questions are daunting enough that we will have a few decades to prepare for neurowars. However, while such debates continue, time is not on our side.

The quest to improve warfighter performance, assess and access human thought, influence well-being, manipulate emotions, and control neurochemical properties is unlikely to end. Instead, risks embedded in linking cybernetic, biotechnical, and neurochemical principles and systems will proceed apace (Suh et al. 2009). The cost of periodic progress, even given the irony of Moore's Law, means we face the risk of deliberately or inadvertently developing a neuroweapon at the same time that we discover a cure for schizophrenia or neurodegenerative diseases.

Modern society entails the risks and benefits of scientific achievement every day. Collective fears about a possible neuroweapon emerging from the midst of neuroscience research still seems unlikely to many. We focus on the salutary effects of continued research and related projects to advance general public health and improve the quality of lives. To a great extent, this will continue largely because it is seen as generally benign. However, brain science research which directly, accidentally, or randomly yields development of neuroweapons will change human history in profound ways. We cannot ignore that weaponization risks are definitely very real.

It is a dilemma that must be balanced, lest we thwart the rate and extent of scientific progress for all. The fact is that unless a globally enforceable mechanism is devised and agreed upon for controlling the conduct and outcomes of neuroscience research itself, we can expect to find no real safeguards and no guarantees. Enhanced brain performance and reductions in battle trauma are welcome. However, we must remain mindful and not be naïve that mechanisms, pathways, and technologies that engage the brain and its properties as weapons are certainly possible. As noted throughout this volume, neuroweaponry, however, we choose to define it, is clearly a case of when—not if.

CONCLUDING COMMENTS

It is far from clear whether any nation that is embarked on a robust neuroscience research effort will have the regard and discipline necessary to be wary of dual-use risks and the latent opportunities for devising neuroweapons. Inherent risk seldom dissuades serious science. During the next 10 years, the genuine risks and trajectories for these speculative outcomes will become much clearer. Several national and global policy questions related to ongoing neuroscience must be addressed as soon as circumstances allow. These include the following:

- Will global neuroscience research be transparent and openly accessible?
- Will global cooperative neuroscientific efforts require security safeguards?
- Will the nature and extent of dual-use neuroscience be globally shared?
- What legitimate neuroscientific "defensive measures" would be encouraged or allowable?
- What can the international community do to limit risks of emerging neuroweaponry?
- What would preclude any nation from covertly developing viable neuroweapons?
- What restrictions, safeguards, and precautions on neuroscience would be globally acceptable?
- Should persons involved in particular aspects of advanced neuroscience be licensed and regulated by government?

These questions and their potential answers will set the stage for addressing the future use(s) of neuroscience and neurotechnology in national security, intelligence, and defense agendas and programs. Opening discussions and serious scientific examinations of existing neuroscience research to assess the possible emergence of neuroweapons is paramount. In many ways, it remains the last undefended frontier and as such necessitates safeguarding against malevolent manipulation.

REFERENCES

Adams, C. 2005. "Empathetic Avionics: Beyond Interactivity." *Aviation Week*. http://www.aviationtoday.com/av/issue/feature/891.html#.UtALumRDssU.

Alexander, J.B. and C.S. Heal. 2003. "Non-lethal and hyper-lethal weaponry." In: *Non-State Threats and Future Wars*, eds. R.J. Bunker. London: Frank Cass.

Art, R.J. et al. 2012. *America's Path: Grand Strategy for the Next Administration*, Washington, DC: Center for National Security. http://www.brandeis.edu/facultyguide/person.html%3Femplid%3D4d0a125256afd15e484fd8bf9819d86.

Boyle, R. 2010. "Air force seeks neuroweapons to enhance US airmen's minds and confuse foes." *Popular Science*. November 3. http://www.popsci.com/technology/article/2010-11/air-force-seeks-neuroweapons-enhance-us-airmens-minds-and-confuse-our-foes.

Bulletin of Atomic Scientists. 2008. *The Military Application of Neuroscience Research*. Bulletin of Atomic Scientists, October 29. http://thebulletin.org/security-impact-neurosciences.

Burgess Jr., R.L. 2012. Annual threat assessment: Statement before the Senate armed services committee, United States Senate, February 16. Defense Intelligence Agency. http://www2.gwu.edu/~nsarchiv/NSAEBB/NSAEBB372/docs/Underground-BurgessThreat.pdf.

Canton, J. 2012. "Foreword." In: *Neurotechnology: Premises, Potential and Promises*, ed. J. Giordano. Boca Raton, FL: CRC Press. http://www2.gwu.edu/~nsarchiv/NSAEBB/NSAEBB372/docs/Underground-BurgessThreat.pdf.

Chen, J. 2011. "The regional impact of Kim Jong-Il's death." *International Policy Digest*, December 20. http://www.uwa.edu.au/people/jie.chen.

Clancy, F. 2006. "At military's behest, DARPA uses neuroscience to harness brain power." *Neurology Today* 6(2):4, 8–10.

Coleman, K. 2008. "Defining the cyber battlespace." *Defense Technology* 8(18):27.

Collina, T.Z. 2010. Chemical and biological weapons status at a glance, Arms Control Association, August.

Dareini, A.A. and L. Jakes. 2012. Nuclear Talks Resume Amid Impasse Worries—Iran Holds Firm, *Newsday*, May 24. http://www.huffingtonpost.com/2012/04/07/iran-nuclear-weapons_n_1409800.html.

Davidson, R. and S. Begley. 2012. *The Emotional Life of Your Brain: How Its Unique Patterns Affect the Way You Think, Feel, and Live—and How You Can Change Them*. New York: Hudson Street Press.

Dockery, K. 2007. *Future Weapons*. New York: Berkeley Caliber.

Durston, S., and B.J. Casey. 2006. "What have we learned about cognitive development from neuroimaging?" *Neuropsychologia* 44(11):2149–2157. http://www.niche-lab.nl/_documents/28_Neuropsychologia%25202006%2520Durston.pdf.

Farre, H. 2012. "Greece, brinkmanship and the euro, again." *Washington Monthly*, May 22, p. 59. http://www.washingtonmonthly.com/ten-miles-square/2012/05/.

Giordano, J. 2012. "Integrative convergence in neuroscience: Trajectories, problems and the need for a progressive neurobioethics." In: *Technological Innovation in Sensing and Detecting Chemical, Biological, Radiological, Nuclear Threats and Ecological Terrorism*, eds. A. Vaseashta, E. Braman, and P. Sussman (NATO Science for Peace and Security Series). New York: Springer.

Giordano, J. and R. Benedikter. 2011. "An early-and necessary-flight of the owl of minerva: Neuroscience, neurotechnology, human social-cultural boundaries, and the importance of neuroethics." *Journal of Evolution and Technology* 22(1):110–115.

Giordano, J. and R. Wurzman. 2011. "Neurotechnologies as weapons in national intelligence and defense—An overview." *Synesis: A Journal of Science, Technology, Ethics, and Policy* 2(1):T55–T71.

Giordano, J., C. Forsythe, and J. Olds. 2010. "Neuroscience, neurotechnology, and national security: The need for preparedness and an ethics of responsible action." *AJOB Neuroscience* 1(2):35–36.

Glannon, W. 2006. *Bioethics and the Brain*. Oxford: Oxford University Press.

Hearns, K. 2008. China and Iran make neuroscience advances. *Washington Times*, October 2, p. 6. http://www.washingtontimes.com/news/2008/oct/02/neuroscience-wake-up-call/%3Fpage%3Dall.

Herbst, J. 2006. *History of Weapons (Major Inventions Through History)*. New York: Twenty-First Century Books.

Hoag, H. 2003. "Neuroengineering: Remote control." *Nature* 423:796–798. doi:10.1038/423796.

Huang, J.Y. and M.E. Kosal. 2008. "The security impact of the neurosciences." *Bulletin of Atomic Scientists*. June 6. http://thebulletin.org/security-impact-neurosciences.

Johnson, S. 2009. Training a Skeptical Eye on Neuroscience. Dana Foundation, November 3. http://dana.org/News/Details.aspx%3Fid%3D43019.

Kagan, R. 2012. *The World America Made*. New York: Knopf Publisher. http://knopfdoubleday.com/2012/02/08/the-world-america-made-by-robert-kagan/.

Michael, A. 2012. "Defense boost ends Tokyo drift—Japan rethinks the value of a 'Peace constitution'." *Wall Street Journal*, January 03. http://online.wsj.com/news/articles/SB10001424052970203550304577138163500771238.

MIT Media Lab. 2010. Synthetic Neurobiology Research programs. http://www.media.mit.edu/research/groups/synthetic-neurobiology.

Morris, S. 2009. "The impact of neuroscience on the free will debate." *Florida Philosophical Review* IX(2):56.

MSNBC. 2012. "U.S. to remain top military power," January 5. http://nbcpolitics.nbcnews.com/_news/2012/01/05/9977091-obama-vows-us-will-stay-worlds-top-military-power.

National Commission on Terrorist Attacks. 2004. *The 9/11 Commission Report: Final Report of the National Commission on Terrorist Attacks Upon the United States* (Authorized Edition). New York: W.W. Norton.

O'Connell, R. 1989. *Of Arms and Men: A History of War, Weapons, and Aggression*. New York: Oxford University Press.

Pickersgilla, M. 2011. "Connecting Neuroscience and Law: Anticipatory discourse and the role of sociotechnical imaginaries." *New Genetics and Society* 30(1):27–40.

Posner, J. and J.A Russell. 2005. The circumplex model of affect: An integrative approach to affective neuroscience, cognitive development, and psychopathology. *Development and Psychopathology* 17(3):715–734. doi:10.1017/S0954579405050340PMCID: PMC2367156.

Rees, D. and S. Rose. 2004. *New Brain Sciences: Perils and Prospects*. Cambridge, MA: Cambridge University Press.

Sheply, J. 1956. "How dulles averted war." *Life*, January 16, pp. 70–72. http://muse.jhu.edu/journals/international_security/summary/v013/13.3.dingman.html.

Smith, D. and J. Bigelow. 2006. *Biomedicine-Revolutionizing Prosthetics*. Baltimore, MD: JHU Press.

Suh, W.H. et al. (2009). "Nanotechnology, nanotoxicology, and neuroscience." *Progress in Neurobiology* 87(3):133–170.

Thomas, T. 1998. "The mind has no firewall." *Parameters*, Spring, pp. 84–92. http://strategic-studiesinstitute.army.mil/pubs/parameters/articles/98spring/thomas.htm.

U.S. Department of Defence. 2012. Secretary of Defense Leon Panetta's speech on emerging naval strength, U.S. Naval Academy Graduation, May 29. http://www.defense.gov/Speeches/Speech.aspx%3FSpeechID%3D1679.

Vaseashta, A. 2012. "The potential utility of advanced sciences convergence (ASC)—analytic methods to depict, assess, and forecast trends in neuroscience and neurotechnologic development(s), and use(s)." In: *Neurotechnology: Premises, Potential, and Problems*, ed. J. Giordano, Boca Raton, FL: CRC Press.

Yokum, D. and F. Rossi. 2009. "A critical perspective on moral neuroscience." *Ethics and Politics* XI(2):18–42.

9 Issues of Law Raised by Developments and Use of Neuroscience and Neurotechnology in National Security and Defense

James P. Farwell

CONTENTS

INTRODUCTION

As noted throughout this volume, neurotechnology is reshaping the practical, tactical, strategic, as well as ethical and legal notions that govern how national security, defense, and military operations are engaged (see also Clancy 2006; Moreno 2006; White 2008; Forsythe and Giordano 2011a, 2011b; Guidorizzi 2013). New technology requires new thinking to shape the legal and ethical standards that govern the use of such technology. Evolving international standards will be important in influencing the precepts that define emerging legal norms.

Neurotechnological developments raise profound legal issues. These include the knowledge and consent required of individuals asked to use—or who volunteer to use—such technologies; the legal duties and responsibilities for those in authority who require subordinates to carry out missions and operations applying neurotechnologies; the duties and responsibilities of designers and manufacturers of neurotechnology and those civilians who support military missions; and more broadly, questions governing the right against self-incrimination when neuroscientific techniques and technologies are used to conduct analyses and queries of human subjects. There are also growing concerns—and questions—as to the admissibility of neuroscientific knowledge and the use of neurotechnological assessments into a court of law.

These questions are not purely legal. They present confluent ethical and policy issues. The interests of the public in advancing and sustaining national security must be weighed in comparison with the private interest in protecting individual rights. This chapter focuses on how the ongoing evolution of neurotechnology currently affects—and will likely influence—both military and public law.

THE IMPACT OF NEUROTECHNOLOGICAL DEVELOPMENTS

It has been said that the world has changed more in the last 30 years than in the past 300 years. It may change once again that quickly in the next 30 years. Commenting on the 2014 Quadrennial Defense Review, Lt. Gen. (Ret.) David Barno of the Center for New American Security has declared:

> The wars of 2034 and beyond will likely be fought and decided with technology, systems, and doctrines that do not exist today. And the stakes may be dramatically higher. (Barno and Bensahel 2013)

Neurotechnology and the neuroscience that fosters and employs these technical developments epitomize the challenge. Organizations like the Defense Advanced Research Projects Agency (DARPA) have funded neurotechnology-based and relevant projects that facilitate and examine advanced signal processing techniques for real-time coding of neural patterns in order to improve decision-making (see Chapter 7; Guidorizzi 2013). New neural interfaces and sensor designs that interact with the central and peripheral nervous systems are being examined for operational applications using nanoneuroscience, neuroimaging, and cyber-neurosystems[1] that can improve the performance of soldiers in operational environments, enhance capabilities to gather strategic intelligence, enable the rapid and accurate detection of visual and auditory information, and to achieve "decision superiority" on the battlefield (see Chapter 4).

Such research seeks new approaches to define, describe, and predict the behavior of individuals and groups. The advantages these developments offer include (1) the ability to process vast volumes of data, (2) facilitating informed, rational decisions rapidly with a high level of success, and (3) optimized crises-based decision-making. Yet, as with any technology, a somewhat "darker" side exists, as these neurotechnological approaches may be employed in ways that impinge upon private rights. The legal implications of such potential intrusions cast into high relief the balancing of public versus private interests when addressing, analyzing, and developing guidelines and governance of the ways that neurotechnology can and should be utilized in national defense and security agendas in real-world scenarios.

THE AUTHORITY ISSUE

Emerging neurotechnology will place new pressure on the need to formulate clear legal rules that govern the authority to act (Giordano 2010). Core questions essential to any genuine consideration of the applicability and relative ethicolegal validity and value of neurotechnology in these circumstances, and the subsequent formulation and execution of action guidelines and regulations include the following:

- Who has the authority to employ neurotechnologies? Although the conventional wisdom in interpreting the Uniform Code of Military Justice requires military personnel to accept medical interventions required under the rubric of fitness for duty, we must query how far such legal requirements extend when the issue involves the use of advanced neurotechnologies that are not fully understood, durably examined, or are invasive.[2]
- What are the implications for a military member to refuse neurotechnological interventions (whether deemed treatment, enablement, or enhancement) that may intrude upon the private realm of cognition and emotion, and in such ways affect, if not control independence of thought, emotion, and action?
- What are the duties of the military to protect—if not enhance—the health and well-being of personnel who are engaged in the exigencies of combat? Do such duties justify, compel, and/or sustain the use of advanced neurotechnologies to affect such goals and ends?
- What are the proximate, intermediate, and distant goals of employing neurotechnology? Legal rules that govern neurotechnological development and use must be consistent with accepted international ethical norms and/or risk assessments. Yet often such parameters are difficult if not impossible to accurately define given the novelty of the technology and/or the information and outcomes that such technology will provide. This makes any postulation of defined goals both speculative and highly contingent upon effects incurred along the trajectories of iteration and use.
- How will unanticipated, side and/or adverse effects be handled, and who (i.e., what individuals, organizations, and entities) will bear responsibility for addressing and militating such effects?
- Is the use of neurotechnology in national security operations offensive or defensive? While there may be general agreement that people have the right to protect themselves against attack, there remains a doubt as to the ways that launching a neurotechnological attack—both intra- as well as internationally—may affect what is considered to be acceptable standards of conduct.

A NEW THREAT ENVIRONMENT

Neurotechnology will create obvious opportunities for benefits in medicine and aspects of daily life—inclusive of medical and daily life parameters of the military and the national security milieu. But such technology will also generate new threat environments that affect critical command decisions. These are addressed in the following sections.

CAPABILITIES IN THE HANDS OF EMPOWERED INDIVIDUALS

In 2000, the creators of the *I Love You Virus* inflicted heavy damage on computers worldwide (Kleinbard and Richtmyer 2000). In 2008, the *Conficker Worm* infected eight million computers (Bowden 2011). The creators of Facebook and Napster offer positive examples of how a single person's creativity and imagination can alter worldviews, use tools, and engage relationships in both the personal and business spheres (Menn 2011).

While these examples pertain to use of the Internet, neurotechnology may have greater impact. But as neuroscience progresses into field use, one should never presume that the capabilities it confers will be limited to state-level activities. States increasingly employ sophisticated, scientifically capable proxies to create plausible deniability and create shields to the imposition of legal responsibility. Individuals, private groups, and commercial entities all seem likely to gain possession of emerging neurotechnologies. These state and nonstate actors will be able to leverage diverse power effects within economic, social, and geopolitical environments that will likely reshape the profile emerging threats and shift focus away from the consequences of purely state action.

THE EFFECT OF NEUROTECHNOLOGY ON NATIONAL SECURITY AND MILITARY OPERATIONS

As noted throughout this volume, neurotechnology can and will be employed to pursue and achieve key national security, intelligence, defense, and military objectives. It may be engaged to affect command and control operations; facilitate intelligence, surveillance, and reconnaissance; optimize force or asset capabilities at tactical and strategic levels; and suppress or negate adversarial activity.

What neurotechnological developments seem plausibly operational? An advanced neurotechnology may use the Internet to self-train, monitor public and nonpublic information, and enhance reasoning abilities (Preden 2010). The legal questions raised by such technology and its use—and potential misuse—are challenging and need to be seriously addressed. Hopefully, the potential impact of such developments will increase cooperation and collaboration among communities of interest, both within and outside of the government, in shaping legal rules.

A threshold question is how such rules must change to protect individual rights against the misuse or abuse of neurotechnological assessment and/or interventions. History teaches that one cannot mandate behavior without repercussions. Current law does not address the more profound implications of neurotechnological use for national security.[3] The law does not reach the intuitive strategic policy and ethical issues that neurotechnology developments pose in a global environment. Yet it seems reasonable to suggest that as the potential impact of using neurotechnology will occur beyond national borders, internationally accepted legal norms (and the ethico-social grounds upon which they might be structured) will need to be acknowledged, respected, and may gain ever wider acceptance (Spranger 2012).[4] The potential power of such technology renders such conclusions prudent and common sense (Vaseashta et al. 2012).

NEW THINKING FOSTERED BY THE REVOLUTIONARY NATURE OF NEUROTECHNOLOGY

Neurotechnological integration of machine and neural substrates enabling mental control of both hardware and software may seem futuristic, yet it is and will become more technically plausible and achievable. Current examples include prosthetic devices to replace limbs and sensory structures and the use of brain–machine interfaces to engage communicative and motoric devices (Lebedev and Nicolelis 2006; Santhanam et al. 2006; Jarosiewica et al. 2008). An artificial arm or leg—although controlled by the brain—hardly seems to pose a threat (Salisbury 2011). But what will future evolutions portend for ethics and law, as neurotechnology enables cognitive processes to become accessible to external assessment or control, and if and when brain functions can be neurotechnogically engineered to affect change in thought, feeling, and actions?

Other interventions may go further to create cyborg capabilities. While the term "cyborg" may conjure science fictional images, as Benanti (2012) and Wurzman and Giordano (Chapter 7) note, integrative neurotechnologies are establishing true cybernetic effects as defined by Cline and Klynes (1995), and more recently Hables-Gray (2001); through the use of biomimetic systems of hardware and software, and physiological enabling techniques and technical implements.

The challenge is sharpened as neurotechnology is used to serve national security, military, or public interests. Consider a sequel to Japan's 2011 Fukushima nuclear incident. A team of 50 volunteers risked their lives by remaining at the plant, braving high levels of radiation, to bring the reactors under control (Wingfield-Hayes 2013). Arguably these volunteers may have sacrificed their lives. Years from now, neurotechnology may alter the capacity of humans to perform tasks in similar or other severely hostile environments.

What happens when individuals are asked—or feel pressured—to volunteer to acquire such neurotechnological implements in order to serve the national or public interests? What confluence of ethical and legal rules should govern recruitment? Such issues need to be fully and soundly addressed, not at some future point, but at present, before an urgent scientific, technological, social, and/or political crisis overrides prudence. The ethical-legal issues, questions, and problems demand attention before these technologies become operational. That will enable a determination as to what legal standards may be viable—or should be developed—to govern (1) the informal or formal pressures placed on individuals to participate in using such technology, (2) the disclosures required for informed consent, (3) the level of care taken to protect individuals from harm, and (4) the liability that parties bear when individuals using such cutting-edge neurotechnologies are harmed.[5]

THE ADVENT OF ARTIFICIAL INTELLIGENCE

There are well-known efforts to exploit the Internet to monitor public and nonpublic information. Neurotechnology will likely play a key role in these advances, similar to traditional forms of cybertechnology. Investment arms of the CIA and Google have backed Recorded Future, which monitors the web in real time and uses the information to predict the future. The company claims that its temporal analytics engine "goes beyond reach" by "looking at the 'invisible links' between documents that talk about the same, or related, entities and events" (Shachtman 2010). "The idea," journalist Noah Schachtman writes, "is to figure out for each incident who was involved, where it happened and when it might go down. Recorded Future then plots that chatter, showing online 'momentum' for any given event" (Shachtman 2010). The company maintains an index with over 100 million events hosted on Amazon.com servers, although it performs analysis on the living web. The technology enables it to spot trends and developments early. Shachtman reports that this technology enabled Israeli President Shimon Peres to corroborate his claim that Hezbollah possessed Scud-like weapons (see Recorded Future's description of its ability to analyze past, present, and future trends using data archives and to anticipate actions, behavior, and intent from the website https://www.recordedfuture.com/.).

Potential cognitive hacking, rooted in neurotechnology, would raise confluent issues of policy, ethics, and law. Overt hacking that spoofs a legitimate website to provide misleading information—for example, to gain access to passwords—could be an effective military tactic. A key offensive objective may be to trick networks into giving away information. One sees a parallel in organized criminal efforts to trick computer users into opening files that lead them to a fraudulent banking site to which confidential financial information is passed (Poulson 2011). Today that would be generally accepted as a valid military tactic.

Active authentication programs employ neuroscience to authenticate user passwords. One aspect uses covert games disguised as computer anomalies to verify unique user features through the user's responses to changes in the games. Another examines cognitive abilities expressed through keystrokes to understand how individuals process information on computer screens in order to validate user identity (Cybenko et al. 2003; Guidorizzi 2013).[6] Will people feel equally comfortable about neurotechnology that provides the ability to covertly manipulate the human brain to extract information from the network?

WHAT LEGAL RULES SHOULD APPLY?

The practical applications and possibilities for neurotechnology in national security scenarios are tantalizing. These applications can be controversial, partly due to their international effects. Hence, informed analysis of the ethical and legal issues requires an integrated approach that looks to include norms embraced in international law. The United Nations Universal Declaration of Human Rights explicitly recognizes "the right to freedom of thought" (Article 18). It observes that all human beings "are endowed with reason and conscience and should act towards one another in a spirit of brotherhood" (Article 1). These are noble words, but future neurotechnology may hinder an individual's ability to think independently.

What rules of law, grounded in what ethical standards should govern the use of technologies that can be used to assess—or affect and control—the cognitive processes, emotions, and neural bases that give rise to intention, belief, and actions? At what point does neurotechnology neutralize or blur the identity of the individual, and status as a person? In the television series *Star Trek: Voyager*, the evil Borg assimilated a character named Seven of Nine. She retained human elements. But neuro-implants controlled most of her functions. Of course, this is just fiction and easily dismissed. But look to near-future reality. What happens when neurotechnology turns today's science fiction into tomorrow's reality? Author Arthur C. Clarke devised a series of amusing laws of prediction. One stated: "Any sufficiently advanced technology is indistinguishable from magic" (Clarke n.d.). Science fiction author Larry Niven presented a corollary set of laws (not all dealing with science fiction). One holds: "Rigorously defined magic is indistinguishable from technology." Another declares: "the ways of being human are bounded but infinite," while a third observes that "ethics change with technology" (Niven 2002). As neurotechnology evolves, expect a call for constraints upon neurotechnology implements.

Concerns over the dangers posed by new technology or weapons is hardly novel. In 1139, Pope Innocent II thought the crossbow so lethal he banned them. Less than a century later, in 1215, Article 51 of the Magna Carta denounced the weapon as a threat to society (Lin 2010). The advent of airpower provoked deep fears. The 1899 Hague Convention banned "the discharge of any kind of projectile or explosive from balloons or by similar means" (Scott 1899). Eight years later, the Hague Convention (1907) restricted the use of airpower. The prospect of terror bombings against civilians initially caused both Germany and Britain to hesitate about launching such missions, although soon enough both carried out strategic air raids. The advent of the nuclear age has prompted numerous calls for a ban on nuclear weapons. Thus, concerns about weaponizing neurotechnology or the use of such technology by militaries should surprise no one.

A number of codes, declarations, and reports have shaped the legal environment, yet may not offer sufficient guidance or regulation of neurotechnologies. Philosopher and ethicist Patrick Lin has framed key issues that reflect the ambiguities raised by the use of neurotechnology for warfighters. These include the duration that neuroenhancement may be engaged and used; the reversibility of such interventions[7]; and possible social disruptions that enhanced warfighters may cause—including examination of potential divisions (in capability, burden, and status) that may arise between the enhanced and the unenhanced (Lin et al. 2009, 2013; Lin 2009; see Chapter 15).

Analysis for developing ethical rules derive largely from the Nuremberg Code. The Code emanated from a 1947 verdict rendered against 16 German physicians, scientists, and administrators who devised, implemented, or supported medical experiments on human subjects in concentration camps without consent.[8] They were charged with war crimes and crimes against humanity. Most of those used for experiments died or sustained crippling injuries. Of great importance, not all of these defendants were members of the National Socialist German Workers Party

(*Nationalsozialistische Deutsche Arbeiterpartei*), i.e., Nazis. The defendants argued that they had operated under a military-duty standard requiring them to carry out orders, that the experiments aimed to protect German fliers and soldiers, and that the "good of the state" took precedence over the individual.[9] The defendants asserted that no law distinguished between legal and illegal experiments.

The Tribunal declined to excuse conduct merely because a defendant was "political". The legal implications resonate powerfully today. German Courts and the North Atlantic Treaty Organization (NATO) divorce the political character or leanings of a party from any consideration of the moral standards of bioscience and medicine that govern conduct. The Nuremberg tribunal recognized the importance of Hippocratic ethics and the maxim of nonharm (*primum non nocere*), requiring physicians to do more than necessary to protect human research subjects, rooted in the precepts that parties must give informed consent to participate in an experiment and retain the right to withdraw from it. Nuremberg merged Hippocratic ethics and the protection of human rights into one code (Shuster 1997).

Working with Dr. Andrew Ivy, Dr. Leo Alexander submitted to the Counsel for War Crimes six points that defined legitimate medical research, to which the Court added an additional four. The ten points that constituted the Nuremberg Code ("The Nuremberg Code"; Germany 1949) hold that the voluntary consent of the human subject is absolutely essential. This means that the person involved should have legal capacity to give consent; should be so situated as to be able to exercise free power of choice, without the intervention of any element of force, fraud, deceit, duress, over-reaching, or other ulterior form of constraint or coercion; and should have sufficient knowledge and comprehension of the elements of the subject matter involved, so as to enable him/her to make an understanding and enlightened decision. As such, neurotechnologies may be used in those circumstances in which a person can be predisposed to believe they exercise free power of choice.

This latter element requires that before the acceptance of an affirmative decision by the experimental subject, there should be made known to him/her the nature, duration, and purpose of the experiment; the method and means by which it is to be conducted; all inconveniences and hazards reasonably to be expected; and the effects upon his/her health or person, which may possibly come from his/her participation in the experiment. The duty and responsibility for ascertaining the quality of the consent rests upon each individual who initiates, directs, or engages in the experiment. It is a personal duty and responsibility which may not be delegated to another with impunity. Moreover, the nature of the research should entail that:

1. The experiment should be such as to yield fruitful results for the good of society, unprocurable by other methods or means of study, and not random and unnecessary in nature.
2. The experiment should be so designed and based on the results of animal experimentation and knowledge of the natural history of the disease or other problem under study, through which the anticipated results will justify the performance of the experiment.
3. The experiment should be so conducted as to avoid all unnecessary physical and mental suffering and injury.

4. No experiment should be conducted, where there is an a priori reason to believe that death or disabling injury will occur, except, perhaps, in those experiments where the experimental physicians also serve as subjects.
5. The degree of risk to be taken should never exceed that determined by the humanitarian importance of the problem to be solved by the experiment.
6. Proper preparations should be made and adequate facilities provided to protect the experimental subject against even remote possibilities of injury, disability, or death.
7. The experiment should be conducted only by scientifically qualified persons. The highest degree of skill and care should be required through all stages of the experiment of those who conduct or engage in the experiment.
8. During the course of the experiment, the human subject should be at liberty to bring the experiment to an end, if he/she has reached the physical or mental state where continuation of the experiment seemed to him/her to be impossible.
9. During the course of the experiment, the scientist in charge must be prepared to terminate the experiment at any stage, if he/she has probable cause to believe, in the exercise of the good faith, superior skill and careful judgment required of him/her that a continuation of the experiment is likely to result in injury, disability, or death to the experimental subject.

The United States elected not to formalize the Code into a binding document to prescribe and proscribe particular activities within the responsible conduct of research,[10] although Secretary of Defense Charles E. Wilson made the Code established policy within the Pentagon.[11] The principles of the Code guided the Declaration of Helsinki in 1964, in which the World Medical Association established recommendations to guide biomedical research that involved human participants. Revised in 1975, 1983, 1989, and 1996, the Declaration is viewed as the basis for Good Clinical Practices used today.[12] The 1948 Universal Declaration of Human Rights (U.N. General Assembly) in principle has global authority but is not enforceable in court.

The current U.S. system of protection for human research subjects is embodied in the Federal Policy for the Protection of Human Subjects, often referred to as the "Common Rule,"[13] published in 1991 (HHS.gov 1991). The rule flowed from the principles enunciated in the Helsinki Declaration (World Medical Association 1964). This regulatory framework binds the Department of Defense (DoD) (Moreno 2006). Rules require that a research ethics committee review proposals and the basis for, and acquisition of, informed consent of volunteers in experiments. It enshrines notions of beneficence, justice, respect for persons, privacy for research participants, the right to withdraw, the return of results, and informed consent.

Although not enforceable in U.S. Courts, certain international protocols or agreements inform policy for evaluating legal rules that should apply to national security and defense uses of neurotechnology. The Geneva Conventions of 1949[14] (ICRC 1949) and the additional Protocols of 1977 (Protocols I and II) stand out as particularly relevant. For example, Article III prohibits "outrages upon personal dignity, in particular humiliating and degrading treatment" (ICRC 1977).

Much of the current discourse about neuroscience is focused upon how that provision applies to neuropharmacological agents (Moreno 2012). It would apply as well to other potentially coercive measures that employ current and future iterations of neurotechnology. A key challenge is that neurotechnology will advance faster than policy or the judiciary system can ascertain and move into domains of potential use and misuse yet to be discovered. The Additional Protocols to the Geneva Convention do not explicitly address neurotechnology. But a reasonable interpretation would impose definable constraints upon the scope of scientific techniques or technologies applied within a clinical or research setting. Here arises a key challenge. As cutting-edge technologies emerge, balancing the anticipated reasonable benefits of a "novel treatment" or approach against the potential harm or burden it may cause can be difficult, especially in ambiguous situations or when managing a crisis.

The 1991 Iraq conflict, Operation Desert Shield, well illustrates the conundrum. United States and allied military advisors were concerned that Saddam Hussein would order the use of sarin nerve gas or biological weapons (e.g., anthrax bacteria). The Iraqi forces also had botulinum toxin, which can quickly cause respiratory paralysis and death. There was no time to conduct clinical trials on available antidotes. The Pentagon needed to ensure that its force was fit for duty, and wanted to give pyridostigmine bromide (PB) pills to counter nerve gas, even though the benefits were uncertain. About 250,000 troops elected to take them. In theory a voluntary act, however, many troops felt pressured to take them (Moreno 1999).[15]

Also administered was a botulism vaccine that the Food and Drug Administration (FDA) considered "investigational" and that was never approved for the military to use against chemical weapons. Record keeping was poor and it is not clear how many were vaccinated. The FDA waived the normal informed consent provisions based upon the need to protect combat troops. The FDA's position has been heavily criticized for violating the precepts adopted into law as a result of the Belmont Report (Levine 1991). The report embraces three principles: respecting people through informed consent for research; beneficence, for example, do no harm while maximizing benefits; and justice, by ensuring that procedures are well considered (Belmont Report 1979).[16] Critics charge that "investigational" is a term applied to any drug the FDA has not approved and is often mistaken for a drug that is the object of research.

Critics charged that the Pentagon administered drugs that failed to conform to federal standards that define research and that characterizing the drugs as "investigational new drugs" was a canard in that they had not been properly researched.[17] Health and Human Services (HHS) rules define research as an activity designed to test a hypothesis, permit conclusions to be drawn, and thus to develop or contribute to knowledge that can be generalized (Belmont Report 1979). Commanders argued that they acted to prevent horrific injuries by administering "the best preventive or therapeutic treatment" available (FDA 1990).[18] A U.S. district court sided with the military, dismissing a suit to declare unlawful the FDA rule that permitted the military to administer the drugs and enjoin their administration without informed consent.

Operation Desert Shield involved combat and an imminent threat of harm to combatants. That situation may not hold true as emerging neurotechnologic devices or approaches are developed and implemented. Very probably, decisions will be taken in ambiguous situations where there are no clear, practical, operating guidelines. What role might customary international law (CIL) play in shaping future standards? Article 11 of the 1977 Additional Protocol I prohibits any procedure against a person in the power of the adverse party which is not consistent with generally accepted medical standards, including medical or scientific experiments, even with the person's consent ("Article 11.2(b)"). Nor can medical personnel be compelled to carry out work contrary to the "rules of ethics" (ICRC 1977). However, the use of neurotechnology is not limited to the confines of the clinical encounter. That presents a central question when addressing and examining legal issues generated by the application of neuroscientific techniques and technical tools in national security agenda.

The U.S. DoD requires all weapons to conform to international law[19] (White 2008), and CIL requires countries to "ensure respect" for international humanitarian law.[20] A U.K. Royal Society report offers cogent illustrations. It has argued:

> degrading the cognitive abilities of an adversary such as they are unable to distinguish between military targets and civilians, which often require a high degree of concentration, will undermine this requirement. This is because such cognitive impairment could easily result in an unintended attack on one's own civilians or other persons or places specifically protected by law. Such attacks could not be prosecuted because the perpetrators will have been rendered mentally incapable of being responsible for the offences. (Brain Waves Module 3 1998)

Corollary issues raise questions about what international legal standards determine intent, culpability, or predispositional states that would enable any court to ascertain intent or culpability. International conventions outlaw "willful killing."[21] However, one must ask how this notion might be aligned with the use of weaponized neuroscience and technology (see Chapters 7 and 8). Analysis must respect the notion of "legality," which accords courts criminal jurisdiction only for acts that have previously been classified as criminal (ICRC 1960).[22]

Arguably, willful killing (murder) is the most egregious crime. The rules for the International Criminal Court (ICC) defining the five elements for the "war crime of willful killing ("Article 8(2) (a)(i)") are set forth in the Report of the Preparation Commission for the ICC. Article 8(2) (a)(i):

- The perpetrator killed[23] one or more persons.
- Such person or persons were protected under one or more of the Geneva Conventions.
- The perpetrator was aware of the factual circumstances that established the protected status.[24]
- The conduct took place in the context of and was associated with an international armed conflict.[25]
- The perpetrator was aware of factual circumstances that established the existence of an armed conflict.

Under most U.S. criminal law, prosecutors must prove the commission of a voluntary act or that some omission by the attacker resulted in one or more victims' death. They must also prove intent to inflict great bodily harm or death. In international law, there is a presumption that a similar rule would apply: prosecutors alleging willful killing must prove (1) *actus reus*—a voluntary act or willful omission,[26] and (2) *mens rea*—intent. Intent under domestic and international law may be inferred objectively by the facts and circumstances of the action. For such proof to be levied in situations that entailed the use of weaponized neurotechnologies, culpability would have to be examined on a case-by-case basis, looking at the facts of a situation and the extent to which an individual was able to exercise independent thought in carrying out an act.

Exculpating an individual from criminal culpability does not relieve commanders who ordered a mission to be executed in accordance with duties or responsibilities. CIL holds commanders and civilian superiors criminally responsible for crimes they direct (Rule 152). This applies to crimes that they knew or should have known would occur.[27] Those who facilitate crimes may face criminal exposure. CIL prohibits means and methods of warfare that cause "superfluous injury or unnecessary suffering," or those that can inflict long-term environmental damage (Customary International Humanitarian Law). That article constrains opposing fighting forces' choice and use of weapons. Article 23(e) of the Hague Convention (1907), to which the United States is a signatory, also endorses that notion.

The War Crimes Act of 1996, passed by the U.S. Congress by overwhelming majorities, and signed into law by then President Clinton, defines a war crime as a "grave breach of the Geneva Conventions" and makes clear that this includes meanings defined in any convention related to the laws of war to which the U.S. is a party. War crimes include the Geneva Convention's language that outlaws acts "committed against persons or property protected by the Convention: willful killing, torture or inhuman treatment, including biological experiments, willfully causing great suffering or serious injury to body or health." It criminalizes as well violations of Articles 23, 25, 27, and 28 of the Annex to the Hague Convention IV.[28]

Language in those international declarations or conventions preceded the advent of weaponizable neurotechnology. Importantly, the use of such weapons does not necessarily cause death or overt bodily harm (see Chapter 7 for detailed discussion). The legal issue of whether such acts that employ weaponized neurotechnologies fall within the scope of the War Crimes Act or, for international law, the rules of the ICC—by which the United States has declined to become legally bound, but which inform international norms[29]—will likely turn on the *effect* and perhaps the *intent* that motivates their use. That raises an additional interesting question: if and when an attack using a neurotechnologic tool is unsuccessful because technology fails or a defense counters its use, what is the attacker's culpability?

Such questions are a red flag for the potential national security employment of neurotechnology. The use of neurotechnology to augment or enhance the ability of warfighters would seem to fall within the ambit of protecting personnel from harm sustained during the conduct of war. That is legally defensible. Offensive uses of weaponized neurotechnology would be more difficult to justify, even under *jus in bellum*.

This important inquiry asks what level of morbidity or lethality must a (neuro) weapon attain to qualify its use as a war crime? One may more often think in terms of chemical or biological weapons. But existing doctrine appears to offer at least initial guidance for governing use of such weapons (Dando 2012). Still, research evokes a gray zone as to the possible effects and countermeasures that arise from the illicit use of chemical/biological neuroweapons, particularly if a national stance of preparation is assumed. Under international treaty and doctrine, would such research be permissible—or even defensible? And as Giordano has noted (Chapter 1), what of the very real potential for dual-use interests and applications? Would such considerations and prohibitions also apply to other forms and types of weaponized neurotechnology? That raises questions as to what rules would apply, and how one might leverage or enforce them.

Competing views exist as to what types of weapons are permissible. The International Committee of the Red Cross has taken a firm position against use of landmines, incendiaries, and nuclear weapons (Henckaerts and Doswald-Beck 2005). But as yet, there is no case law or authoritative commentary that governs the use of weaponized neurotechnology, outside the rubric, noted above, of microbes and chemicals. This generates further query as to what this would mean for assessment of neurotechnologies, or use of electrical, magnetic, and or surgical approaches to modify brain function.

EVIDENTIARY STANDARDS

Under U.S. law, attempting to examine evidence that may enable a court to draw a distinction between actions that are voluntary and satisfy appropriate standards of intent, or sufficient awareness of action that triggers a legal duty of responsibility for action raises significant issues. Where does the constitutional right against self-incrimination under the Fifth Amendment, or the Fourth Amendment right against unreasonable searches or seizures come into play? What standards of evidence apply to the operation or effect of neurotechnology in such circumstances?

Issues arise as to the use of brain scans or other evidence obtained using neurotechnology against a defendant. In the United States, the Fifth Amendment right against self-incrimination would seem to offer protection in such cases. But it remains to be seen how it will be applied as support grows for the validity, viability, and value of neurotechnology in assessing brain structure and function(s) to be involved in predispositions, cognitions, and actions of criminality. A corollary question is how legal rules apply to both the use of neurotechnology and to the designers and developers of that technology.[30] One must discern how neurotechnology affects individual behavior and determine if and how neurotechnology can and should be used to meet explicit calls for approaches to assess, predict, and even alter human cognition that is instrumental to the acts of individual and/or social violence.

At issue are uses of assessment neurotechnologies (e.g., combinatory applications of neuroimaging, encephalography, neurogenomics/genetics, and complex neuro-cognitive and behavioral analyses effected through use and analysis of large-scale data portals and banks [see, e.g., Chapters 2 and 7]) that have been posited for use in these approaches.

The U.S. federal courts have long entertained division over what legal standard can and should be employed to guide the admissibility of expert testimony of scientific evidence. The leading case, *Daubert v. Merrell Dow Pharmaceuticals, Inc.*, 509 U.S. 579, 113 S.Ct. 2786 (1993), overturned a prior, narrower standard (see below). The case turned on the Supreme Court's interpretation of Federal Rule of Evidence (FRE) 702, which provides:

> If scientific, technical or other specialized knowledge will assist the trier of fact to understand the evidence or to determine a fact in issue, a witness qualified as an expert by knowledge, skill, experience, training or education may testify thereto in the form of an opinion or otherwise …

The decision makes the District Judge the gatekeeper for deciding the admissibility of experts offering testimony on scientific evidence and the techniques and technologies they employ. The Court set forth a three-part test for admissibility, which asserts the following:

- The testimony must be about *scientific knowledge*. It must be grounded in knowledge and achieved through the use of the scientific method.
- The scientific knowledge must assist the trier of fact in understanding the evidence or determining a fact that is at issue. The trier of fact may be the judge or a jury. Being "helpful" requires establishing a valid scientific connection to the pertinent inquiry before the Court.
- The judge is entitled to make the initial determination as to whether certain scientific knowledge would assist the trier of fact as contemplated by FRCP 702. Critically, and much influenced by the sophisticated "friend of the court" briefing in the case from a diverse group of scientists, scholars, physicians, historians, and sociologists of science that relied on philosophical notions as much as legal precedent, the Court drew a distinction between evidence rooted in the judicial process and aimed at resolving a dispute that aimed at and the search for scientific truth.[31]

The *Daubert* standard varies sharply from a competing view advanced in a 1923 case, *Frye v. United States*, 293 F. 1013 (D.C. Cir. 1923), in which the Circuit Court held that expert testimony as to scientific evidence was admissible only where "the thing from which the deduction is made … is sufficiently established to have gained general acceptance in the particular field in which it belongs." The *Frye* court declined to allow the testimony of an expert as to the results of a systolic blood pressure test, a predecessor to the modern polygraph.

Daubert ruled that its standard overruled *Frye*'s restricted view for federal cases. As an interpretation of FRE 702, it does not bind state courts. Although a products' liability case involving the combination drug pyridoxine/doxylamine, marketed under the brand name *Bendectin*, it raised larger issues about the relationship of science and law that are highly relevant to evaluating what standards of evidence govern neuroscientifically and neurotechnologically based and/or derived evidence.

The proposed use of functional magnetic resonance imaging (fMRI) lie detection tests illustrates this legal challenge (Chen 2009; Brown and Murphy 2010). This technology has threatened to up-end the very function of a jury in assessing the credibility of a witness. Only a few cases have considered the admissibility of evidence derived from fMRI technology for lie detection.[32] The New York State Court considered the issue in a civil case involving sexual harassment, *Wilson v. Corestaff Servs.*, 900 N.Y.S. 2d 639 (N.Y. Sup. Ct. May 14, 2010). (The Opinion can be accessed at http://blogs.law.stanford.edu/lawandbiosciences/files/2010/06/ CorestaffOpin1.pdf.) Applying *Frye*, the Court excluded the proferred testimony as to fMRI evidence of Dr. Steven Laken of Cephos Corporation[33] on two grounds: (1) it went to the credibility of a fact witness, a matter solely reserved for the jury, and (2) the inability "to establish that the use of the fMRI test to determine truthfulness or deceit is accepted as reliable in the relevant scientific community" (opinion in *Wilson* 2010).

In 2012, the United States Sixth Circuit excluded fMRI evidence in *United States v. Semrau*, 693 F.2d 510 (6 Cir. 2012). Dr. Laken tested Dr. Semrau, the owner and CEO of two firms accused of criminal fraud in billing Medicare and Medicaid for psychiatric services that the firm supplied to nursing homes. Laken concluded that the accused was "generally truthful" in all of his answers, collectively. The trial judge directed a magistrate to conduct a two-day *Daubert* hearing to make recommendations as to whether to admit fMRI evidence.

The magistrate recommended against admission.[34] The District Court concurred and applying *Daubert*, the Sixth Circuit affirmed, on the grounds that in evaluating the probability of detecting untruthful testimony, the fMRI assessment lacked sufficient probative value under FRE 403, which permits a court to exclude relevant evidence if its probative value is substantially outweighed by a danger of confusing the issues or misleading the jury.[35] Rule 403 provides a basis for excluding evidence independent of Rule 702 and *Daubert*, even though consideration of Rule 403 is included in the *Daubert* analysis. For more details see *United States v. Smiths*, 212 F.2d 306, 322 (6 Cir. 2000); *United States v. Hawkins*, 969 F.2d 169, 174 (6 Cir. 1992); and *United States v. Ramirez-Robles*, 386 F.2d 1234, 1246 (9 Cir. 2004).

Truthful Brain Corporation (TB) (2013) (truthfulbrain.com) claims to employ fMRI for lie detection using an approach that differs from that used by Cephos. TB uses a magnetic resonance imaging (MRI) machine with a magnetic field of 3 Tesla, compared to Cephos' 1.5 Tesla. Lakens refutes that, saying "that we have used 3 Tesla and 1.5 Tesla and there are no substantial differences." For his part, TB CEO Joel Huizenga argues that its technology, procedures, and analysis render its approach a "different science" compared to that used by Cephos (Huizenga 2013). TB's website asserts that its technology has been studied in 31 original peer-reviewed scientific journals that included researchers from 13 different countries having tested 723 subjects (truthbrain.com).[36] There is a competing view that the science has not evolved to the point that evidence derived from this technology is sufficiently reliable to be admitted into court (Belcher and Sinnott-Armstrong 2009; Schauer 2010a, 2010b; Meixner 2012).

Lakens work now focuses on DNA evidence, which courts today accept as scientific fact. He points out that

> it took over ten years for DNA evidence to be accepted in court. Partially that was because Judges were concerned that jurors would hear scientific data that might overwhelm their independent judgment as to the credibility of testimony. The fMRI technology, which our testing showed had a 97% reliability, faces the same resistance. It's unfortunate, because fMRI does work and is a useful, relevant, scientifically valid way to ascertain whether a person is telling the truth.[37]

The case in which TB has been most visibly engaged is a murder proceeding involving the second-time conviction of Gary Smith in the 2006 shooting death of fellow Army Ranger Michael McQueen, whom Smith says committed suicide.[38] Employing the *Frye* standard, the District Court declined to admit evidence derived from fMRI testing that concluded that Smith was innocent. The Court stated that it was "not swayed" by 25 peer-reviewed scientific journal articles that have studied fMRI for lie detection and truth verification.[39]

The Court stated:

> [The] Defendant offers that none of these twenty-five articles 'conclude that the technology does *not* work.' The Court is not persuaded that the fact that there is no evidence a scientific method does 'not work' is evidence that it *is* reliable and valid. The standard required for admissibility in a court of law is higher than the method simply working. There must be evidence of the method's reliability and validity as determined by its general acceptance in the relevance scientific community.[39]

Smith is appealing his conviction to the Maryland Supreme Court. Upon this writing, the case is pending.

The fMRI cases illustrate that admitting evidence derived from emerging neurotechnology may prove challenging, especially where the evidence addresses credibility issues as to a witness's thoughts or intentions. Clearly, what concerns courts, beyond whether the evidence is reliable, is whether it intrudes upon the jury's traditional role in assessing the credibility of witnesses. The fMRI technology is only one emerging approach for lie detection. New nonintrusive neurotechnology that will become operational in the next twelve-to-eighteen months will help ascertain, for example, whether an interviewee is telling the truth. Current polygraph technology is considered to have a reliability factor of about 72% and is not admissible in Court. The developers of this new neurotechnology believe it may have a 90% reliability.[40] Whether the evidence derived from use of such technology would be admissible in court remains to be seen.

At present, the broad issue of legal admissibility of evidence derived from neurotechnology remains incipient. However, continued progress in neuroscience and neurotechnology, recurrent calls for applications of neurotechnology to be used—or not used—to foster public safety, national security and defense, a sustained penetration of neurotechnology into the legal sphere, and the dedication of disciplines of neuroethics and neurolaw to such issues and questions all contribute to both the expansion of these controversies and the need for prompt and ongoing thinking to surmount the legal challenges posed.

CONCLUSION

Neurotechnology developments will require new thinking that may challenge international norms and standards. Key observations are as follows:

1. *Embrace, do not fear neurotechnology.*

New technology that confronts traditional beliefs can provoke strong resistance. The use(s) of fMRI technology aptly illustrate the point. The technology has aroused disagreement among experts, just as debate over the admission of a polygraph has done. A key issue is that judges tend to be wary of the use of technology to determine the credibility of witnesses in a trial. Courts worry that technology may replace individual assessments. That concern clearly influenced the judicial thinking in the few cases that have considered the admissability of fMRI technology.

The principles enunciated in *United States v. Scheffer*, 521 U.S. 346 (1997), offer the most complete and incisive analysis of both sides of the issue. It merits discussion as it underscores how sharply new technology divides opinion (in even the nation's highest court) over whether evidence derived from new technology is admissible. Airman Edward Scheffer passed a polygraph examination in which he denied using methamphetamine. Court-martialed for using illicit drugs, he sought to introduce that evidence to exculpate himself. Military Rule of Evidence 707 mandated its exclusion. The legal issue was whether the military judge's decision to exclude based upon that rule violated Scheffer's Sixth Amendment to present his case.

A divided Supreme Court affirmed his conviction. Although the case dealt with polygraphs, the reasoning almost certainly applies to the use and admissabilty of fMRI technology. Speaking for a divided majority, Justice Thomas held that a defendant's right to present relevant evidence

> is not unlimited but subject to reasonable restrictions that "do not abridge an accused's right to present a defense so long as they are not 'arbitrary' or 'disproportionate to the purposes they are designed to serve. Moreover, we have found the exclusion of evidence to be unconstitutionally arbitrary or disproportionate only where it has infringed upon a weighty interest of the accused.[41]

Upholding the constitutionality of Rule 707 and the conviction, Justice Thomas held that the rule

> serves legitimate interests in the criminal trial process. These interests include ensuring that only reliable interest is introduced at trial, preserving the jury's role in determining credibility, and avoiding litigation that is collateral to the primary purpose of the trial. The rule is neither arbitrary nor disproportionate in promoting these ends. Nor does it implicate a sufficiently weighty interest of the defendant to raise a constitutional concern under our precedents.[42]

Echoing arguments raised against the admissablity of fMRI, the Court pointed out that the "scientific community remains extremely polarized about the reliability of polygraph techniques."[43] Accordingly, applying the *Daubert* standard, the Court declined to find that the military judge had acted arbitrarily or disportionately in promulgating a *per se* rule excluding all polygraph evidence.

The debate does not end there. Justice Stevens offered an incisive dissent that eviscerated the reasoning of the majority.[44] It will be interesting to see if his views, especially in a military setting, ultimately prevail. First he disposed of *Frye*, dismissing its general acceptance test as "now discredited" and "repudiated" by the Supreme Court and pointing out that trial courts have "broad discretion" under *Daubert* to evaluate the admissibility of scientific evidence. For the military, the issue was especially sensitive, given the widespread use it makes of polygraphs. Justice Stevens acknowledged the point, noting that

> use of the lie detector plays a special role in the military establishment ... because the military carefully regulates the administration of polygraph tests to ensure reliable results.... The military has administered hundreds of thousands of such tests and routinely uses their results for a wide variety of official decisions.[45]

In contrast to the courts in *Semrau* and *Smith*, Justice Stevens expressed confidence in a jury's ability to sift through conflicting evidence and testimony and to properly evaluate expert opinion—however disputed—while allowing a defendant to use expert testimony to bolster his credibility and present a complete defense in accordance with his right under the Sixth Amendment. Brushing aside concerns that a jury might get confused by expert opinion, he declared:

> Vigorous cross-examination, presentation of contrary evidence, and careful instruction on the burden of proof are the traditional and appropriate means of attacking shaky but admissible evidence.[46]

While acknowledging some risk that

> juries will give excessive weight to the opinions of a polygrapher, clothed as they are in scientific expertise.... it is much more likely that juries will be guided by the instructions of the trial judge concerning the credibility of expert as well as lay witnesses.[47]

As to the potential unreliability of such evidence, he declared: "[T]he reliance on a fear that the average jury is not able to assess the weight of this testimony reflects a distressing lack of confidence in the intelligence of the average American."[48]

When the life, career, or freedom of an individual is at stake, should not a defendant have every reasonable right to present a complete defense under the Sixth Amendment to establish credibility? Specifically, under what circumstances should a jury be allowed to hear testimony derived from and/ or based upon the use of neurotechnology? The challenge will intensify as new questions arise as to what brain–machine interfaces and other emerging technology may strive to reveal about human thought processes.

Open minds and confidence in the judicial system, which has withstood many severe tests, are essential. Of course, one must assess the admissibility of evidence derived from such technology in light of the Fifth and Fourth amendment rights against self-incrimination and unreasonable searches and seizures. Although some have wondered whether a defendant might be forced to submit a scientifically validated lie detection test (Greely 2004),[49] or have a refusal used against him/her, there is no basis for believing that would hold true in a criminal case, although it would in a civil case.

Courtrooms are not the only venue in which neurotechnology might be employed for detecting lies. What about military interrogations of detainees accused of terrorism? fMRI employs a noninvasive approach. A detainee may refuse to voluntarily answer questions, raising familiar issues of how aggressively the military can legally and ethically conduct interrogation (see, e.g., Chapter 12). Still, as General Stanley McChrystal has pointed out, the most successful interrogations employ noncoercive techniques that yield voluntary cooperation (McChrystal 2013). The point is that neurotechnology may have broad utility in detecting lies, and the law for the admissibility of evidence it yields is still evolving. *Frye* employs an unreasonable test, and *Daubert*'s broader standard seems likely to promote the admission, not exclusion, of such evidence as neurotechnology evolves, subject to constitutional constrains against self-incrimination.

2. *What limits constrain the use of neuroscience as a predictive mechanism for criminal behavior?*

Although courts have not yet found fMRI neurotechnology sufficiently developed so as to warrant admission into evidence, Stanford University's Henry Greely has suggested that some might employ neuroscience to predict a person's future dangerousness, by showing that he/she has poor control of anger, aggressiveness, or sexual urges (Greely 2004). In *Kansas v. Crane*, 534 U.S. 407 (2002), the U.S. Supreme Court upheld by a 7-2 vote the constitutionality of a Kansas statute that authorizes commitment in a civil case of a "sexually violent predator" where there is a determination that the defendant lacks the ability to control dangerous behavior.

What will happen as neurotechnology is able to provide means to predict societally disruptive behavior that sits on the borderline between free expression and an accepted definition of criminality? At what point does the Fifth Amendment right against self-incrimination or the Fourth Amendment right against unreasonable searches or seizures prevail over the State's interest in preventing a future crime?

Cases will be fact specific, but the prospect should raise concerns. We should never forget that the Soviets invoked a very broad concept of mental illness to repress political dissent. One should not presume that prosecutors or government officials will hold beneficent, just views about what constitutes lawful behavior. Similarly, Greely (2004) observes that there is potential for abuse of nuerotechnologically-derived information by insurers and/or employers. States have imposed legal restrictions on the use of genetic data, but neurotechnology may offer an alternative way for parties to reach prejudicial conclusions about an individual. Rules to address this concern need to be forged.

3. *The Nuremberg principles that put moral virtue ahead of politics remain valid and should guide rule-making for neurotechnology.*

Neurotechnology offers remarkable opportunities to enhance physical and mental capacities. They may provide us with a longer, healthier life. They offer the military new potential to achieve effects or end-states without using kinetic means, and may be operationalized for command-and-control; intelligence; surveillance and reconnaissance; intelligence collection, processing, and analysis; and other activities that may save lives and reduce casualties. Yet as neurotechnology evolves, challenging legal issues will arise as we compare benefits to risks, and balance public against private interest. National security interests must not outweigh ethical and legal responsibility. We must respect the principles of informed and voluntary consent, transparency, and the right to withdraw or cease to use a technology.

As neurotechnology develops, what additional rules are needed to ensure reliability and safety? Neurotechnology policies should be consistent with the broader principles of CIL so as to be relevant to the broad international community. Yet, that goal may prove elusive. Legal standards of conduct that guide and govern developing and using neurotechnology in the United States and Western nations may deviate substantially from those employed in places like China or India. That would challenge both efforts to establish international norms and the stability of a global economy and perhaps power.

Given the cultural, philosophical, and values' diversity of the current international environment, developing philosophical and ethical bases upon which to ground or develop international legal frameworks remains challenging (Benedikter and Giordano, 2012; Giordano and Benedikter, 2012; Spranger, 2012). This has prompted more overt calls and efforts to both develop internationally relevant neuroethical approaches, and to address how existing and newly devised neuroethical principles and standards might be employed in international relations and the social and legal issues generated in and by national security concerns (Lanzilao, Shook, Benedikter and Giordano, 2014; Shook and Giordano, 2014).

4. *Commanders bear a special responsibility to avoid placing undue pressure on their subordinates to use neurotechnology that may (1) interfere with independent thinking, (2) harm them, or (3) create irreversible effects.*

Today's science fiction could be tomorrow's science.[50] The Quadriennel Defense Review of 2014 looks to 2034. But what weaponized neurotechnology will tomorrow bring? What legal duties and responsibilities—consistent with international norms as well as cultural values—should guide the use of these future technologies? The formal and informal pressures that commanders or others in authority place upon subordinates can be compelling. We need to develop relevant, clear, fair, practical legal rules that bind the military scope of authority to use neurotechnology as these technologies become operational.

5. *Legal rules governing neurotechnology must achieve fair play and justice in ways that avoid social disruption, invasion of privacy, or coercion.*

Neurotechnology offers the potential to change the way we see ourselves and think about society. That may prove beneficial. It may also produce outcomes that disrupt society and raise questions about discrimination and fair play by empowering a new elite who may have abilities unavailable to others. This is not far-fetched.

The military is testing exoskeleton suits that will vastly increase the capacity of soldiers to function. These include Cyberdyne's HAL-5, which interprets faint electrical signals in the skin around damaged muscles and moves motorized joints in response (Ponsford 2013). That technology is likely to have civilian applications. Zac Vawter used a thought-controlled bionic leg to climb all 103 floors of Chicago's Willis Tower. Over 220,000 people have cochlear implants. Deep brain stimulators used for Parkinson's disease are being tested to treat a number of other neurological conditions, as well as psychiatric disorders and states. Artificial limbs have already provoked controversy in Olympic competition, as some have complained that bionic limbs provide an unfair competitive edge (Naam 2013). Similar complaints may arise from those who argue than inequitable access to neurotechnology provides unfair enhancements for performance—perhaps to enhance memory recall or to accelerate learning—in the workplace (Lynch 2004).

How do we apportion the benefits of neurotechnology equitably so that broader goals of fair play and justice are realized? What rules must be put in place to ensure that individuals remain free from its use in coercion? These raise legal and policy questions, to achieve justice and avoid creating classes of victims suffering from new forms of invidious discrimination.

6. *At what point does weaponized neurotechnology become so lethal as to constitute a prohibited weapon under international norms?*

Open-source data do not reveal the current existence of such neurotechnology. But conceptually, it might be developed. Now is the time to think through what standards might define illicit levels of development, testing and use, and what rights states have to protect their populations against the use of that technology.

At present no treaty specifically applies to neurotechnology. Efforts to forge controls of cybermalware have proven—and are likely, outside of narrow criminal activities such as banning spam or child pornography—to be fruitless. We should not deceive ourselves into believing that forging an international accord on weaponized neurotechnology would be any easier. The issues are complicated and nuanced. On the one hand, the use of such technology should conform to the Law of Armed Conflict (LOAC).

Equally, how would LOAC considerations address the challenge posed by the notion of "unrestricted warfare," (Lang and Xiangsui 2007) in which there are no limits, no rules, no boundaries, battlefields, and sovereign places? Victory in engagement and conflict occurs through the "principle of addition" of five dominant means. Liang and Xiangsu posit that nations should use "all means, including armed force or nonarmed force, military and nonmilitary, and lethal and nonlethal means to compel an enemy to accept one's interests" (Lang and Xiangsui 2007). Their approach envisions a world in which "a pasty-faced scholar wearing thick eyeglasses is better suited to be a modern solider than a strong young-lowbrow with bulging biceps" (Lang and Xiangsui 2007). In short, they envision the use of scientists, technicians, mathematicians, and other nonmilitary parties to achieve military goals—including, presumably, developers and users of neurotechnology.

That vision arguably ignores or rejects international norms for the use of weaponized neurotechnology. It raises profound questions about (1) what precautions other nations might or might not take in developing weaponized neurotechnology (at least for passive or active defense), (2) military doctrines, and (3) how LOAC might be interpreted in addressing nations who embrace the notion of unrestricted warfare. Indeed, avoiding the catastrophic consequences of war, especially to noncombatants, is a principal reason that most nations respect the LOAC.

A new set of rules governing weaponized neurotechnology for offensive or defensive purposes must be established that can achieve some international consensus in order to minimize, if possible, the risks of engagement or conflict. That will not be easy to accomplish.

7. *The corollary to the above question is at what point does use of weaponized neurotechnology constitute a "use of force" under Article 2 of the United Nations Charter?*

The use of the Stuxnet malware put into question the existing discourse over U.S. doctrines of active offense versus defense of weaponized malware (Farwell and Rohozinski 2011, 2012). Although dealing with weaponized code, the strategic and legal considerations under international law parallel those for neurotechnology. Stuxnet has shown that the United States (and Israel) will use cyber weapons offensively.[51]

The United States' cyber strategies respect international law and there is no reason to believe that such a stance would not apply to neurotechnology. The key normative standards are set forth in the United Nations Charter Articles 2(4) and 51. Article 2(4) prohibits the "threat or use of force against the territorial integrity of independence of any state". Article 51 states that nothing in the present Charter shall impair the inherent right of individual or collective self-defense if an armed attack occurs against a member of the United Nations (Charter of the United Nations 1945).

But force is not defined. No international convention defines whether the employment of weaponized neurotechnology constitutes a use of force. Probably, the term covers attacks that injure persons or irreparably damage property. Apparently, the U.S. government viewed employment of Stuxnet as a use of force. The intent was to irreparably damage critical infrastructure used to develop nuclear technology and, presumably, weapons. The tenor of the operation and strategic intent—and statements from CIA Director Mike Hayden, strongly imply that White House and DoD lawyers considered the operation a use of force.

The implications for neurotechnology are significant and offer lessons for evaluating what legal rules might apply. The precedent that Stuxnet set is that a nation could use weaponized neurotechnology when and where it is deemed a sufficient national security interest is at stake. Stuxnet and a comparable employment of weaponized neurotechnology raise important questions of authority. In deploying Stuxnet, did the White House exceed its jurisdiction under the Constitution, which reserves to Congress the right to declare war? Did it exceed its authority under the War Powers Resolution of 1973?[52]

One would have a hard time characterizing the operation—*Olympic Games*—as an act of war. The U.S. statute defines it as armed conflict, (between two or more nations or between military forces of any origin.)[53] Weaponized neurotechnology that aims to paralyze command and control, destroy an enemy's warfighting capacity, or otherwise substantially impair an enemy's warfighting capacity may or may not fall under the ambit of any current U.S. statute. Little by way of example provides guidances.

One analogy might be the air war in Libya, which at least offers insight into the mindsets of policy-makers. The Obama administration declined to ask Congress for authorization to act, arguing that "US operations do not involve sustained fighting or active exchanges of fire with hostile forces, nor do they involve ground troops" (Savange and Landler 2011). Subsequently, there were suggestions that the White House was responding to an unfolding crisis and there was no time to ask for authorization. By contrast, when the administration pondered whether to launch strikes at the regime of Bashar al-Assad in Syria over the use of chemical weapons, it adopted a more conservative, if ambiguous, approach and sought Congressional approval while maintaining that the White House had the authority to strike even if Congress disapproved ("Statement on Syria" 2013). Whether launching a strike in the face of Congressional disapproval would have led to the President's impeachment and potential conviction by the Senate remains an open question.

Whether deployment of weaponized malware was a use of force raises collateral issues. One must consider the larger implications of an individual event. Does a pattern of use convert the employment of weaponized neurotechnology into a use of force? The answer is not clear. The unpredictable nature of damage that weaponized neurotechnology may inflict may require new definitions of "use of force" and of "wafare".

Intent may come to bear in determining whether an engagement using neurotechnology constitutes use of force. Open-source reporting for Stuxnet indicates that damage inflicted was temporary. But that was not the intent. Deciphering intent may pose a challenge, but it may be inferred objectively. Finally, at what point might Article 51 come into play? Like "force," "armed attack" is undefined, even where force is clearly employed. The implications for Article 51 or other international conventions of using neurotechnologies remain unclear.

8. *What legal rules should govern the use of weaponized neurotechnology?*

The key principles of war include distinctions of combatants from non-combatants; military necessity; proportionality in tactics; avoiding superfluous injury; banning indiscriminate weapons, notably biological and chemical weapons; perfidy or providing visual and electronic symbols like the Red Cross to identify persons or property protected from attack; and neutrality. These would almost certainly apply to the use of weaponized neurotechnology. But other issues also arise.

One emerging issue is whether there should be "an unambiguous standard of conduct" for the use of neurotechnology in warfare that will be universally recognized *jus in bello* (Brown 2006). The challenge lies in forging an international consensus. That may prove highly elusive.

Other significant legal issues for weaponized neurotechnology include:

The neutrality doctrine. Important to bear in mind is that emerging weaponized neurotechnology may create effects that cross international boundaries. As one evaluates the impact of these effects, the neutrality doctrine comes into play. The Hague Convention offers rules that govern neutrality and dictates that the territory of a neutral state is inviolable (Hague Convention 1907; Kelsey 2008). Article 8 of Hague Convention V provides an exception for telecommunications, permitting a neutral country to allow belligerents to use communications equipment (telegraph or cable wires or wireless telegraphy apparatus), on an impartial basis. The United States has taken the position that "the plain language of this agreement would appear to apply to communication satellites as well as to ground-based facilities" (Department of Defense 1999).

The use of neurotechnology that create effects in one State that affect a neutral State would seem to violate the neutrality doctrine. This has broad implications in the current global environment, where weaponized neurotechnology potentially may be used in ways—for example, cognitive hacking—that may be routed through a number of nonbelligerent states. There is no formula, no hard cast prediction that can be made. What is important in thinking about the legal implications is that if such technology creates effects that cross borders of nonbelligerents, legal issues arise under the neutrality doctrine.

The rule of distinction. The Geneva Convention and its Additional Protocols were written prior to the advent of neurotechnology. One can argue that they apply to weaponized neurotechnology by analogy (Department of Defense 1999). This raises the question of what "distinction" between military and noncombatants means under the Geneva Convention. The 1977 Additional Protocol requires parties to an armed conflict to distinguish between the two.[54] In concept, states "must never use weapons that are incapable of distinguishing between civilian and military targets" (International Court of Justice 1996). Dual-use targets further complicate the issue. The DoD would presumably deal with these challenges on a case-by-case basis.

What rules define the opponent? Confirming an attacker's identity or intention does not solve the attribution problem if an unauthorized person gains access to neurotechnology and employs it. A key issue is whether an attack that cannot be shown to be state-sponsored justifies an act of self-defense that has an effect in another nation's territory. And what if incursions are perpetrated by allied nations or parties within them? (Bamford 2008).

Who will formulate the weaponized neurotechnology regulations that govern DoD strategy and tactics? Weaponized neurotechnology is still at an embryonic stage, and while current doctrine and treaties may regulate neurobiochemical agents, other forms of weaponizable neurotechnology do not fall within the scope of those agreements. What international bodies will develop and oversee these governances? Domestically would such governance fall to the DoD, or the Department of Justice, and how might Congress regulate a permissible scope of action?

Should the use of weaponized neurotechnology apply only to persons in uniform? As laws and conventions of war require uniforms, a similar rule arguably applies to the use of weaponized neurotechnology, to distinguish civilian from military individuals or assets. It is doubtful that such a rule has been applied in practice and there is no evidence the DoD has considered it.

Is new legal authority needed to meet the challenge of using or defending against weaponized neurotechnology? In sum, there are no specific U.S. laws that govern the use of neurotechnology in national security and defense. Similarly, international law offers no concrete answers. What international bodies will develop and oversee these governances? In conclusion, the rise of neurotechnology poses new challenges and opportunities in law, that are heightened by complexities of national security and defense on a global scale.

NOTES

1. It also includes pharmacology, but that topic falls outside the scope of this chapter.
2. During Operation Desert Storm, the Federal Drug Administration adopted Rule 23d, 21 C.F.R. Section 50.25(d) to permit the military to administer certain medications, including pyridostigmine bromide—an antidote to nerve gas—to protect against Saddam Hussein's chemical or biological weapons. The issue is discussed below. For a broader discussion, see Annas G. (1992). While that case pertains to pharmacology, the consent of wavier requirements sought by the DoD is analogous to issues that neurotechnology could present. DoD's position was that the military could not tolerate refusals to provide certain drugs because of "military combat exigencies." As noted throughout this volume, there is a real possibility—if not likelihood—that neurotechnology will be even more studied, developed, and considered for use in national defense, and that such use might involve both national and nonstate actors. Appreciating this reality must be accompanied by equal recognition of the legal issues that may be incurred, and the need to address these issues of law with forethought, prudence, and expedience.

3. Mostly these focus on cyber issues that fall outside the scope of neurotechnology developments.

4. A finely grained evaluation of the distinctions and nuances of international standards of neurolaw lie beyond the scope of this chapter. But it offers one of the most complete and careful discussions on comparative aspects of international law relevant to the applications of neuroscience and neurotechnology.

5. The U.S. Government's record in using care in conducting experiments that affect individuals is questionable. See Moreno (1999) for an excellent history and discussion on this topic. The legal duties include the relevant scope of foreseeability for harm.

6. See Cybenko et al. (2003) and Guidorizzi (2013). The latter offers illustrations of cognitive hacking that could easily be applied in a military operation. Such hacking raises legal issues under the Lanham Act, 15 U.S.C. 1125(a), which outlaws false advertising on the Web. The Lanham Act, copyright and trademark law, are being used to decide cases related to cognitive hacking. For example, a company that uses another's trademark to divert web searches to itself arguably violates the Lanham act. A competing view cautions against overly aggressive application of the Lanham Act as violating the constitutional right to free expression. (See also Thompson, P. "Cognitive hacking and intelligence and security informatics." Thayer School of Engineering and Department of Computer Science, Dartmouth College. http://www .ists.dartmouth.edu/library/75.pdf.

7. For example, should enhancements be reversed routinely upon discharge, the impact of an individual's ability to return to civilian life, and whether the individual may have the right to refuse to reverse an enhancement become an issue.

8. Seven received death sentences, five life imprisonments, two 25-year prison term, and two others to prison for 10 and 15 years, respectively. The military governor confirmed the sentences. The U.S. Supreme Court declined to hear the appeals.

9. Transcript of the Nuremberg Medical Trial. See *United States v. Karl Brandt et al.*, (Case 1). Washington, DC: National Archives, November 21, 1946—August 20, 1947 (Microfilm publication no. M887).

10. States like California have recognized that the Code is not enforceable but in effect codified it. (see CA Health & Safety Code, Section 24172).

11. Pentagon Policy TS-01188. Referred to as the "Wilson Memorandum," it created a Nuremberg Code-based policy to govern the DoD's human experiments in atomic, biological, and chemical warfare for defensive purposes. (See "Radiation" http:// www2.gwu.edu/~nsarchiv/radiation/dir/mstreet/commeet/meet8/brief8/tab_k/ br8k1.txt.)

12. For discussion, see "History of ethics," Claremont Graduate University, http://www.cgu .edu/pages/1722.asp.

13. The rule is codified in separate regulations by 15 Federal departments and agencies. See especially HHS regulations, 45 CFR part 46. The Common Rule was influenced by the Belmont Report, issued in 1979 by the National Commission for the Protection of Human Subjects of Biomedical and Behavioral Research. This report stated basic ethical principles and guidelines to help resolve ethical problems that surround research on humans. They include respect for persons, application of informed consent, not harming people while maximizing benefits, distributing the benefits and risks of research fairly, systematically assessing risks and benefits, and adopting fair procedures and outcomes in selecting research participants. Claremont's excellent summary, supra notes that the Belmont Report established three basic ethical principles: autonomy/ respect for persons, beneficence, and justice as the cornerstone for regulations involving human participants.

14. The text of the Geneva Conventions and the Additional Protocols. http://www.icrc.org/ eng/war-and-law/treaties-customary-law/geneva-conventions/index.jsp.

15. Moreno (2006) offers a good history of the basic facts surrounding the administration of vaccines during Operation Desert Shield.
16. Issued on September 30, 1978, and published in the Federal Register on April 18, 1979 (45 CFR 46.102e), the Report summarizes the ethical principles and guidelines for research involving human subjects that the federal government imposes through the Department of Health and Human Services.
17. *John Doe and Mary Doe v. Louis W. Sullivan and Richard Cheney*, Civil Action No. 91–51, SSH, U.S. District Court for the District of Columbia.
18. DoD's letter is published as part of the supplementary information.
19. U.S. DoD, The Defense Acquisition System, E1.1.15 (2003).
20. See also Customary International Humanitarian Law, Vol. 1, Rules: International Committee of the Red Cross (ICRC). Cambridge University Press, 2005.
21. *Prosecutor v. Kordic*, Case No. IT-95-14/2-T, Judgment, Paragraph 233, February 26, 2001.
22. See also The Rome Statute, which established the International Criminal Court, Articles 22(1) and 24(1),(2). Article 22(1) declares: "A person shall not be criminally responsible under this Statute unless the conduct in question constitutes, or at the time it takes places, a crime within the jurisdiction of the Court." Article 22(2) limits the authority of the Court to extend the definition of a crime: "The definition of a crime shall be strictly construed and shall not be extended by analogy. In case of ambiguity, the definition shall be interpreted in favor of the person being investigated, prosecuted, or convicted." See also White (2008) supra, p. 194.
23. In this context, "killed" is interchangeable with "caused death."
24. This requires a mental element of awareness about the status of the victims.
25. "International armed conflict" includes military occupation.
26. *Actus reus* is not a constituent element of the Rome Statute but criminal law generally requires its proof.
27. The International Criminal Tribunal for Yugoslavia set forth a three-pronged test for determining the legal responsibility of commanders: (1) was the defendant a superior? (2) did the commander know or possess information that would give rise to a suspicion that subordinates were breaching the laws of war? (3) did the commander take measures to prevent subordinates from committing crimes? (See *"Prosecutor v. Delali"* 1998, Case No. IT-96-21-T, Judgment, pp. 64–73, and notably p. 65, (November 16, 1998).
28. United States War Crimes Act of 1996, 18 U.S.C. 2401 (1996).
29. Currently 122 states are parties to the Statute of the Court. A further 31 countries, including Russia, have ratified but not signed the Rome Statute, and 41 United Nations Member States have neither ratified nor signed. The United States signed but has informed the U.N. Secretary General that it no longer will be a party and will not be legally obligated by its rules or jurisdiction.
30. The Royal Society notes that industrialists who supplied Zyklon B to German concentration camps were found guilty of war crimes. See Brain Waves Module 3.
31. Justice Blackmun went out of his way to point out that the Court sought a "balance struck by Rules of Evidence designed not for the exhaustive search for cosmic understanding but for the particularized resolution of legal disputes." *Daubert* at 113 S.Ct. 2799.
32. Other courts have reviewed attempts to use fMRI tests to demonstrate impaired brain functionality. *Turner v. Epps*, 460 F. App'x 322, 323–24 (5th Cir. 2012); *Hooks v. Thomas*, No. 2:10CV268-WKW, 2011 WL 4542901, at *2–3 (M.D. Ala. July 1, 2011) (Report and Recommendation), adopted 2011 WL 4542675 (M.D. Ala. September 30, 2011); see also *State v. Andrews*, 329 S.W.3d 369, 383–84 and no.12 (Mo. 2010) (reviewing fMRI research on juvenile brain development); *Entm't Software Ass'n v. Granholm*, 404 F. Supp. 2d 978, 982 (E.D. Mich. 2005) (reviewing fMRI research on media violence exposure and brain activation).

33. Laken reports that he closed Cephos in 2012 and in 2013 reopened it as Cephos LLC, focusing primarily on DNA forensics. Phone interview with Steven Laken, September 23, 2013, and e-mail on September 20, 2013.

34. The Magistrate excluded the evidence under Rule 403 for three reasons: (1) the fMRI tests were undertaken independently, without the Government's knowledge or testing supervision; (2) use of lie detection tests are deemed highly prejudicial, especially as to issues central to the verdict; and (3) the jury would not be assisted by hearing that Dr. Semrau's answers were truthful "overall" without learning which specific questions he answered truthfully or deceptively. The Sixth Circuit concurred. *Semrau*, published decision, pp. 18–19. In the *Gary Smith case*, discussed below, the defense team issued a point-by-point comparison to show that all objections that the *Semrau* Magistrate had cited were overcome.

35. Among the problems was that "testing indicates that a positive test result in a person reporting to tell the truth is only accurate 6% of the time and may be affected by fatigue." *Semrau*, published decision, p. 11. That did not necessarily mean that those found to lie were telling the truth. The Court did not find that the evidence satisfied the requirements for reliability articulated in *Daubert* and FRE 702 as the error rates for lie detection could not be reliably established. The Court found no way to detect whether Dr. Semrau was lying or not deceptive.

36. A review of the commentaries shows that they do lend support, in different degrees, to Huizenga's arguments. Huizenga highlights the following as strongly supporting the scientific validity of his approach: McPherson, B., K. McMahon, W. Wilson, and D. Copland. 2012. "'I know you can hear me': Neural correlates of feigned hearing loss." *Human Brain Mapping* 33:1964–1972; Weixiong, J., L. Jian, L. Huasheng, T. Yan, W. Wei (org. capitalization), 2012. "Function MRI analysis of deception among people with anti-social personality disorders." *Journal of Central South University* (Med. Sci.) 37(11):1141–1146; Jiang, W., H. Liu, J. Liao, X. Ma, P. Rong, Y. Tang, and W. Wang. 2013. "A function MRI study of deception among offenders with antisocial personality disorders." *Neuroscience* 244:90–98; Liang, C.-Y., Z.-Y. Xu, W. Mei, L.-L. Wang, I. Xue, J. de Lu, and H. Zhao. 2012. "Neural correlates of feigned memory impairment are distinguishable from answering randomly and answering incorrectly: an fMRI and behavioral study," *Brain and Cognition* 79:70–77; Marchewka, A., K. Jednorog, M. Falkiewicz, W. Szeszkowski, A. Grabowska, and I. Szatkowska, "Sex, lies and MRI—Gender differences in neural basis of deception." *PLoS One* 7(8):e43076; Sip, K.E., D. Carmel, J.L. Marchant, J. Li, P. Petrovic, A. Roepstorff, W.B. McGregor, and C.D. Frith. 2013. "When pinnochios nose does not grow: Belief regarding lie-detectability modulates production of deception." *Frontiers in Neuroscience* 7:16; and Ito, A., N. Abe, T. Fujii, A. Ueno, Y. Koseki, R. Hashimoto, S. Mugikura, S. Takashi, and E. Mori. 2011. "The role of the dorsolateral prefrontal cortex in deception when remembering neural and emotional events." *Neuroscience Research* 69:121–128.

37. Telephone interview with Steven Lakens, September 23, 2013.

38. See *Gary James Smith v. State of Maryland*, Circuit Court for Montgomery County Criminal Case No. 106589C. http://statecasefiles.justia.com/documents/maryland/court-of-appeals/10-11.pdf?ts=1323898022. The Court of Appeals judgment that reversed the original conviction and remanded for retrial did not address the fMRI issue. See also: "Gary Smith sentenced to 28 years in Michael McQueen's murder," *WJLA.com*, October 15, 2012.

39. Memorandum Opinion and Order, *State of Maryland v. Gary Smith*, Case No. 106589C, In the Circuit Court for Montgomery County, Maryland (the Smith Opinion).

40. Interview with Robert Cutlie and John Deaver of the National Center for Credibility Assessment, Washington, DC, August 21, 2013. They are leads in developing this technology for the Defense Intelligence Agency.

41. The Opinion. http://www.law.cornell.edu/supct/pdf/96-1133P.ZO. Page citations in this chapter are from the Opinion and Dissent published on that cite by Cornell University.

42. The Opinion. http://www.law.cornell.edu/supct/pdf/96-1133P.ZO. Page citations in this chapter are from the Opinion and Dissent published on that cite by Cornell University, p. 5.

43. The Opinion. http://www.law.cornell.edu/supct/pdf/96-1133P.ZO. Page citations in this chapter are from the Opinion and Dissent published on that cite by Cornell University, pp. 5, 6. The Court discussed that debate at length.

44. Justice Stevens' dissent. http://www.law.cornell.edu/supct/pdf/96-1133P.ZD. This chapter cites from the Dissent pubished by Cornell and the page numbers referenced come from there.

45. Dissent, pp. 4–5. The fact is, access to this nation's highest secrets generally requires passing a polygraph.

46. Dissent, pp. 4–5; 16–18 citing *Daubert*, 509 U.S. 596.

47. The Opinion. http://www.law.cornell.edu/supct/pdf/96-1133P.ZO.

48. The Opinion. http://www.law.cornell.edu/supct/pdf/96-1133P.ZO.

49. Greely's paper was, by its own terms, a speculation, not a prediction.

50. In the mid-1970s, as this author commenced the practice of law, his secretary used a manual typewriter and copies of legal briefs were as often made on carbon paper. That was not that long ago. The concept of email never crossed the minds of most people. We have come a long way. We have farther to go.

51. The use of Stuxnet is classified, but its use as operation OLYMPIC GAMES and the considerations surrounding its use are reported by journalist David Sanger. Analysis hereof legal issues arising out of, or the facts as to OLYMPIC GAMES, presumes *arguendo* the accuracy of Sanger's reporting and is based solely upon that and not on classified sources. (See Sanger "Obama Order" 2012; see also Sanger "Confront and Conceal" 2012.)

52. 50 U.S.C. 1541–1548.

53. 18 U.S.C. 2331.

54. See generally "Convention Respecting the Rights and Duties of Neutral Powers and Persons In Case of War on Land," October 18, 1907, 36 Stat. 2310, T.S. 540 [1907 Hague Convention IV]; Convention Concerning the Rights and Duties of Neutral Powers in Naval War, October 18, 1907, 36 Stat. 2415, T.S. 545 (1907 Hague Convention XIII).

REFERENCES

Annas, G.J. 1992. "Legal Issues in Medicine: Changing the Consent for Operation Desert Storm," *New England Journal of Medicine* 326.

Bamford, J. 2008. The Shadow Factory: The Ultra-Secret NSA from 9/11 to the Eavesdropping of America. New York: Doubleday.

Barno, D. and N. Bensahel. 2013. "Decisions deferred: Balancing risks for today and tomorrow." Center for New America Security, August 5. www.cnas.org/sites/default/files/publications-pdf/scme_BarnoBensahel_commentary_0_1.pdf.

Belcher, A. and W. Sinnott-Armstrong. 2009. "Neurolaw." *Wiley Interdisciplinary Reviews: Cognitive Science* 1:1.

The Belmont Report. 1979. The National Commission for the Protection of Human Subjects of Biomedical and Behavioral Research: Ethical Principles and Guidelines for the Protection for Human Subjects of Research. Washington, DC: DHEW Publications No. (OS) 78-0012:3. http://www.hhs.gov/ohrp/humansubjects/guidance/belmont.html.

Benanti, P. 2012. "The cyborg: A core concept for answers and questions of neurotechnology." *Current Pharmaceutical Biotechnology* 12.

Benedikter, R. and J. Giordano. 2012. "Neurotechnology: New frontiers for European policy." *Pan European Networks Science and Technology* 3:203–207.

Bowden, M. 2011. *Worm: The First Digital World War.* Washington, DC: Atlantic Monthly Press.

Brain Waves Module 3: Neuroscience, Conflict and Security. 1998. The Royal Society, supra, p. 20.

Brown, D. 2006. "A proposal for an international convention to regulate the use of information systems in armed conflict." *Harvard International Law Journal* 47:179–181.

Brown, T. and E.R. Murphy. 2010. "Through a scanner darkly: Functional neuroimaging as evidence of a criminal defendant's past mental state." *Stanford Law Review* 62(119):1139.

Charter of the United Nations. 1945. http://www.un.org/en/documents/charter/index.shtml.

Chen, I. 2009. "The court will now call its expert witness: The brain." *Stanford Lawyer* 81. http://news.stanford.edu/news/2009/november19/greely-neurolaw-issues-111909.html.

Clancy, F. 2006. "At military's behest, DARPA uses neuroscience to harness brain power." *Neurology Today* 6(2):4, 8–10.

Clarke, A. n.d. "Clarke's three laws." http://www.princeton.edu/~achaney/tmve/wiki100k/docs/Clarke_s_three_laws.html.

Clynes, M.E., and N.S. Kline. 1995. "Cyborgs in Space." In *The Cyborg Handbook*, eds. C.H. Gray, J.H. Figueroa-Sarriera, and S. Mentor, pp. 3–42. New York: Routledge.

Code of Federal Regulations—Title 45, 46. 2009. HHS.gov.

Customary International Humanitarian Law, Vol. 1, Rules: International Committee of the Red Cross, supra, Rule 70.

Cybenko, G., A. Giani, and P. Thompson. 2003. "Cognitive hacking." *Advances in Computers.*

Dando, M. 2012. *Bioterror & Biowarfare: A Beginner's Guide.* Oxford: Oneworld Publications, Kindle Edition.

Department of Defense Office of General Counsel. 1999. An Assessment of International Legal Issues in Information Operations. http://www.au.af.mil/au/awc/awcgate/dod-io-legal/dod-io-legal.pdf.

Farwell, J. and R. Rohozinski. 2011. Stuxnet and the future of cyberwar. *Survival: Global Politics and Strategy* 52(6):127–150.

Farwell, J. and R. Rohozinski. 2012. "The new reality of cyber war." *Survival: Global Politics and Strategy* 54(4):17–120.

Forsythe, C. and J. Giordano. 2011a. "Neurotechnology in national security, intelligence, and defense." *Synesis: A Journal of Science, Technology, Ethics, and Policy* 2:T1–T2.

Forsythe, C. and J. Giordano. 2011b. "On the need for neurotechnology in the national intelligence and defense agenda: Scope and trajectory." *Synesis: A Journal of Science, Technology, Ethics, and Policy* 2(1):T5–T8.

Food and Drug Administration. 1990. "Informed consent for human drugs and biologics; determination that informed consent is not feasible; interim rule and opportunity for public comment." *Federal Register* 55(246):52814–52817.

Germany (Territory under Allied occupation, 1945–1955: US Zone). 1949. *Trials of War Criminals before the Nuremberg Military Tribunals under Control Council Law No. 10.* Vol. 2. Washington, DC: U.S. Government Printing Office, pp. 181–182. http://www.loc.gov/rr/frd/Military_Law/pdf/NT_war-criminals_Vol-II.pdf.

Giordano, J. 2010. "The mechanistic paradox." *Synesis: A Journal of Science, Technology, Ethics, and Policy* 1:T1–T4.

Giordano, J. and Benedikter, R. 2012. "An early—and necessary—flight of the Owl of Minerva: Neuroscience, neurotechnology, human socio-cultural boundaries, and the importance of neuroethics." *Journal of Evolution and Technology* 22(1):14–25.

Giordano, J. and R. Wurzman. 2011. "Neurotechnologies as weapons in national intelligence and defense—An overview." *Synesis: A Journal of Science, Technology, Ethics, and Policy* 2:T55–T71.

Greely, H. 2004. "Neuroethics: The neuroscience revolution, ethics, and the law." Marjula Center for Applied Ethics. Santa Clara University. www.scu.edu/ethics/publications/submitted/greely/neuroscience_ethics_law.html.

Guidorizzi, R. 2013. "DARPA's active authentication: Moving beyond passwords." IDGA video, December 21, 2012. http://www.idga.org/communications-engineering-and-it/videos/moving-beyond-passwords-darpa-s-active-authentica/.

Hables-Gray, C. 2001. *Cyborg Citizen: Politics in the Posthuman Age.* New York: Routledge.

Hague Convention. 1907. "Convention (IV) respecting the laws and customs of war on land and its annex: Regulations concerning the laws and customs of war on land." The Hague, October 18. http://www.opbw.org/int_inst/sec_docs/1907HC-TEXT.pdf.

Henckaerts, J. and Doswald-Beck, L. 2005. "Customary International Humanitarian Law, 1, Rules: Rule 70; Rule 152" International Committee of the Red Cross, Cambridge. www.icrc.org/eng/assets/files/other/customary-international-humanitarianlaw-i-icrc-eng.pdf.

HHS.gov. 1991. "Federal policy for the protection of human subjects ('Common Rule')." http://www.HHS.gov.

Huizenga, J. 2013. Telephone interview with Joel Huizenga, September 20, 2013, and email from Huizenga.

International Committee of the Red Cross (ICRC) 1949. "Geneva Convention Relative to the Protection of Civilian Persons in Time of War, Fourth Geneva Convention Article 67." Geneva, Switzerland, August 12. www.icrc.org/ihl.nsf/.

ICRC. 1960. "Essential rules. I. General principles (Third Geneva Convention), Article 99." www.icrc.org/ihl.nsf/.

ICRC. 1977. "Protocol addition to the Geneva Conventions of 12 August 1949, and relating to the protection of victims of international armed conflicts (Protocol I), 8 June 1977." Article 1, 10.2. www.icrc.org/ihl.nsf/.

International Court of Justice. 1996. "Legality of the threat or use of nuclear weapons, advisory opinion." July 8, pp. 226, 257. www.icj-cij.org/docket/files/95/7495.pdf.

Ito, A., N. Abe, T. Fujii, A. Ueno, Y. Koseki, R. Hashimoto, S. Mugikura, S. Takashi, and E. Mori. 2011. "The role of the dorsolateral prefrontal cortex in deception when remembering neural and emotional events." *Neuroscience Research* 69:121–128.

Jarosiewica, B. et al. 2008. "Functional network reorganization during learning in a Brain–Computer interface paradigm." *Proceedings of the National Academy of Sciences of the United States of America* 105:49.

Kelsey, J.T.G. 2008. "Hacking into International Humanitarian Law: The principles of distinction and neutrality in the age of cyber warfare." *Michigan Law Review*, May, Part II. http://www.michiganlawreview.org/articles/hacking-into-international-humanitarian-law-the-principles-of-distinction-and-neutrality-in-the-age-of-cyber-warfare.

Kleinbard, D. and R. Richtmyer. 2000. "U.S. catches 'Love Virus'." *CNN Money*, May 5. http://money.cnn.com/2000/05/05/technology/loveyou/.

Lang, Q. and W. Xiangsui. 2007. *Unrestricted Warfare.* Dehradun, India: Natraj Publishers.

Lanzilao, E., J. Shook, R. Benedikter, J. Giordano. 2013. "Advancing neuroscience on the 21st century world stage: The need for—and proposed structure of—an internationally-relevant neuroethics." *Ethics, Biology, Engineering and Medicine* 4(3):211–229.

Lebedev, M. and M. Nicolelis. 2006. "Brain-machine interfaces: Past and future." *Trends in Neuroscience* 29(9):536–546.

Levine, R. 1991. "Commentary on E. Howe's and E. Martin's 'treating the troops'." *The Hastings Center Report* 21(2):27–29.

Lin, P. 2010. "Robots, Ethics, and War." Center for Internet and Society, Stanford Law School.

Lin, P., G. Bekey, and K. Abney. 2009. "Robots in war: Issues of risk and ethics." *Ethics and Robotics* 49–67. http://works.bepress.com/palin/3.

Lynch, Z. 2004. "Neurotechnology and Society (2010–2060)." Lifeboat Foundation.

McChrystal, S. 2013. *My Share of the Task.* New York: Penguin.

Meixner, J. 2012. "Liar, Liar, Jury's The Trier? The Future of Neuroscience-Based Credibility Assessment and the Court." *Northwestern University Law Review* 106:3.

Menn, J. 2011. "Napster creates reunite for video start-up." *Financial Times*, October 7, 2011.

Miller, R.J. 2010. *Wilson v. Corestaff Services. New York Law Journal* 3. http://blogs.law.stanford .edu/lawandbiosciences/wp-content/uploads/sites/8/2010/06/CorestaffOpin.pdf.

Moreno, J. 1999. *Undue Risk: Secret State Experiments on Humans.* New York: W.H. Freeman & Co., pp. 267–271.

Moreno, J. 2006. *Mind Wars: Brain Research and National Defense.* Washington, DC. Dana Press.

Naam, R. 2013. "Are Bionic Superhumans on the Horizon?" *CNN.com* [Internet].

National Institute of Health. "Nuremberg code." http://history.nih.gov/research/downloads/ nuremberg.pdf.

Niven, L. 2002. "Niven's laws 2002 edition." http://www.larryniven.net/stories/nivens _laws_2002.shtml.

The Nuremberg Code. 1949. "Trials of War Criminals before the Nuremberg Military Tribunals under Control Council Law No. 10", Vol. 2, pp. 181–182. Germany. Washington, DC: U.S. Government Printing Office. http://www.loc.gov/rr/frd/Military_Law/pdf/NT _war-criminals_Vol-II.pdf.

Oie, K. and K. McDowell. 2011. "Neuroconitive engineering for systems development." *Synesis: A Journal of Science, Technology, Ethics, and Policy* 2:T27.

Opinion in Wilson, p. 3 (published opinion) http://blogs.law.stanford.edu/lawandbiosciences/ wp-content/uploads/sites/8/2010/06/CorestaffOpin.pdf.

Ponsford, M. 2013. "Robot exoskeleton suits that could make us superhuman." *CNN.com*, May 22. http://www.cnn.com/2013/05/22/tech/innovation/exoskeleton-robot-suit.

Poulson, K. 2011. *Kingpin.* New York: Crown Publishers.

Preden, J.S. 2010. *Enhancing Situation-Awareness, Cognition, and Reasoning of Ad-Hoc Network Agents.* Tallinn, Estonia: TUT Press.

Preparatory Commission of the International Criminal Court. 2000. Report of the Preparatory Commission of the International Criminal Court. "Article 8 (2) (a)." New York. http:// www.refworld.org/docid/46a5fd2e2.html.

Salisbury, D. 2011. "New 'bionic' leg gives amputees a natural gait," Research News. Vanderbitt, August 17. http://news.vanderbitt.edu/2011/08/bionic-leg.

Santhanam, G. et al. 2006. "A high-performance brain-computer interface." *Nature* 442:195–198.

Savange, C. and M. Landler. 2011. "White House defends continuing US role in Libya opera-tion." *New York Times*, June 15. www.nytimes.com/2011/06/16/us/politics/16powers .html?pagewanted=all&_r=0.

Schauer, F. 2010a. "Can Bad Science Be Good Evidence? Neuroscience, Lie Detection and Beyond," *Cornell Law Review* 95: 6. http://www.lawschool.cornell.edu/research/cornell-law-review/upload/Schauer-final.pdf.

Schauer, F. 2010b. "Neuroscience, lie-detection, and the law: Contrary to the prevailing view, the suitability of brain-based lie-detection for courtroom or forensic use should be determined according to legal and not scientific standards." *Trends in Cognitive Sciences* 14:3. http://www.sciencedirect.com/science?_ob=ArticleURL&_udi=B6VH9-4Y40THJ-2 &_user=10&_coverDate=03%2F31%2F2010&_rdoc=1&_fmt=high&_orig=browse&_ sort=d&view=c&_acct=C000050221&_version=1&_urlVersion=0&_userid=10&md5 =ebfa54ff4aef7dd255697aa08b35e609.

Scott, J.B. 1899. "Laws of war: Declaration on the launching of projectiles and explosives from balloons (Fourth Hague Convention)." July 29. www.nytimes.com/2011/06/16/us/ politics/16powers.html?pagewanted=all&_r=0.

Shachtman, N. 2010. "Google, CIA invest in 'future' of web monitoring." *Wired*. www.wired .com/2010/07/exclusive-google-cia/.

Shook, J.R. and J. Giordano. 2014. "A principled, cosmopolitan neuroethics: Considerations for international relevance." *Philosophy Ethics and Humanities in Medicine* 9 (1).

Shuster, E. 1997. "Fifty years later: The significance of the Nuremberg code." *The New England Journal of Medicine* 337:1426–1440. www.wired.com/2010/07/exclusive-google-cia/.

Spranger, T. 2012. *International Neurolaw: A Comparative Analysis.* London: Springer.

Truthful Brain Corporation. www.truthfulbrain.com. Accessed September 10, 2013.

Terry, J. 2001. "The lawfulness of attacking computer networks in armed conflict and in self-defense in periods short of armed conflict: What are the targeting constraints?" *Military Law Review* 169:70–91.

U.N. General Assembly. 1945. "Charter of the United Nations, 24 October, 1 UNTS XVI." http://www.un.org/en/documents/charter/index.shtml.

U.N. General Asssembly. 1948. *Universal Declaration of Human Rights.* 217A (III). Article 1, 18. www.un.org/en/documents/udhr.

United States War Crimes Act of 1996. 18 U.S.C 2401. www.gpo.gov/fdsys/pkg/PLAW-104publ192/pdf/PLAW-104publ192.pdf.

Vaseashta, A., E. Braman, and P. Sussman, eds. 2012. *Technological Innovation and Ecological Terrorism* (NATO Science for Peace and Security Series). New York: Springer.

White, S. 2008. "Brave new world: Neurowarfare and the limits of international humanitarian law." *Cornell International Law Journal* 41(2):177–210.

Whitehouse.gov. 2013. "Statement by the President on Syria, August 31, 2013." http://www.whitehouse.gov/the-press-office/2013/08/31/statement-president-syria.

Wingfield-Hayes, R. 2013. "Why Japan's 'Fukushina 50' remain unknown." *BBC News.* www.bbc.com/news/world-asia-20707753.

World Medical Association. 1964. "Declaration of Helsinki: Ethical principles for medical research involving human subjects." Adopted by the 18th World Medical Association General Assembly. Helsinki, Finland. www.wma.net/en/30publications/10policies/b3/.

10 Neuroscience, National Security, and the Reverse Dual-Use Dilemma*

Gary E. Marchant and Lyn M. Gaudet

CONTENTS

INTRODUCTION

The "dual-use" problem is a well-known, yet unresolved, problem vexing much of the cutting-edge research and development in science and technology. Simply stated, the dual-use problem is the risk that scientific and technological innovations created for legitimate civilian or defensive military purposes will be diverted for offensive or malicious uses. For example, research on virulent flu virus strains intended to develop preventive or therapeutic measures could provide knowledge that could be misused to spread the virus for terrorist purposes (Hale et al. 2012).

We address here the opposite problem, for which we have coined the term "reverse dual-use" problem (Marchant and Gulley 2010). It occurs when technologies developed for legitimate military objectives spread to the civilian sector and have largely unanticipated disruptive or detrimental effects. This is a type of rebound effect, where a technology developed for one beneficial purpose has an unintended detrimental secondary effect.

Although much less recognized than the more familiar dual-use problem, the reverse dual-use problem is becoming an increasingly prevalent and serious problem. Many technologies developed by the military for legitimate national security objectives, such as unmanned aerial vehicles (UAVs) (or "drones"), various surveillance technologies, and cyber-attack capabilities, are likely to have destabilizing or disruptive consequences if allowed to proliferate unrestricted in the civilian sector.

---

* This chapter is an expanded and updated version of Marchant and Gulley (2010), adapted with permission from *AJOB Neuroscience* 1(2):20–22.

Perhaps no technology is more prone to the reverse dual-use problem than military neuroscience. This chapter explores the reverse dual-use problem focusing on military neuroscience as a case study. Part I explains in greater detail the reverse dual-use problem and how it differs from the more traditional dual-use problem. Part II summarizes the field of military neuroscience, at least how it appears from a nonclassified perspective. Part III then assesses the reverse dual-use problem in the context of military neuroscience. Finally, Part IV provides preliminary thoughts on how the reverse dual-use problem might be addressed.

THE REVERSE DUAL-USE PROBLEM

Over the years, many technologies and products developed by military research and development have valuable subsequent applications in the civilian economy. Examples include aviation technologies, global positioning system (GPS) surveillance, and the Internet. Yet, as the power of new emerging technologies in fields such as genomics, synthetic biology, nanotechnology, neuroscience, information and communication technologies, and robotics surge forward at an unprecedented pace, the spread of powerful technologies that may have legitimate national security applications may have consequences that are not beneficial or even benign when spread to the civilian sector. We refer to this problem as the "reverse dual-use dilemma" (Marchant and Gulley 2010). While the dual-use problem primarily focuses on the (mis)use of beneficial, civilian technology for nefarious or military applications (National Research Council 2004), the reverse dual-use dilemma is concerned with the disruptive impact of spillover into the civilian sector of technology developed by the military for legitimate national security purposes.

Two things are different today that are bringing the reverse dual-use problem to the forefront. First, the immense power and pervasiveness of emerging technologies such as neuroscience create a much stronger potential for adverse and potentially catastrophic societal impacts when such civilian spillovers occur. Indeed, there is likely to be a strong negative correlation between the value and utility of an emerging technology to the military and its potential disruptive consequences for the civilian sector. Second, the time lag between military development and civilian application is shrinking—what would take decades to make its way into the mainstream in the early part of the century may take only a handful of years to reach general society today. For example, UAV developed by Honeywell with Defense Advanced Research Projects Agency (DARPA) funding was recently licensed to the Miami-Dade Country Police almost simultaneously with its military debut (Brown 2008).

In fact, this development of miniature UAVs by various national security agencies (Weiss 2007) is a powerful illustration of the reverse dual-use dilemma. These micro- or even nanoscale miniaturized UAVs would provide unprecedented capabilities for covert surveillance or stealth assassination of adversaries. From a military perspective, such a capability would be enormously attractive, vastly increasing operational effectiveness and perhaps even reducing net casualties. Yet, once such devices spill into the civilian sector, how can there be any privacy if tiny, undetectable "insects" can covertly enter and videotape any bedroom, boardroom, or other once-private space? And what if a "fly" buzzing around the president is a robotically controlled assassin "insect"? Of course, the power of these UAVs need not stop with surveillance, but can and is being extended to create weaponized vehicles capable of targeted killings or mass murder (Marchant et al. 2011; Wan and Finn 2011). Again, not the kind of technology that one would want to be easily and widely available.

There are many other emerging examples of the reverse dual-use problem. Another example is the recent development and deployment of the malware viruses nicknamed "Stuxnet" and "Flame" to spy on and disrupt the Iranian nuclear weapon program. A string national security case can be made in support of these efforts to at least delay the development and potential use of a nuclear bomb by Iran that could kill millions of innocent lives. At the same time, if this same technology was to become available to civilians and terrorist groups, the damage could be equally if not more severe. According to one media description of the Flame program, it is "designed to replicate across even highly secure networks, then control everyday computer functions to send secrets back to its creators. The code could activate computer microphones and cameras, log keyboard strokes, take screen shots, extract geolocation data from images, and send and receive commands and data through Bluetooth wireless technology" (Nakashima et al. 2012). It is easy to imagine how such a powerful technology could wreak havoc if it could get into the wrong hands.

The first rebound effect would be if other nations use the technology and precedent set by Stuxnet and Flame to launch their own cyber-attacks against U.S. resources— these programs "will invite imitation and retaliation in kind, and it has established new and disturbing norms for state aggression on the Internet and in its side channels. American and Israeli official action now stands as a justification for others. In national security as in much, what goes around often comes around" (Coll 2012). An even more disruptive effect would occur if or when individual citizens can obtain and use similar cyber-attack technologies, whether it be the thrill-seeking teenage hacker, spurned or jealous lovers, industrial espionage agents, or full-blown terrorists intent on inflicting unparalleled economic and physical harm to the U.S. infrastructure.

MILITARY NEUROSCIENCE

While the development of applied neuroscience is in its infancy, the long-term applications and impacts of neuroscience are likely to be both powerful and profound (National Research Council 2009). Military and intelligence agencies, with the most at stake from such applications in terms of both benefits and risks, recognize the potential of neuroscience to revolutionize intelligence-gathering and warfare (National Research Council 2009). According to one estimate, the U.S. Department of Defense expenditures on neuroscience exceeded $350 million in 2011 (Tennison and Moreno 2012). The national security establishment has the time, money, and mission needs to explore the cutting edge of neuroscience research and applications. With budgets running to billions of dollars per year, the luxury of no hard deadlines, and the ever-pressing need to innovate technology to maintain the U.S. global edge in military technology and capability, U.S. military research entities such as the DARPA can pursue particular lines of research for as long as they are inclined (Moreno 2004; Royal Society 2012).

The military has set their sights on neurotechnology for some time—DARPA has been "dabbling" in neuroscience research since the early 1990s (Hoag 2003). While many of the potential applications discussed may be over-hyped today, there is no reason to think that persistent research will not bring at least some to fruition. Research typically takes longer to implement and assimilate than initially anticipated, and there is a real danger of complacency when the negative impacts are not immediately apparent (Brooks 1983). What may be exaggerated fears today may be too real tomorrow.

Most military interest in neuroscience can be grouped into two categories: performance enhancement and performance degradation (Royal Society 2012). Neuroscience for performance enhancement includes pharmaceutical and other measures for improving the cognitive capabilities, alertness, endurance, and communication of warfighters, as well as less exotic applications such as better training and rehabilitation of soldiers (National Research Council 2009). Neuroscience for performance degradation includes various types of nonlethal weapons that can impair the cognitive functioning of adversaries on the battlefield. In addition to these enhancement and de-enhancement applications, neuroscience can also be used for interrogating and interviewing both enemy and friendly actors for their veracity and intentions (Tennison and Moreno 2012). All of these actual or potential military uses of neuroscience, in many (but at least theoretically not all) cases directed to pursue legitimate and appropriate national security objectives, could have widespread applications in the civilian sector, many with dubious ethical or consequential impacts.

THE REVERSE DUAL-USE PROBLEM APPLIED TO MILITARY NEUROSCIENCE

From a national security perspective, there are many longer-term applications of neuroscience that may be justified (National Research Council 2009). Such advancements will be compelled, again from purely a military perspective, by a need to maintain military superiority, perform operations more effectively and safely, and be able to anticipate and counter potential offensive applications by adversaries. Of course, such applications should be developed in an ethical manner, including appropriate human subject protections, but the point is that the benefits of these applications to the military might clearly outweigh the risks. Yet, when these same technologies are evaluated in a broader context that includes the impacts of spillover into the civilian sector, the overall cost–benefit balance may shift into the negative.

Several military applications of neuroscience, still primarily in the R&D stage, provide possible examples of this reverse dual-use dilemma. The first is remote brain scanning. National security agencies are attempting to develop remote brain scanning technologies that involve focusing some form of optical or other beam on a person's head that is capable of detecting veracity, anxiety, or malevolence (Silberman 2006). Such a technology, if viable, could be used to "interrogate" individuals without their consent or even knowledge. If remote brain scanning is going to become a reality, the most likely candidate is near-infrared spectroscopy (NIRS). NIRS measures changes in blood flow in the brain and is based on the fact that the transmission and absorption of near-infrared light provides information about change in oxygenation levels

(Villringer et al. 1993). NIRS is currently being used to investigate a wide variety of scientific questions such as the effects of exercise and alcohol, semantic processing in learning disorders, and the measurement of oxygen saturation in preterm infants (Tsujii et al. 2011; Sela et al. 2012; Thompson et al. 2013; Tsujii et al. 2013).

NIRS can only be used on cortical tissue in the brain so it is unable to provide the same type, quantity, or quality of data as other scanning technologies such as functional magnetic resonance imaging (fMRI), but it has other advantages that make it the best candidate for remote scanning. It is far more portable and it can be used to gather data from patients that are not amenable to functional scans like premature infants or patients engaged in physical activity such as exercise. NIRS can also be a wireless system, increasing its convenience and ease of use. In theory, once the technology becomes advanced enough, NIRS could be administered to an individual without their knowledge because the equipment could be in a neighboring room.

With access to an individual's brain activity comes the potential to control. It may become possible to manipulate individuals' brain activity and consequently their thoughts, behaviors, or both. Two techniques are relevant in this area, transcranial magnetic stimulation (TMS) and transcranial direct current stimulation (tDCS). Both techniques are becoming increasingly popular methods to study—including by the military—because they are noninvasive and inexpensive compared to other treatment interventions and they have amazing potential to affect change in the brain and therefore behavior. The possible military applications of these techniques are to enhance soldiers' performance by increasing memory and learning, altering mood states, or by reducing fatigue as well as inducing different types of behavioral change in individuals besides our own soldiers (Tennison and Moreno 2012).

TMS is currently used in research and to treat depression and other neurological conditions (Welberg 2007). TMS causes depolarization or hyperpolarization of neurons in the brain by placing an electricity-generating magnetic coil above the head that induces currents in the cortex of the brain (Allen et al. 2007). TMS can be used to induce activity in specific or large areas of the brain. How TMS currents actually affect neuronal processing has been a question under study for a number of years (Allen et al. 2007). TMS lasts for anywhere from seconds to minutes, and it has the potential to be used as a type of mind, or more aptly, behavioral control. The military benefits of such a technology, especially one that can be applied remotely, are significant. It could allow for the peaceful control of hostile individuals, allowing them to be subdued and disarmed, as well as potential crowd control during riots or other situations where predictably violent or unpredictable crowds may compromise safety of all involved in a particular mission. The civilian applications of such a technology are numerous as well, although the motivation behind some applications may be suspect as it is not hard to imagine uses driven solely by the prospect of financial gain. What if TMS could be used to reduce inhibition for a short period of time? What retailer—from grocery stores to car dealerships—would not like to be able to apply this to customers walking into and around their stores? By reducing their inhibition, it may increase the likelihood that they will not only purchase something but perhaps purchase far more than they normally would. More sinister uses can easily be conjured up as well.

Like TMS, tDCS is being pursued in research and in the military for its ability to enhance learning, memory, and other domains of cognitive function and as a potential treatment in various psychiatric conditions (Bullard et al. 2011; Brunelin et al. 2012). tDCS involves neurostimulation (or inhibition) by means of constant, low-intensity current delivered directly to the brain area of interest through small electrodes placed on either side of the scalp. It requires only small amounts of energy, is portable, and is easy to administer. Even though the exact mechanisms by which tDCS influences behavior are presently unknown, studies in humans are proceeding and producing fascinating results (Bikson et al. 2009; Clark 2012).

For example, research sponsored by DARPA has discovered that tDCS has an amazingly robust effect on learning. A 2012 study investigated whether tDCS would increase the learning rate of identifying concealed objects placed in naturalistic surroundings (Clark et al. 2012). Two levels of tDCS were applied, low current and high current. The low-current tDCS resulted in a 10.5% increase in accuracy between pretraining and delaying posttraining (one hour after training concluded), and high-current tDCS resulted in a 21.3% increase in accuracy after delayed posttesting. There was a 104% increase in overall performance accuracy between the non-tDCS and tDCS groups, one of the largest effects reported in the learning literature (Clark et al. 2012).

Additional studies have found tDCS to be significantly more effective when applied to novice rather than experienced learners and that the effects of tDCS-enhanced training will carry over into subsequent learning sessions (Bullard et al. 2011). Future work will investigate whether applying tDCS to different areas of the brain at different stages of will maximize training effectiveness (Bullard et al. 2011). The results to date suggest that the benefits of tDCS are significant in terms of education and training. It would not only reduce the amount of time required to obtain levels of expertise but could also result in individuals being able to acquire skills that they may not have had the time (or the aptitude) to acquire before.

While the research continues, ethicists have begun to ask questions and consider the implications of this work, including concerns regarding safety, justice, and autonomy, particularly as the technology migrates into widespread use in the commercial and civilian sectors (Hamilton et al. 2011). From a national security perspective, such remote mind scanning technologies may be seen as having many benefits, in particular the more effective and less intrusive screening of airline passengers, prospective intelligence agents, or suspected terrorists. While such applications have the potential to be beneficial for national security, what would happen if (or more poignantly when) such technology becomes widely available? How will it affect police–suspect, employer–employee, business–customer, spouse–spouse, parent–child, and many other interactions? While there may be many benefits to some of the civilian applications as well, such a technology would also have the potential for many disruptive and disconcerting consequences. The point being, should we not consider those impacts before making a societal decision to develop such a technology?

A second example is behavior-modifying agents that can placate, calm, incapacitate, or otherwise control individuals (National Research Council 2008; Royal Society 2012). There is substantial evidence that militaries are investigating and developing such behavior-altering agents, as evidenced by the Russian military's use of a fentanyl opiate during the Moscow theater siege in 2002 (Wheelis and Dando 2005; Dando 2009). From a purely military perspective, there are applications of such technologies that would be both beneficial and arguably ethical, such as peacefully controlling a potentially hostile local population during a military operation in order to reduce both military and civilian casualties. Again, though, what would happen if such technologies could be acquired by the general public? A recent report by the Royal Society in the United Kingdom warned that "[t]he development of incapacitating chemical agents also increases the proliferation of these weapons and the risk of acquisition by rogue states, terrorists or criminals" (Royal Society 2012). Do we want to give terrorists, aspiring bank robbers, and rebellious teenagers the power to cheaply and relatively easily alter the behavior and perhaps wellbeing of other people? Obviously not, but are those not probable scenarios if we develop such agents?

There are no doubt many other examples of military neuroscience research that present reverse dual-use concerns. For example, the use of noninvasive brain stimulation technologies to interrogate unwilling participants may be justifiable in certain military contexts (although even that is debatable), whereas such uses for involuntary interrogations in the civilian sector would undoubtedly be seen as objectionable (Heinrichs 2012).

A PROPOSED SOLUTION

The examples provided above are merely illustrative. They are intended to demonstrate that emerging technologies in the neurosciences and other scientific fields being developed by the military for legitimate national security objectives could have disruptive and destructive impacts if allowed to flow freely into the civilian sector. There is no indication that the military is considering the potential future civilian implications of these technologies, as this has never been part of their mission or responsibility up until now. Our national security agencies have been given the challenging task of protecting the nation from external threats, which include pursuing technologies that will help them carry out that mission.

Of course, the civilian applications of military neuroscience innovations will not be all negative, there will be some positive benefits as well, such as the potential assistance to quadriplegics from brain–machine interface technologies being developed by the military (Weinberger 2012). Yet, as the power of emerging technologies continue to grow, the potential detrimental impacts also grow, even if the positive impacts likewise grow. If the longer-term civilian implications of military technologies are not addressed at the formative stages when such technologies are first being considered, shaped, and developed, we run the very real danger of blindly committing ourselves to a future we do not want (Moreno 2004). How can this problem be addressed? No obvious and simple solutions present themselves. The most logical approaches would be a review mechanism either inside or outside the military establishment. Either approach would have its limitations. An internal review mechanism would have the benefit of better access to classified information and the technology developers, but may not have the independence and freedom to raise tough questions about technological developments that might be beneficial for the military but disruptive for the general society. Alternatively, some form of civilian review mechanism would likely offer greater independence and objectivity, but would almost certainly be limited (even with security clearances) in its access to information and key personnel.

Moreno (2006) has recommended the formation of an advisory committee to consider the ethical and policy aspects of military development and utilization of neurotechnologies. Such a committee, if constituted, could potentially take on the related function of assessing the civilian spillover implications of such military technologies.

Alternatively, a model that could be used, with some important modifications, is the environmental impact statement (EIS) required under the National Environmental Policy Act (NEPA) for major federal actions that will significantly affect the human environment (NEPA 1969). The agency sponsor of such a project must undertake an analysis of the potential environmental impacts of the project and indentify potential alternatives that might reduce any such adverse impacts. The agency first undertakes an initial screening assessment called an environmental statement (ES) to determine whether the project is likely to have significant environmental impacts; a full EIS is only required for the relatively small subset of projects that are likely to have substantial impacts. The EIS has been interpreted to be a procedural requirement, with the sponsor agency under no obligation to adopt any substantive alternatives to its proposed project. Nevertheless, the procedural requirement to study and report environmental impacts of, and alternative to, a project is seen as having two major benefits. First, it creates awareness within the sponsoring agency of the potential environmental impacts of its actions, by requiring the hiring of staff with environmental expertise who will then conduct an analysis that will be shared with the key decision makers and staff within the agency. In addition to this internal educational and awareness benefit, a second benefit of the EIS is to inform the public of the potential impacts of proposed projects, thereby triggering the political process to debate the wisdom of the project and its possible alternatives (Rodgers 1990).

The impact assessment model created for environmental effects by NEPA has now been cloned into a number of other contexts, such as (1) social impact assessment statements for many development and other projects (Dietz 1987), (2) a privacy impact assessment statements implemented by many private companies and mandated by all federal agencies by Section 208 of the E-Government Act of 2002 (Clarke 2009), (3) health impact assessments "to examine the effects that a policy, program, or project may have on the health of a population" (Collins and Koplan 2009), and (4) ethical technology assessments (Palm and Hansson 2006). In 2007, the National Science Advisory Board for Biosecurity (NSABB 2007) proposed an impact-assessment type model for dual-use problems in which investigators would identify potential substantive dual-use concerns associated with their research and confer with local oversight boards about those concerns.

Military and intelligence agencies that are developing new technologies that could have significant detrimental civilian implications could be required to conduct an impact assessment, perhaps called a civilian impact assessment statement (CIAS). Just as NEPA screens out projects with minor environmental impacts from having to conduct a full EIS, the military entity responsible for a new technology would conduct an initial assessment and only proceed to a full CIAS for technologies that are reasonably likely to have a significant adverse civilian impact. Findings of no significant impact would be reviewed by one office within the Pentagon to ensure that offices are not ducking their CIAS responsibility by unreasonably downplaying societal impacts. For projects that are subject to the full CIAS, the sponsoring entity would be required to identify potential impacts on the civilian sector and to identify potential alternatives for mitigating, managing, or avoiding those impacts.

This proposed CIAS requirement may seem like an unduly burdensome and bureaucratic imposition on first impression, but given the transformative impacts some of the emerging military-developed technologies in neuroscience and others may have, this new burden may be necessary to avoid much greater hardship. While not perfect, this proposal is likely to improve the reverse dual-use dilemma. The internal education/awareness function may be most important. It is likely that the scientists, engineers, funders, project managers, and implementation decision makers give very little consideration to civilian sector impacts. Experience with programs intended to increase awareness of dual-use concerns among civilian scientists finds that the greater awareness that the researchers have about the potential risks, the more likely they are to think through risk reduction and avoidance (Davidson et al. 2007). Presumably, the requirement to identify, and possible alternatives to reduce or avoid, adverse civilian impacts of the technologies they are working on will spur military scientists to also be more sensitive to potential problems associated with their work.

The external benefit of a NEPA-like assessment process in informing the public and triggering the political process where appropriate is likely to be more difficult for the CIAS process. For one thing, the CIAS will often involve classified or sensitive information, since it involves the assessment for risks associated with powerful technologies with military or national security applications that are still in the development stage. Thus, full public disclosure may not be warranted. Moreover, the biggest criticism of NEPA is that it provides an opportunity for judicial review that is used to delay many projects. Such a mechanism would be inappropriate for important military technologies as it could jeopardize important national security objectives, and so CIAS should not be judicially reviewable. Of course, the cost of eliminating judicial enforcement is reduction in the incentives to take seriously the CIAS process and requirements. To provide some enforcement, there would need to be some independent review process for the CIAS. Perhaps a committee with appropriate security clearances set up under the auspices of the National Academy of Sciences or similar institution could serve as the CIAS review mechanism.

Regardless of what institutional form is adopted, the substantive criteria and actions that might be considered as a less-damaging alternative are probably the most difficult aspect of the reverse dual-use problem. It would be difficult to convince or require the military to forgo most militarily beneficial technologies in order to avoid future civilian impacts, especially if military adversaries are likely to develop similar technologies themselves. While relinquishment may be an appropriate remedy in extreme cases, most examples will likely require a different policy approach such as trying to restrict the technology to legitimate military applications through classification or other measures, preparing and educating civilian policy makers, law enforcement officials, and other relevant stakeholders about the potential disruptive technologies on the horizon, and developing potential mitigation strategies. While the details and implementation of these various options need to be elaborated, what is clear is that the reverse dual-use dilemma is real and increasingly urgent and that the policy response to this problem is undeveloped, inadequately recognized, and in need of greater attention.

REFERENCES

Allen, E., B.N. Pasley, T. Duong, and R.D. Freeman. 2007. "Transcranial magnetic stimulation elicits coupled neural and hemodynamic consequences." *Science* 317(5846):1918–1921.

Bikson, M., A. Datta, and M. Elwassif. 2009. "Establishing safety limits for transcranial direct current stimulation." *Clinical Neurophysiology* 120(6):1033–1034.

Brooks, H. 1983. "Technology, competition, and employment." *Annals of the American Academy of Political and Social Science* 470(1):115–122.

Brown, T. 2008. "Miami police could soon be the first in the United States to use cutting-edge, spy-in-the-sky technology to beef up their fight against crime." *Reuters*, March 26. Accessed on September 1, 2012, http://www.reuters.com/article/2008/03/26/us-usa-security-drones-idUSN1929797920080326.

Brunelin, J., M. Mondino, L. Gassab, F. Haesebaer, L. Gaha, M. Suaud-Chagny, M. Saoud et al. 2012. "Examining transcranial direct-current stimulation (tDCS) as a treatment for hallucinations in schizophrenia." *American Journal of Psychiatry* 169(7):719–724.

Bullard, L., E. Browning, V. Clark, B. Coffman, C. Garcia, R. Jung, A. van der Merwe et al. 2011. "Transcranial direct current stimulation's effect on novice versus experienced learning." *Experimental Brain Research* 213(1):9–14.

Clark, V., B. Coffman, A. Mayer, M. Weisand, T. Lane, V. Calhoun, E. Raybourn, C. Garcia, and E. Wassermann. 2012. "TDCS guided using fMRI significantly accelerates learning to identify concealed objects." *Neuroimage* 59(1):117–128.

Clarke, R. 2009. "Privacy impact assessment: Its origins and development." *Computer Law and Security Review* 25(2):123–135.

Coll, S. 2012. "The rewards (and risks) of cyber war." *The New Yorker*, http://www.newyorker.com/online/blogs/comment/2012/06/the-rewards-and-risks-of-cyberwar.html. Accessed on May 11, 2014.

Collins, J. and J.P. Koplan. 2009. "Health impact assessment." *Journal of the American Medical Association* 302(3):315–317.

Dando, M. 2009. "Biologists napping while work militarized." *Nature* 460(7258):950–951.

Davidson, E.M., R. Frothingam, and R. Cook-Deegan. 2007. "Practical experiences in dual-use review." *Science* 316(5830):1432–1433.

Dietz, T. 1987. "Theory and method in social impact assessment." *Sociological Inquiry* 57(1):54–69.

E-Government Act. 2002. Public Law No. 107-347, 107th Cong, 1st Sess., December 17.

Hale, P., S. Wain-Hobson, and R. May. 2012. "The folly of resuming avian flu research." *Financial Times*, http://www.gpo.gov/fdsys/pkg/PLAW-107publ347/content-detail.html. Accessed on May 11, 2014.

Hamilton, R., S. Messing, and A. Chatterjee. 2011. "Rethinking the thinking cap: Ethics of neural enhancement using noninvasive brain stimulation." *Neurology* 76(2):187–193.

Heinrichs, J.-H. 2012. "The promises and perils of non-invasive brain stimulation." *International Journal of Law and Psychiatry* 35(2):121–129.

Hoag, H. 2003. "Remote control." *Nature* 423(6942):796–798.

Marchant, G., B. Allenby, R. Arkin, E. Barrett, J. Borenstein, L. Gaudet, O. Kittrie et al. 2011. "International governance of autonomous military robots." *Columbia Science and Technology Law Review* 12:272–315.

Marchant, G. and L. Gulley. 2010. "National security neuroscience and the reverse dual-use dilemma." *American Journal of Bioethics—Neuroscience* 1(2):20–22.

Moreno, J.D. 2004. "DARPA on your mind." *Cerebrum* 6(4):1–9.

Moreno, J.D. 2006. *Mind Wars: Brain Research and National Defense.* Washington, DC: Dana Press.

Nakashima, E., G. Miller, and J. Tate. 2012. "U.S., Israel developed flame computer virus to slow Iranian nuclear efforts, officials say." *Washington Post*, http://www.washingtonpost .com/world/national-security/us-israel-developed-computer-virus-to-slow-iranian-nuclear-efforts-officials-say/2012/06/19/gJQA6xBPoV_story.html. Accessed on May 11, 2014.

National Environmental Policy Act (NEPA). 1969. Public Law No. 91–190, 91st Cong, 1st Sess., https://www.govtrack.us/congress/bills/91/s1075/text. Accessed on May 11, 2014.

National Research Council. 2004. *Biotechnology Research in an Age of Terrorism: Confronting the Dual Use Dilemma.* Washington, DC: National Academies Press.

National Research Council. 2008. *Emerging Cognitive Neuroscience and Related Technologies.* Washington, DC: National Academies Press.

National Research Council. 2009. *Opportunities in Neuroscience for Future Army Applications.* Washington, DC: National Academies Press.

National Science Advisory Board for Biosecurity (NSABB). 2007. *Proposed Framework for the Oversight of Dual Use Life Sciences Research: Strategies for Minimizing the Potential Misuse of Research Information.* Bethesda, MD: Office of Biotechnology Activities, National Institutes of Health.

Palm, E. and S.O. Hansson. 2006. "The case for ethical technology assessment (eTA)." *Technological Forecasting and Social Change* 73(5):543–558.

Rodgers, W.H. 1990. "NEPA at twenty: Mimicry and recruitment in environmental law." *Environmental Law* 20(485):485–504.

Royal Society. 2012. *Brain Waves Module 3: Neuroscience, Conflict and Security.* London: Royal Society.

Sela, I., M. Izzetoglu, K. Izzetoglu, and B. Onaral. 2012. "A functional near-infrared spectroscopy study of lexical decision task supports the dual route model and the phonological deficit theory of dyslexia." *Journal of Learning Disabilities* 47(3):279–288. doi:10.1177/0022219412451998.

Silberman, S. 2006. "The cortex cop." *Wired* 14(1):149.

Tennison, M. and J. Moreno. 2012. "Neuroscience, ethics, and national security: The state of the art." *PLoS Biology* 10(3):e1001289. doi:10.1371/journal.pbio.1001289.

Thompson, A., P. Benni, S. Sevhan, and R. Ehrenkranz. 2013. "Meconium and transitional stools may cause interference with near-infrared spectroscopy measurements of intestinal oxygen saturation in preterm infants." *Advances in Experimental Medicine and Biology* 765:287–292.

Tsujii T., K. Komatsu, and K. Sakatani. 2013. "Acute effects of physical exercise on prefrontal cortex activity in older adults: A functional near-infrared spectroscopy study." *Advances in Experimental Medicine and Biology* 765:293–298.

Tsujii T., K. Sakatani, E. Nakashima, T. Igarashi, and Y. Katayama. 2011. "Characterization of the acute effects of alcohol on asymmetry of inferior frontal cortex activity during a go/no-go task using functional near-infrared spectroscopy." *Psychopharmacology* 217(4):595–603.

Villringer, A., J. Planck, C. Hock, L. Schleinkofer, and U. Dirnagi. 1993. "Near infrared spectroscopy (NIRS): A new tool to study hemodynamic changes during activation in human adults." *Neuroscience Letters* 154(1/2):101–104.

Wan, W. and P. Finn. 2011. "Global race on to match U.S. drone capabilities." *Washington Post*, http://www.washingtonpost.com/world/national-security/global-race-on-to-match-us-drone-capabilities/2011/06/30/gHQACWdmxH_story.html. Accessed on May 11, 2014.

Weinberger, S. 2012. "Mind control moves into battle." *BBC News*, http://www.bbc.com/future/story/20120704-mind-control-moves-into-battle. Accessed on May 11, 2014.

Weiss, R. 2007. "Dragonfly or insect spy? Scientists at work on robobugs." *Washington Post*, October 9. Accessed on January 30, 2010, http://www.washingtonpost.com/wp-dyn/content/article/2007/10/08/AR2007100801434.html.

Welberg, L. 2007. "Technology: TMS reveals its workings." *Nature Reviews Neuroscience* 8(11):813.

Wheelis, M. and M. Dando. 2005. "Neurobiology: A case study of the imminent militarization of biology." *International Review of the Red Cross* 87(859):553–571.

11 Neuroskepticism
*Rethinking the Ethics of Neuroscience and National Security**

Jonathan H. Marks

CONTENTS

INTRODUCTION

It is hard to imagine anywhere darker, more esoteric, and—to be frank—more thrilling than the domain of national security neuroscience. In this chapter,[1] I explore that intriguing place where neuroscience and national security intersect, each enchanting to the initiates of the other, and both somewhat mysterious to the rest of us. I confess that my aim here is to puncture that aura of mystery and enchantment, to defuse the understandable thrill, and to offer some words of caution—in particular to scientists, ethicists, research funding bodies, policy makers, and anyone else who may play a significant role in shaping the kinds of neuroscience research that will be conducted in the years ahead. Before proceeding, however, I should make two things clear.

---

* This chapter combines and adapts two articles with permission, Marks (2010) and Marks, J. "Neuroconcerns: Some responses to my critics." *AJOB Neuroscience* 1(2):W1–W3 (2010).

First, I readily acknowledge that neuroscience offers unparalleled opportunities to transform our lives, and (for some) it has already done so. Few of these opportunities are more dramatic than the potential use of functional magnetic resonance imaging (fMRI) to identify patients with impaired consciousness who might be candidates for rehabilitation (Owen and Coleman 2008), and of deep brain stimulation to release them from imprisonment in hitherto unresponsive bodies (Schiff et al. 2007). However, my thesis here is premised on what might be called *neuroskepticism*—that is, a perspective informed by science studies scholarship that views with some healthy skepticism claims about the practical implications and real-world applications of recent developments in neuroscience. The need to probe and question is, I contend, especially acute in the context of national security neuroscience— where the translation from research laboratory to real life may involve great leaps, among them the troubling jump from brain scanning to terrorist screening.

The approach I adopt here is consonant with and sympathetic to the goals of "critical neuroscience"—a multidisciplinary project defined by some scholars as "a reflexive scientific practice that responds to the social and cultural challenges posed both to the field of science and to society in general by recent advances in the behavioral and brain sciences" (Choudhury et al. 2009). The proponents of critical neuroscience aim to bridge the gap between science studies and empirical neuroscience by engaging scholars and practitioners from the social sciences, humanities, and empirical neuroscience to explore neglected issues: among them, the economic and political drivers of neuroscience research, the limitations of the methodological approaches employed in neuroscience, and the manner in which findings are disseminated. The project's avowed and worthy goals include "maintaining good neuroscience, improving representations of neuroscience, and ... creating an awareness of its social and historical context in order to assess its implications" (Choudhury et al. 2009, 66).

Second, I acknowledge the legitimate aims and objectives of the national security enterprise and of the officials solemnly charged with its pursuit. However, sometimes national security threats may be overstated or invoked for political ends, and the means employed in the pursuit of these objectives are often fundamentally violative of the human rights of others (for a more detailed explication, see Marks 2006). In addition, as I outline later, there are many examples from the Bush administration's "war on terror" of medicine, other health sciences (including behavioral psychology), and polygraphy being abused in the name of national security. So, while there are risks that the national security community may be misled about what neuroscience can offer, I am also concerned about the ways in which national security may pervert neuroscience.

NEUROSCIENCE NARRATIVES AND SECURITY SEMANTICS

Neuroscience and national security both jealously guard their own argot. In the case of neuroscience, the lexicon is replete with Latin and Greek and innumerable portmanteau constructions that fuse (or confuse) both classical languages. Consider, for example, the subthalamic nucleus, a neuroanatomical term that sandwiches a Greek derivative between two Latin ones and is (infelicitously) susceptible to the translation "the nut under the bedroom." For some cognoscenti, there may be a familiar poetry in the classical language of the brain's anatomy—the *sulci* and *gyri* becoming, perhaps, the neuro-topographical analogs of William Wordsworth's "vales and hills." But those who do not possess either training in neuroscience or an anatomical dictionary are lost.

National security too has its special (albeit less colorful) language, consisting—for the greater part—of cryptic and somewhat intimidating initials, acronyms, and not-quite-acronyms, such as human intelligence (HUMINT) and behavioral science consultation team (BSCT, pronounced "biscuit"). For readers unfamiliar with these terms, HUMINT is commonly defined as a category of intelligence derived from information collected and provided by human sources. This includes interrogations (as well as other forms of overt or clandestine conversations with sources that may be considered "neutral," "friendly," or "hostile"). BSCTs are teams of psychologists and/or psychiatrists and their assistants (often mental health technicians), who have been tasked by the Department of Defense (DoD) with advising interrogators how to ramp up interrogation stressors at Guantanamo Bay, Abu Ghraib, and elsewhere (see, e.g., Mayer 2008).

My intention here is not to take an easy shot at the neuroscience and national security communities, and the linguistic practices in each of these domains. Rather, I wish to express concern about the naming of things in these contexts and, in particular, about the hazards—practical and ethical—that arise from the deployment of opaque terminology. It is easy for outsiders to the world of neuroscience to believe that, because there is a polysyllabic name for some part of our brain, we have a deep understanding of what it does and how it does it. This is, of course, not necessarily the case. To give just one example, physicians can sometimes achieve dramatic improvements in the motor symptoms of some parkinsonian patients by using small electrical pulses to stimulate the subthalamic nucleus (the "nut" mentioned earlier). However, the mechanism by which this effect is achieved is still being explored. In addition, as one neurosurgeon colleague recently made clear to me, our stimulators are not "smart." They do not monitor what is occurring in the subthalamic nucleus and respond to it, nor do they monitor the response of the nucleus to their stimuli.

Some might argue that, in this clinical example, the intervention works, and while we should seek to refine the technique and our understanding of the efficacy of the intervention, the limited nature of our current understanding should not bar its use. I do not intend to address that claim here. However, the nonclinical application of neuroscience in the murkier national security context creates serious potential hazards, and these risks are amplified in the absence of solid theoretical models and robust empirical data. Not least, there is a real danger that pseudoneuroscience will become a vehicle for the abuse of those who are perceived as a threat to national security.

Although I will substantiate this point shortly, allow me to briefly explore the foundations of neuroscience's linguistic hazards. When the language of neuroscience is used to construct explanatory narratives, the results can be unduly persuasive due to a phenomenon the philosopher J. D. Trout (2008) has termed "explanatory neurophilia." A recent study indicates that nonexperts—including college students taking a cognitive neuroscience class—are not very good at critiquing neuroscience narratives. Deborah Weisberg and colleagues (2008) explored the hypothesis that even irrelevant neuroscience information in an explanation of a psychological phenomenon may interfere with a person's ability to consider critically the underlying logic of that explanation. Weisberg found that nonexperts judged explanations with logically irrelevant neuroscience information to be more satisfying, particularly in the case of bad explanations. The authors try to explain the results in a number of ways. They suggest that the *seductive details effect* may be in play. According to this theory, seductive details that are related—but logically irrelevant—make it more difficult for subjects to code and later recall the main argument of a text. They also hypothesize that lower level explanations, in particular, may make bad explanations seem connected to a larger explanatory system and therefore more insightful.

Contemplating the implications of the Weisberg study, J.D. Trout (2008) has argued that placebic neuroscientific information may "promote the feeling of intellectual fluency" and that "all too often humans interpret the positive hedonic experience of fluency as a mark of genuine understanding." Trout suggests that "neurophilic fluency flourishes wherever heuristics in psychology are reductionist," that is, where they focus on a small number of local factors with apparent causal significance in order to explain a complex problem. Although more work may be required to provide a full account of explanatory neurophilia—that is, our blind (or at the very least blinkered) love for neuroscientific explanations—there is little doubt that the phenomenon has been persuasively demonstrated.

In the national security domain, there is also a temptation to believe that a claim is true because it carries the label HUMINT or is similarly packaged in the specialist language of national security. Many senior administration officials appear to have believed (or wanted to believe) that "EITs"—so-called enhanced interrogation techniques—were, as the name suggested, "enhanced." But, as experienced interrogators have repeatedly asserted, the products of aggressive interrogation tactics such as waterboarding, exposure to temperature extremes, stress positions, and the like tend not to be reliable, whatever one calls them (see, e.g. Bennett 2007; Soufan 2009). This is because interrogatees under pressure tend to say whatever it is they believe their captors want to hear, or anything just to stop their abuse. Numerous detainees in the Bush administration's "war on terror" retracted claims they had made during torturous interrogations, once they were removed from their high-pressure interrogation environments—most notably, Khalid Sheikh Mohammed, who was waterboarded 183 times in March 2003 (CIA Inspector General 2004, 90–91) and later told the Red Cross:

> During the harshest period of my interrogation I gave a lot of false information in order to satisfy what I believed the interrogators wished to hear in order to make the ill-treatment stop. ... I'm sure that the false information I was forced to invent in order to make the ill-treatment stop wasted a lot of their time. (International Committee of the Red Cross 2007)

Psychologists James Mitchell and Bruce Jessen were the principal architects of the post-9/11 interrogation regime to which Khalid Sheikh Mohammed and others were exposed (Mayer 2008; SASC 2008). But as Scott Shane observed in the *New York Times*,

[Mitchell and Jessen] had never carried out a real interrogation, only mock sessions in the military training they had overseen. They had no relevant scholarship; their Ph.D. dissertations were on high blood pressure and family therapy. They had no language skills and no expertise on Al Qaeda.

But they had psychology credentials and an intimate knowledge of a brutal treatment regimen used decades ago by Chinese Communists. For an administration eager to get tough on those who had killed 3,000 Americans, that was enough. (Shane 2009)

Mitchell and Jessen drew on their experience of the survival, evasion, resistance, escape (SERE) training program—a program designed to inoculate the U.S. service personnel against abusive treatment at the hands of enemy captors by exposing them to the kinds of treatment that they had historically received, for example, during the Korean War. Mitchell and Jessen reverse-engineered those techniques and used them as the basis for a new aggressive interrogation regime in the "war on terror" (Mayer 2008). But the reverse engineering of SERE tactics not only violated fundamental human rights norms and the baseline protections for detainees found in Common Article III of the Geneva Conventions (Marks 2007a), it was also premised on a fundamental strategic error. As several experienced interrogators have repeatedly made clear (see, e.g., Bennett 2007; Soufan 2009) and as some psychologists tried to warn the Bush administration (Fink 2009), these techniques are not reliable methods for the extraction of intelligence. On the contrary, as the North Koreans demonstrated in the 1950s (Margulies 2006) and the British government discovered in the wake of several questionable convictions for terrorism in the 1970s (Gudjonsson 2003), these techniques tend to be excellent methods for extracting sham confessions and getting detainees to say whatever they believe their captors want to hear. While this was the intended effect in the former case, in the latter it was not. As a result, several Irish Republican Army (IRA) suspects were falsely convicted, while those responsible for the mainland terror attacks in Britain continued to roam free. However, this vital element was lost in the translation of stress tactics from the SERE training program to the U.S. detention and interrogation operations.

In my view, this account demonstrates the perils arising from a lack of critical engagement with purported scientific expertise, in this case behavioral psychology, in a national security context. These perils arise, in part, from the seductive nature of national security terminology—such as "enhanced interrogation techniques"—that exaggerates or misrepresents the scientific foundations of a particular practice. Two related examples from the interrogation context are "truth serum" and, its even more troublesome cousin, "lie detector." These terms reflect a profound lack of precision and tend to reinforce the operation of mental heuristics that deprive us of the opportunity to think critically.

The U.S. military and intelligence communities have long had a fascination for psychoactive drugs as interrogation aids (see, e.g., Intelligence Studies Board 2006, 73–74). The term "truth serum" is loosely used to describe a variety of psychoactive drugs including scopolamine, sodium pentothal, and sodium amytal. The colloquialism seems to promise the Holy Grail—detainees in a drug-induced state of compliance were unable to resist imparting explosive nuggets of actionable intelligence. However, there is, to date, no drug that can live up to this title. These drugs may make some people more talkative, but there is no guarantee that what they say is either accurate or useful. It appears that this may not have prevented the use of psychoactive drugs as interrogation aids in the "war on terror." Several detainees have claimed that they were drugged prior to interrogation (Warrick 2008). The Central Intelligence Agency (CIA) and DoD dispute these claims. However, the Bush administration commissioned and received legal opinions that took a permissive approach to the use of drugs in interrogation (see Bybee 2002; Yoo 2003). The CIA has acknowledged that detainee Abu Zubaydah was waterboarded 83 times in August 2002 (CIA Inspector General 2004, 90–91). It has also been reported that Zubaydah was drugged with sodium pentothal (see, e.g., Follmann 2003). If this allegation is true, Abu Zubaydah's abusive interrogation may speak as much to the efficacy of so-called truth serums as to the utility of waterboarding.

Not surprisingly, the search for the Holy Grail continues, and neuroscience cannot resist stepping up to the plate. There has been much discussion in both academic journals and the media recently about oxytocin. This hormone is released in the bodies of pregnant women during labor (see, e.g., Lee et al. 2009) and there is some evidence to suggest that "intranasal administration of oxytocin causes a substantial increase in trusting behavior" (Kosfeld et al. 2005; see also Baumgartner et al. 2008). As a recent report of the National Research Council (NRC 2009) acknowledged, the drug is of particular interest to the defense and national security communities, not simply because of its implications for soldier performance, but also because it might allow for "new insights into adversary response." Although the report does not expressly discuss this, one potential use is its administration as an aid to interrogation—perhaps covertly prior to interrogation in aerosolized form. This may sound fanciful. However, aerosolized oxytocin is already being marketed by one corporation, Verolabs, as "Liquid Trust," promising to deliver "the world at your fingertips, whether you are single, in sales, an unhappy employee who wants to get ahead"—or perhaps all three of these. So we should expect interrogators to be tempted to use it if they have not already done so.

If there were such a thing as a "truth serum," of course, there would be little need to direct intelligence efforts toward the detection of lies. But that too is an enterprise with a long and colorful history—discussed in more detail than is possible here in a report of the NRC (2003). For the greater part of the last century, "lie detector" was the moniker associated with the polygraph, although the device does nothing of the kind suggested by the term. The polygraph does not detect lies; on the contrary, it only measures physiological changes that tend to be associated with anxiety. This is problematic because for many polygraph subjects the experience of being tested itself is sufficient to cause considerable anxiety. As a result, the NRC (2003, 2) concluded, the polygraph is "intrinsically susceptible to producing erroneous results"—in particular, false positives. In addition, "countermeasures" are possible: People can be trained to beat the polygraph by reducing external manifestations of anxiety, creating false negatives. In spite of these limitations, the "lie detector" label has stuck, reinforced by countless television series and movies, and figurative labels, like their literal counterparts, are often hard to peel away.

There is strong evidence that polygraphy was abused in the "war on terror" and, in my view, this misuse is attributable to misunderstandings of the technology reinforced by the "lie detector" monicker. Documents obtained by M. Gregg Bloche and me pursuant to Freedom of Information Act (FOIA) requests reveal that between August 2004 and October 2006, the U.S. Air Force Polygraph Program conducted 768 polygraphs in Iraq.[2] According to an internal summary, 47% of the polygraph tests indicated no deception, while 46% purported to indicate some form of deception. This was interpreted by the drafter of the summary in the following way: "Detainee personnel are just as likely to have committed the suspected act as not." (Of course, this might equally have been interpreted to mean: "Detainee personnel are just as likely not to have committed the suspected act.") But further reading suggests that the attribution of guilt—defined as "involvement in multiple acts of anti-coalition force activities"—on the basis of these results requires an unjustifiable leap of faith. The summary itself offers one important reason why the report's conclusion is unwarranted. It states (without acknowledging the implications of this) that "only 10% of requests for polygraph support contain sufficiently detailed information for specific issue exams." The remainder of the polygraph tests were (according to the summary) "by definition screening examinations wherein the examiner is called to resolve numerous and divergent issues based on extremely generic, anonymous and perishable reporting." In polygraphers' feedback forms accompanying this summary, many respondents complained that the polygraph technology was either "over utilized or not utilized properly." Two described the use of a failed polygraph test as "a hammer to be used against the detainee." One said this never resulted in anything positive, while another said that, having participated in 240 polygraph examinations in Iraq, on only one occasion did he witness this approach produce "anything of value." Despite this, detainees were "regularly" hammered with polygraph results—even when deception was not indicated (i.e., even when they had "passed" the polygraph test). In such cases, the only clear evidence of dishonesty was on the part of the interrogators.

Not surprisingly, many polygraphers complained about the way their services had been deployed. One noted that interrogators "did not fully understand how to use our services despite multiple briefings and pretest coordination discussions." The polygraph was often used as a "crutch" to avoid unnecessary interrogations, one polygrapher claimed. Another complained about the use of what s/he considered to be worthless questions and estimated that in 70% of cases interrogators asked: "Have you ever been involved in attacking coalition forces?" Others described larger issues that the military failed to address. Most notably, one concluded:

> 'I encountered nothing but difficulties with the exams and have no reason to have any confidence the results were valid. I attribute these problems to a host of reasons: bad environment, problems with interpreters [who were used in most interrogations], and cultural differences.'

Even in its traditional use in the United States, it is clear that the polygraph does not merit the moniker, "lie detector." The NRC (2003) has noted that while the technology performs better than chance, it is far from perfect. But in the national security context, where most interrogations are mediated through an interpreter and cultural issues are often ignored, the label is surely even more problematic.

The language of lie detection is more worrisome still when it is deployed to describe the use of brain imaging and related technologies that do not patently rely on external manifestations of anxiety. Newer technologies purport to show us what is going on in someone else's brain and—if one were to believe much of the press coverage—in their mind. As the British experimental psychologist Richard Henson (2005, 228) has succinctly observed, "There is a real danger that pictures of blobs on brains seduce one into thinking that we can now directly observe psychological processes." It is to the seductive power of brain images that I now turn.

NEUROIMAGING AND NEUROSCIENTIFIC IMAGINARIES

Brain images are ubiquitous. They are no longer the sole province of medical and scientific journals. Viewers of cable news and print media alike are frequently shown brain images. They usually accompany features breathlessly reporting that brain imaging has heralded the end of lies or finally lifted the shroud to reveal "how the brain handles love and pain" (Kane 2004). But functional neuroimages are not images insofar as that word is used to connote optical counterparts of an object produced by an optical device. Brain activity does not have an optical component. We cannot literally see people think—although these images may suggest as much to those with little or no understanding of how they are produced. Rather, neuroimages are carefully constructed representations of the brain and its functions. When the results of fMRI-based cognitive neuroimaging studies are presented to us in image form (as is almost invariably the case), tiny changes in blood oxygenation levels (less than 3%) are represented by bright colors (usually reds, yellows, and blues). These changes are the product of a comparison between levels of blood oxygenation for a chosen cognitive task and those for an activity considered a suitable baseline. These changes are interpreted as markers of local activation or inhibition in the regions of the brain in which they occur.

Many science studies scholars and bioethicists have critiqued the manner in which brain images are produced, constructed, and interpreted (see, e.g., Dumit 2003; Wolpe et al. 2005; Marks 2007b; Joyce 2008). I do not review all these critiques here. Instead, I wish to highlight a simple methodological point that is often not appreciated. An fMRI—the kind so frequently reproduced in glossy magazines for lay readers—is usually not a single image of one person's brain. There are two reasons for this. First, the changes represented in brain images are often not those of a single experimental subject. More commonly, they are representations of composite data from a small experimental cohort. Second, these color images are superimposed on a higher resolution structural brain image, just as Doppler weather radar images are superimposed on geographic maps. The higher resolution image is intended to reveal the topography of the brain—just as the geographic map (on to which the constructed Doppler image is superimposed) is intended to show that, for example, the latest hurricane is 50 miles off the coast of Georgia. However, the structural image of the brain need not be taken (and is often not taken) from any of the subjects of the experiment. This is important because the neurological analog of the state of Georgia in my brain is (like the lyrical Georgia on my mind) not necessarily the same as yours. Put another way, there is considerable variation in the anatomical structure of the human brain—variations that occur even as between identical twins, and the right and left hemispheres of the same brain (Weiss and Aldridge 2003). So the colored area of activation or inhibition may not correlate precisely with the area represented in the structural image.

This reinforces two points recently made by the neuroscientist and philosopher Adina Roskies. First, as Roskies (2008) notes, "the conventions of the brain image are representational translations of certain nonvisual properties related to neural activity in the brain"—that is, the comparative magnetic properties of oxygenated and deoxygenated hemoglobin; and second, "the choices that are made [in the construction of a brain image] are not visible in or recoverable from the image itself." In my view, functional brain images might properly be considered the product of "neuroscientific imaginaries" comprising the values, beliefs, and practices of neuro-imaging communities.[3] This is not to say that functional neuroimages are complete fabrications, entirely divorced from reality (in this case, brain functions). Rather, brain images are the product of decisions about scanning parameters and the criteria for statistical significance, in conjunction with acts of interpretation and representation that may tell us as much about the imagers as they do about those who are imaged (although we must look beyond the images to learn much about either of them).

This analysis is important because it highlights the chasm between the manner in which brain images are constructed and the way lay viewers in particular comprehend them. Many of us appear to assume that visually arresting brain images share the evidential characteristics and epistemic status of photographs (see Roskies 2008). Empirical support for this view has been provided by David McCabe and Alan Castel, who have shown that readers attribute greater scientific value to articles summarizing cognitive neuroscience research when those articles include brain images than they do when the articles include no image, a bar graph, or a topographical map of the brain. They found that this effect (albeit not large) was demonstrable regardless of whether the article included errors in reasoning (McCabe and Castel 2008). One explanation for this may be, as Roskies contends, that neuroimages are "inferentially distant from brain activity, yet they appear not to be." Put another way, she argues, brain images are "seemingly revelatory." In my view, this latter claim needs to be probed a little further. To whom is what revealed, and by what means? The answer to this question (one I next endeavor to provide) is vital to a nuanced understanding of some of the potential hazards of brain imaging in the arsenal of national security neuroscience.

In spite of their ubiquity, brain images are essentially meaningless to the uninitiated. When we look at these images in isolation, there is no "aha!" moment, no epiphany. The images require the explanation of experts. However, at the same time, the images also tend to reinforce the expert narrative. It is hard not to be impressed by the expert (real or apparent) who can guide us through the images—who can bring us to the point of revelation. And if the expert is compelling enough, it can be hard to look at the image again without relying heavily on the expert's interpretive framework. In this way, brain images and neuroscientific narratives rely on each other to work their persuasive and pervasive magic. This point may be illustrated by a notable recent analog regarding the use of a different kind of image in a national security context, the run-up to the invasion of Iraq in 2003. I have chosen this example not simply to illustrate how images and narratives work together, but also to demonstrate the potential hazards when science is purportedly deployed in a national security context.

In February 2003, the U.S. Secretary of State Colin Powell made a presentation to the United Nations Security Council that was intended to substantiate the U.S. argument that Iraq presented an imminent threat to regional and global security. Most people recall Powell's presentation—which he subsequently described as a permanent "blot" on his record (Weisman 2005)—because of a theatrical flourish. He held up a small vial of white powder while describing the threat that a similar quantity of weaponized anthrax might present. Fewer may now recall some of the other visual elements of his presentation—in particular, his reliance on image intelligence (or IMINT—the term used in "national securitese " to denote aerial and satellite photography). Powell showed time-sequenced satellite images of buildings whose functions were described in yellow text boxes—among them a chemical munitions bunker. He also showed computer-generated images of trucks and railway carriages that were described as "mobile production facilities for biological agents." Before displaying the images, Powell warned that we could not understand them, but that imaging experts had shed light where, otherwise, there would only be darkness:

> The photos that I am about to show you are sometimes hard for the average person to interpret, hard for me. The painstaking work of photo analysis takes experts with years and years of experience, poring for hours and hours over light tables. But as I show you these images, I will try to capture and explain what they mean, what they indicate, to our imagery specialists.

As the filmmaker Errol Morris subsequently noted in a *New York Times* blog:

> I don't know what these buildings were *really* used for. I don't know whether they were used for chemical weapons at one time, and then transformed into something relatively innocuous, in order to hide the reality of what was going on from weapons inspectors. But I *do* know that the yellow captions influence how we *see* the pictures. 'Chemical Munitions Bunker' is different from 'Empty Warehouse' which is different from 'International House of Pancakes.' The image remains the same but we *see* it differently. . . . (Morris 2008)

The interpretations of these photographs offered so convincingly by Powell have, of course, failed to stand up to scrutiny. Embarrassingly for the Bush administration, the Iraq Survey Group failed to find any evidence of an active program for the development of weapons of mass destruction. But, by that time, the images had done their work. According to Morris, the captions did the "heavy lifting," while the "pictures merely provide[d] the window dressing." But this account does not give full credit to the powerful way in which the images and the narrative work together, each reinforcing the other. The image, requiring interpretive expertise, validates the expert interpreter. And the act of interpretation gives meaning to the image that is otherwise incomprehensible to the lay viewer. In this deceptively attractive circularity (that I call the "image-expert bootstrap"), the image tells us how important the expert is, while the expert tells us how important the image is. Since nonexperts are hardly well placed to provide an alternative interpretation of the image, or to challenge the interpreter's expertise (or his expert interpretation), it is extremely hard for them to break the circle.

NATIONAL SECURITY NEUROSCIENCE: PERILOUS TO THE VULNERABLE?

An NRC (2008, p. 19) report on "Emerging Cognitive Neuroscience and Related Technologies" issued in August 2008 cautioned that when using neurophysiology data to determine psychological states, "it is important to recognize that acceptable levels of error depend on the differential consequences of a false positive or a misidentification." The authors expressed particular concern that "the neural correlates of deception could be construed as inconvertible evidence of deception and therefore (mistakenly) used as the sole evidence for making critical legal decisions with lasting consequences" and noted that "insufficient high-quality research using an appropriate research model and controls has been conducted on new modalities of credibility assessment to make a firm, data-driven decision on their accuracy" (NRC 2008, p. 34).

These concerns were exacerbated the following month when reports emerged that a woman named Aditi Sharma had been convicted in India of killing her former fiancé with arsenic and that her conviction relied principally on "evidence" from brain electrical oscillation signature testing (EEOS)—purportedly a variation on the electroencephalograph (EEG)-based technique of so-called brain fingerprinting developed and aggressively marketed by Lawrence Farwell (Giridharadas 2008). Many commentators have been understandably appalled that a judge would permit such a travesty of justice, noting hopefully that such an outcome should not be possible in the United States. Judicial gatekeeping may well prevent such an incident from recurring in Europe or North America (for a discussion of the admissibility of evidence derived from neurotechnology, see, e.g., Chapter 9). A more immediate concern in the United States, however, is the use of neurotechnologies in the national security context, where there is no judicial gatekeeper.

In particular, an fMRI "test result" could be used to label a detainee as a terrorist—a troubling prospect described more fully elsewhere (see Marks 2007b). One can imagine without great difficulty that an intelligence operative—seduced by colorful brain scans, pseudoscientific explanatory narrative, and media hype—might say "the fMRI picked him out as a liar"—or, worse still, "as a terrorist." It is not difficult to see how that could influence the subsequent treatment and interrogation of such a detainee. Labels such as the "worst of the worst" (used to described the detainees at Guantanamo Bay), and—a fortiori—specific labels such as the "mastermind of 9/11" and the "20th hijacker" (ascribed to detainees Khalid Sheikh Mohammed and Mohamed Al Qahtani, respectively) led to abusive and, in the most extreme cases, life-threatening treatments. But a label invoking a much-hyped and little-understood technology, such as fMRI, is likely to be all the more powerful. These concerns are highlighted by two factors.

The first is the aggressive marketing of these technologies, which may further exacerbate the undue confidence in and uncritical assessment of these technologies fueled by the specialist language and images I have already described. We have seen this in the case of so-called fMRI-based "lie detection"—the most conspicuous in the field being *No Lie MRI*, a corporation that claims accuracy rates of 90% or more and asserts that its proprietary technology is "insensitive to countermeasures" (No Lie MRI 2006). A similar point can be made about aerosolized oxytocin too, and the colorful marketing strategies described earlier. The second factor is the interest of the national security community in neuroscience. Although it has been difficult to corroborate the claim that fMRI has already been used—in conjunction with EEG— to screen suspected terrorists (see Marks 2007b), there is clearly interest in its use for this purpose (see Intelligence Studies Board 2006). In January 2007, the DoD's Polygraph Institute was renamed the Defense Academy for Credibility Assessment. There appear to have been two rationales for the new title—first, an expansion of the portfolio of the institute to encompass the use of newer technologies including but not limited to fMRI and, second, shift in the institute's mandate to address counterterrorism.

The NRC's (2009, 96) recent report "Opportunities in Neuroscience for Future Army Applications" acknowledged that there were challenges to the detection of deception due to individual variability and "cultural differences in attitudes to deception." Despite this, the report continues to pose the question "... is there some kind of monitoring that could detect if a subject being interrogated is responding in a 'contrary to truth' manner?" (NRC 2009, 96). This should serve as a potent reminder. Lysergic acid diethylamide (LSD) and, increasingly, the polygraph may seem consigned to the annals of the U.S. national security (for a discussion of some of this history, see Moreno 2006, 2012). But no one should doubt that the concerns that have motivated their use are alive and well in the neuroscience and national security communities.

THE WAY FORWARD

In the space available, I cannot attempt a comprehensive ethical critique of national security neuroscience—a project to which Canli and colleagues (2007), among others, make important contributions. Nor can I address all the potential perils of national security neuroscience. I have focused on the population I consider most vulnerable, detainees, and I have discussed two core examples in relation to that population. But there are clearly other vulnerable populations, most notably, soldiers (or warfighters, as they are now called), whose freedom is constrained not simply by their enlisted status, but also by their natural desire for recognition and advancement in the military. A fuller discussion of the potential hazards in relation to that population can be found in Moreno's book *Mind Wars* (Moreno 2012).

Moreno calls for the creation of a national advisory committee on neurosecurity, staffed by professionals who possess the relevant scientific, ethical, and legal expertise. The committee would be analogous to the National Science Advisory Board for Biosecurity established in 2004, which is administered by the National Institutes of Health (NIH) but advises all cabinet departments on how to minimize the misuse of biological research. Such a committee could certainly play a role in the oversight of national security neuroscience if it had the authority to monitor the misuse of neuroscience within—as well as outside—government. But in my view, the time has also come for a broader public debate about the legitimate nonclinical applications of neuroscience—one that takes into account the concerns addressed here and seeks to learn from the abuse of medicine, behavioral psychology, and polygraphy in the national security context.

If we are to have a meaningful discussion, we will have to ask ourselves some difficult questions that explore the kinds of neuroscience research that are being funded and address the broader context in which that research takes place. For example, many of us do not blink at the use of brain imaging to detect lies in detainees. In contrast, no one advocates the use of the technology during U.S. Senate hearings for nominees to the federal judiciary. Surely, the detection of an answer "contrary to truth" in such a context might be of some interest to the senators charged with the confirmation of federal judges!

I am, of course, not the first to make this kind of point. Commenting on the fMRI-based research to screen children for potentially violent behavior, the sociologist Troy Duster (2008) has noted that studies are "not designed to capture the kind of diffuse, anonymous violence reflected in the behavior of unscrupulous executives, traders, subprime lenders and so on." It is tempting to add to the list the arm's-length architects of torturous interrogation, and the legal and health professionals who—purportedly exercising their professional skill and judgment—approved or facilitated their use.

Duster continues with more than a hint of sarcasm:

> But for the sake of argument, suppose we could monitor children and determine that greater activity in the prefrontal cortex means that they are likely to exhibit violent behavior. Surely, then, we should scan preteens to intervene in the lives of potential Enron-style sociopaths before they gut the pensions of the elderly, right? Oops, I guess I have the wrong target group in mind. (Duster 2008, B4)

In this piece, Duster applies to recent neuroscience research some of the criticism that social activist Martin Nicolaus directed at sociologists and criminologists in 1968, roughly paraphrased as "you people have your eyes down and your hands up, while you should have your eyes up and your hands down." He explains:

> 'Eyes down' meant that almost all the research on deviance and crime was focused on the poor and their behavior, while 'hands up' meant that the support for such research was coming from the rich and powerful—from foundations, the government, and corporations. Conversely, of course, 'eyes up' meant turning one's research focus to the study of the pathological behavior of the elite and privileged, and 'hands down' meant giving more of a helping hand to the excluded, impoverished, and disenfranchised. (Duster 2008, B4–B5)

Although Duster does not address how neuroscience might help the disenfranchised, it is not difficult to conjure other uses of magnetic resonance imaging (MRI) technologies that might redound to their benefit (including perhaps the provision of government-funded diagnostic services to the uninsured and the underinsured or a broader use of the technology to rescue and rehabilitate patients trapped in minimally conscious states). But how we use these technologies, "the wicked problems"—to use the language of Frank Fischer (2003, 128–129)—that we call on them to address, should be a matter of informed debate in which experts engage with and listen closely to the public.

ADDENDUM: RESPONSES TO MY CRITICS

Neuroscience has its limits. Perhaps oxytocin can diminish neuroskepticism. But no psychoactive drug or neurotechnological innovation can enable me to engage meaningfully in around a thousand words with more than a dozen commentaries—embracing the work of Foucault (Thomsen 2010) and Spinoza (Bentwich 2010), in addition to the latest scholarship in neuroscience and neuroethics, some of which appeared after the text above was originally submitted for publication (e.g., Illes et al. 2010). Although my responses are shaped by these thoughtful commentaries, the latter raise many important points to which I cannot do justice here.

NEUROSKEPTICISM

Let me begin by saying something about my view of neuroskepticism. It is certainly not an ideology, as one of my harshest critics has alleged (Keane 2010). Nor is it meant to be construed as some kind of ethical theory. I am even hesitant to elevate it to the status of "method" as the members of the Stanford Interdisciplinary Group in Neuroscience and Law (SIGNAL) suggest (Lowenberg et al. 2010). Rather, neuroskepticism is a "perspective." One might also call it a sensibility or an orientation. I am glad to acknowledge, as the SIGNAL authors have suggested, that I am also "neuroconcerned: concerned about whether even scientifically reliable neuroscience could be used to cause harm or to further harmful ends" (Lowenberg et al. 2010). This acknowledgment should help address the views of some commentators who asserted that I placed too great an emphasis on bad science as the source of concern (Strous 2010; Thomsen 2010). There are clearly serious ethical issues, whether the science is sound or, as Fisher (2010) puts it, "absolute junk."

(NEURO)ETHICS AND (NEURO)SCIENCE

My purpose is not to provide a comprehensive ethical assessment of neuroscience and national security. Rather, it is to offer an explanatory account that might inform neuroethical debate. Such a debate must obviously address the legitimacy of the means by which and the ends for which neuroscience is applied (Lowenberg et al. 2010).

However, I do not (as Benanti [2010] has suggested) consider that "ethical questions remain, in a certain way, external to neuroscience, because they are related only to practical use of neuroscience." Ethics must infuse science from inception—from the first flicker of an idea—through testing and development, to myriad actual and potential applications. As Schienke and colleagues (2009) argue, a comprehensive account of scientific research ethics needs to address three distinct but related spheres of concern that those authors describe as "procedural ethics" (i.e. the responsible conduct of research, which includes issues of falsification, fabrication, plagiarism, care of human and animal subjects, and conflicts of interest), "intrinsic ethics" (comprising issues internal to or embedded in the production of a given inquiry or mode of analysis), and "extrinsic ethics" (i.e., issues arising from the application of science to policy or from the impact of science and technology on society).

I agree that the slippage from military into civilian applications of neurotechnologies presents ethical issues (see Chapter 10; Marchant and Gulley 2010). But, of course, applications that remain within the national security domain still raise important ethical concerns. For example, as Kirmayer et al. (2010) have noted, there is a risk that neuroscience can become "a screen on which to project our prejudices and stereotypes" and that it may shift the focus away from the origins of terrorism as "social and political processes." This has implications for national security too. As John Horgan (2008) has argued, recognition of these social processes is vital to our understanding of terrorism and to the crafting of effective counterterrorism policies.

Ethics and Neurohype

The discussion above does not focus on the problem of "neurohype" and the related ethical responsibilities of the scientific community. However, Rippon and Senior (2010) have taken "… umbrage with the impression [my work] gives that the neuroscientific community is not 'policing its own house'"—a phrase I do not employ. At the same time, Rippon and Senior acknowledge that "neuroscientists need to be more aware of the potential dangers of 'overselling our wares'" and that "the brain imaging community needs to be more vigorous in communicating its concerns to the public at large and, more importantly to the policy makers and funders." I am grateful for these acknowledgments and would commend to Rippon and Senior the discussion of "neurohype" in Caulfield et al. (2010; and the works cited therein). I would also endorse Nagel's (2010) "plea for responsibility in science"—a plea that recognizes that some scientists live up to such responsibilities better than others and, more importantly, that there are systemic factors that can promote problematic behaviors.

Emotion, Cognitive Biases, and Counterterrorism

Bentwich (2010), drawing on the work of Spinoza, argues that I tend to pay too little attention to the "pernicious linkage among fear, superstition, and prejudice" in the national security context. To the contrary, I have done so at substantial length elsewhere. In Marks (2006), I argue that our emotional responses to counterterrorism and associated cognitive biases have an impact not just on individual behavior but also (through a variety of social mechanisms) on counterterrorism policy and practice. Resulting policies tend to violate human rights and often fail to address real threats: see Marks (2006), where my debt to Spinoza is mediated by the work of neuroscientist Antonio Damasio, upon which I draw (Damasio 2003).

ETHICS AND HUMAN RIGHTS

Bentwich (2010) posits that the answers to these problems are "not found in the jurisdiction of neuroethics, but rather by reasserting human rights and constitutional civil liberties." In other work on neuroscience and national security, I too have emphasized the importance of human rights (see, e.g., Marks 2007b). More broadly, I have also advocated human rights impact assessments of counterterrorism policies (Marks 2006) and used international human rights law to critique health professionals who are complicit in detainee abuses (Marks 2007a). However, I do not believe that professional ethics is an entirely autonomous enterprise. In other recent work, I elaborate more fully on the relationship between human rights and professional ethics (Marks 2012). I also note that the American Association for the Advancement of Science (AAAS) and Human Rights Program recognizes the centrality of human rights to the ethics of science, and its coalition explores how greater attention to human rights might also result in improvements in scientific process and practice (AAAS 2010).

THE PROSPECTS FOR AND LIMITS OF FURTHER DISCOURSE

I am heartened by the lively debate that my work has provoked. I encourage my colleagues to work with me to take this discussion beyond the pages of this volume and into the larger public domain. I acknowledge, as Lowenberg and colleagues (2010) and Giordano (2010b) point out, that national security considerations may place limits on what may be discussed in public. But the presumption should be in favor of openness, and there is—in any event—more than enough information in the public domain about which to conduct meaningful discussions.

In these discussions, the neuroscience and neuroethics communities must be frank with the public about the potential, the limitations, and the perils of neuroscience. We should empower the public to challenge decisions regarding the development and application of neuroscience (see Dickson 2000) and engage with them in figuring out the road ahead. The educational and communication challenges in this exercise should not be underestimated. But we should rise to them. If we fail to reconsider and refocus the gaze of neuroscience, we risk abandoning or—worse still—imperiling the vulnerable. And if we do that, tomorrow's historians and science studies scholars will, rightly, not look kindly on us.

NOTES

1. This chapter is based on a plenary lecture of the same title delivered at the Novel Tech Ethics Conference in Halifax, Nova Scotia, in September 2009. The author is indebted to the conference organizers—in particular, Jocelyn Downie and Francoise Baylis—for extending this invitation and providing him with the opportunity to develop his views.
2. Documents on file with author.
3. The term is my own, but it draws some inspiration from the notion of "technoscientific imaginaries." See Marcus (2005).

REFERENCES

American Association for the Advancement of Science (AAAS). 2010. Science and human rights coalition. Accessed on February 9, 2010, http://shr.aaas.org/coalition/index.shtml.

Baumgartner, T., M. Heinrichs, A. Vonlanthen, U. Fischbacher, and E. Fehr. 2008. "Oxytocin shapes the neural circuitry of trust and trust adaptation in humans." *Neuron* 58(4): 639–650.

Benanti, P. 2010. "From neuroskepticism to neuroethics: Role morality in neuroscience that becomes neurotechnology." *American Journal of Bioethics: Neuroscience* 1(2):39–40.

Bennett, R. 2007. "Endorsement by seminar military interrogators." *Peace and Conflict: Journal of Peace Psychology* 13(4):391–392.

Bentwich, M. 2010. "Is it really about neuroethics? On national security, fear, and superstition." *American Journal of Bioethics Neuroscience* 1(2):30–32.

Bybee, J. 2002. Memorandum for Alberto Gonzales. *Washington Post*, August 1. Accessed on January 4, 2010, http://www.washingtonpost.com/wp-srv/politics/documents/cheney/torture_memo_aug2002.pdf.

Canli, T., S. Brandon, and W. Casebeer. 2007. "Neuroethics and national security." *American Journal of Bioethics* 7(5):3–13.

Caulfield, T., C. Rachul, and A. Zarzeczny. 2010. "Neurohype and the name game: Who's to blame?" *American Journal of Bioethics: Neuroscience* 1(2):13–15.

Central Intelligence Agency Inspector General. 2004. Counterterrorism, detention and interrogation activities (September 2001–October 2003). http://media.washingtonpost.com/wp-srv/nation/documents/cia_report.pdf.

Choudhury, S., S.K. Nagel, and J. Slaby. 2009. "Critical neuroscience: Linking neuroscience and society through critical practice." *BioSocieties* 4(1):61–77.

Committee on Armed Services of the U.S. Senate (SASC). 2008. Inquiry into the treatment of detainees in US custody. November 20, http://documents.nytimes.com/report-by-the-senatearmed-services-committee-on-detainee-treatment. Accessed December 21, 2009.

Damasio, A. 2003 *Looking for Spinoza: Joy, Sorrow and the Feeling Brain*. Orlando, FL: Harcourt Press.

Dickson, D. 2000. "The science and its public: The need for a third way." *Social Studies of Science* 30(6):917–923.

Dumit, J. 2003. *Picturing Personhood: Brain Scans and Biomedical Identity*. Princeton, NJ: Princeton University Press.

Duster, T. 2008. "What were you thinking? The ethical hazards of brain imaging studies." *Chronicle Review*, October 10, B4–B5.

Fink, S. 2009. Tortured profession: Psychologists warned of abusive interrogations, then helped craft them. *Propublica*, May 5, http://www.propublica.org/article/tortured-professionpsychologists-warned-of-abusive-interrogations-505. Accessed December 21, 2009.

Fischer, F. 2003. *Citizens, Experts, and the Environment: The Politics of Local Knowledge*. Durham, NC: Duke University Press.

Fisher, C.E. 2010. "Brain stimulation and national security: Considering the narratives of neuromodulation." *American Journal of Bioethics: Neuroscience* 1(2):22–24.

Follmann, M. 2003. Did the Saudis know about 9/11? *Salon*, October 18. Accessed on January 3, 2010, http://dir.salon.com/story/news/feature/2003/10/18/saudis/print.html.

Giordano, J. 2010a. "Neuroethics-sharp ideas about neurohormones, love, and drugs." *Oxford Medical Gazette* 60(1):26–28.

Giordano, J. 2010b. "Neuroscience, neurotechnology and national security: The need for preparedness and an ethics of responsible action." *American Journal of Bioethics: Neuroscience* 1(2):35–36.

Giridharadas, A. 2008. India's novel use of brain scans in courts is debated. *New York Times*, September 15. Accessed on January 4, 2010, http://www.nytimes.com/2008/09/15/world/asia/15brainscan.html?_r = 3&pagewanted = print.

Gudjonsson, G. 2003. *The Psychology of Interrogations and Confessions: A Handbook*. Chichester: John Wiley & Sons.

Henson, R. 2005. "What can functional imaging tell the experimental psychologist?" *Quarterly Journal of Experimental Psychology* 58A(2):193–233.

Horgan, J. 2008. "From profiles to *Pathways* and roots to *Routes*: Perspectives from psychology on radicalization into terrorism." *The ANNALS of the American Academy of Political and Social Science* 618(1):80–94.

Illes, J., M.A. Moser, J.B. McCormick, E. Racine, S. Blakeslee, A. Caplan, E.C. Hayden et al. 2010. "Neurotalk: Improving the communication of neuroscience research." *Nature Reviews Neuroscience* 11(1):51–69.

Intelligence Studies Board. 2006. *Educing Information: Interrogation: Science and Art Foundations for the Future (Phase I Report)*. Washington, DC: National Defense Intelligence College.

International Committee of the Red Cross. 2007. *Report on the Treatment of Fourteen "High-Value Detainees" in C.I.A. Custody*. Washington, DC: International Committee of the Red Cross.

Joyce, K. 2008. *Magnetic Appeal: MRI and the Myth of Transparency*. Ithaca, NY: Cornell University Press.

Kane, D. 2004. How your brain handles love and pain. MSNBC, February 29. http://www.msnbc.msn.com/id/4313263/ns/technologyandscience-science/.

Keane, J. 2010. "Neuroskeptic or neuro-ideologue?" *American Journal of Bioethics: Neuroscience* 1(2):33–34.

Kirmayer, L., S. Choudhury, and I. Gold. 2010. "From brain image to the bush doctrine: Critical neuroscience and the political uses of neurotechnology." *American Journal of Bioethics: Neuroscience* 1(2):17–19.

Kosfeld, M., M. Heinrichs, P.J. Zak, U. Fischbacker, and E. Fehr. 2005. "Oxytocin increases trust in humans." *Nature*, 435(7042):673–676.

Lee, H., A.H. Macbeth, J.H. Paganil, and W.S. Young III. 2009. "Oxytocin: The great facilitator of life." *Progress in Neurobiology* 88(2):127–151.

Lowenberg, K., B.M. Simon, A. Burns, L. Greismann, J.M. Halbleib, J.M.G. Persad, G.D.L.M. Preston, H. Rhodes, and E.R. Murphy. 2010. "Misuse made plain: Evaluating concerns about neuroscience in national security." *American Journal of Bioethics Neuroscience*, 1(2): 15–17.

Marchant, G. and L. Gulley. 2010. "National security neuroscience and the reverse dual-use dilemma." *American Journal of Bioethics: Neuroscience* 1(2):20–22.

Marcus, G. 2005. *Technoscientific Imaginaries: Conversations, Profiles, and Memoirs*. Chicago, IL: University of Chicago Press.

Margulies, J. 2006. The more subtle kind of torment. *Washington Post*, October 2, A19. http://www.washingtonpost.com/wp-dyn/content/article/2006/10/01/AR2006100100873.html. Accessed December 21, 2009.

Marks, J.H. 2006. "9/11 + 3/11 + 7/7 = ? What counts in counterterrorism." *Columbia Human Rights Law Review* 37(3):559–626.

Marks, J.H. 2007a. "Doctors as pawns? Law and medical ethics at Guantanamo Bay." *Seton Hall Law Review* 37(3):711–731.

Marks, J.H. 2007b. "Interrogational neuroimaging in counterterrorism: A "No-brainer" or a human rights hazard?" *American Journal of Law & Medicine* 33:483–500.

Marks, J.H. 2010. A neuroskeptic's guide to neuroethics and national security. *American Journal of Bioethics: Neuroscience* 1(2):4–12.

Marks, J.H. 2012. "Toward a unified theory of professional ethics and human rights." *Michigan Journal of International Law* 33:215–263.

Mayer, J. 2008. *The Dark Side: The Inside Story of How the War on Terror Turned into a War on American Ideals*. New York: Doubleday.

McCabe, D. and A. Castel. 2008. "Seeing is believing: The effect of brain images on judgments of scientific reasoning." *Cognition* 107(1):343–352.

Moreno, J.D. 2006. *Mind Wars: Brain Research and National Defense*. Washington, DC: Dana Press.

Moreno, J.D. 2012. *Mind Wars: Brain Science and the Military in the 21st Century*. New York: Bellevue Literary Press.

Morris, E. 2008. Photography as a weapon. *New York Times*, August 11. Accessed on September 20, 2009, http://morris.blogs.nytimes.com/2008/08/11/photography-as-a-weapon/?ref = opinion.

Nagel, S. 2010. "Critical perspective on dual-use technologies and a plea for responsibility in science." *American Journal of Bioethics: Neuroscience* 1(2):27–28.

National Research Council (NRC). 2003. *The Polygraph and Lie Detection*. Washington, DC: National Academies Press.

National Research Council. 2008. *Emerging Cognitive Neuroscience and Related Technologies*. Washington, DC: National Academies Press.

National Research Council. 2009. *Opportunities in Neuroscience for Future Army Applications*. Washington, DC: National Academies Press.

No Lie MRI. 2006. Product Overview. Accessed on September 20, 2009, http://noliemri.com/products/Overview.htm.

Owen, A.M. and M.R. Coleman. 2008. "Functional neuroimaging of the vegetative state." *Nature Reviews Neuroscience* 9(3):235–243.

Rippon, G.C. and Senior. 2010. "Neuroscience has no place in national security." *American Journal of Bioethics Neuroscience* 1(2):37–38.

Roskies, A. 2008. "Neuroimaging and inferential distance." *Neuroethics* 1(1):19–30.

Schienke, E.W., N. Tuana, and D.A. Brown. 2009. "The role of the NSF broader impacts criterion in enhancing research ethics pedagogy." *Social Epistemology* 23(3/4):317–336.

Schiff, N.D., J.T. Giacino, K. Kalmar, J.D. Victor, K. Baker, M. Gerber, B. Fritz et al. 2007. "Behavioural improvements with thalamic stimulation after severe traumatic brain injury." *Nature* 448(7153):600–603.

Shane, S. 2009. Interrogation Inc., 2 U.S. Architects of harsh tactics in 9/11's Wake. *New York Times*, August 12. Accessed on September 20, 2009, http://www.nytimes.com/2009/08/12/us/12psychs.html?_r = 5&hp = &pagewanted = print.

Soufan, A. 2009. What torture never told us. *New York Times*, September 5. Accessed on December 21, 2009, http://www.nytimes.com/2009/09/06/opinion/06soufan.html?_r=2.

Strous, R. 2010. Response to a neuroskeptic: Neuroscientific endeavor versus clinical boundary violation. *American Journal of Bioethics: Neuroscience* 12:24–26.

Thomsen, K. 2010. "A Foucauldian analysis of 'A neuroskeptic's guide to neuroethics and national security.'" *American Journal of Bioethics: Neuroscience* 1(2):29–30.

Trout, J.D. 2008. "Seduction without cause: Uncovering explanatory neurophilia." *Trends in Cognitive Sciences* 12(8):281–282.

Warrick, J. 2008. Detainees allege being drugged, questioned. *Washington Post*, April 22. Accessed on December 21, 2009, http://www.washingtonpost.com/wp-dyn/content/article/2008/04/21/AR2008042103399_pf.html.

Weisberg, D.S., F.C. Keil, J. Goodstein, E. Rawson, and J.R. Gray. 2008. "The seductive allure of neuroscience explanations." *Journal of Cognitive Neuroscience* 20(3):470–477.

Weisman, S.R. 2005. Powell calls his U.N. speech a lasting blot on his record. *New York Times*, September 9. Accessed on January 4, 2010, http://www.nytimes.com/2005/09/09/politics/09powell.html.

Weiss, K. and K. Aldridge. 2003. "What stamps the wrinkle deeper on the brow?" *Evolutionary Anthropology* 12(5):205–210.

Wolpe, P., K.R. Foster, and D.D. Langleben. 2005. "Emerging technologies for lie-detection: Promises and perils." *American Journal of Bioethics: Neuroscience* 10(10):40–48.

Yoo, J. 2003. Memorandum for William Haynes. *American Civil Liberties Union*, March 14. Accessed on January 4, 2010, http://www.aclu.org/files/pdfs/safefree/yooarmytorture-memo.pdf.

12 Prison Camp or "Prison Clinic?"
Biopolitics, Neuroethics, and National Security[*]

Kyle Thomsen

CONTENTS

INTRODUCTION

Within the larger neuroethical debate regarding functional magnetic resonance imaging (fMRI) scanning, the nonclinical application of medical technology is one of the more intriguing issues. One such application focuses on the use of this medical tool in a national security context. Authors such as Jonathan Marks (2010; see also Chapter 11) and Apoorva Mandavilli (2006) have voiced clear concern regarding the nonmedical usage of this technology. After all, medical technology is designed to heal by directly treating an illness, by assisting medical staff in diagnosing a particular problem, or by furthering medical research. This is the only truly legitimate way to use medical technology, one could claim. To use fMRI technology outside of the clinical context stands in opposition to the intended use of this neuroscientific tool. Those such as Marks and Mandavilli can strengthen this claim by pointing to the fact that in a national security context, the targets of these scans are vulnerable populations whose rights have already been violated by the State. fMRI technology

[*] This chapter is adapted and expanded with permission from Thomsen, K. 2010. *AJOB Neuroscience* 1(2):29–30.

becomes a tool of oppression, rather than a tool of healing. As such it is illegitimate to use neuroimaging for the purposes of national security.

In what follows, I will challenge some of the background assumptions made by the likes of Marks and Mandavilli. This is not to say that I disagree with them insofar as they say nonclinical applications of brain scanning are unethical in certain contexts, such as the environment of national security. Far from it. It seems quite reasonable to cast a dubious eye on the fMRI as a lie detector, and the systematic violation of human rights associated with some aspects of national security are clearly matters of ethical concern. What I will challenge is the background assumption that we can call fMRI scans in the national security context "nonclinical" in the first place, and the connected claim that these uses of fMRI technology are necessarily illegitimate from a normative point of view. On the contrary, certain interpretations of contemporary power structures seem to indicate that this use of neuroimaging has a profoundly clinical character, and as such may be medically legitimate (in a nonnormative sense). Michel Foucault's account of biopower and the clinical character of the State will serve as a launching point for this discussion.

Throughout this work, I will focus on Guantanamo Bay as a concrete national security space (Thomsen 2010).[1] I chose Guantanamo for several reasons. Over time it has become synonymous with the darker aspects of post 9/11 national security, given the compromised rights status of detainees and the violation of anything that resembles due process. It is also a space where the issue of lie detection is quite prevalent. One of the functions of this prison camp is to extract information from high-profile "enemy combatants." Lie detectors are a favorite tool of interrogators, in spite of their frequent ineffectiveness. This chapter is structured as follows. First I will introduce fMRI and show how it relates to lie detection in a national security context. Next, I will describe an objection (derived from Jonathan Marks) to the use of fMRI as a lie detector. Following this description, I will offer a Foucauldian analysis of State power and show how this analysis complicates the claim that fMRI lie detecting technology is illegitimate and nonclinical. I will conclude with a call to action that takes this analysis into account.

fMRI LIE DETECTORS AND NATIONAL SECURITY

In the interest of framing our larger discussion, it is necessary to offer a brief description of what fMRI technology measures and how these measurements affect the national security landscape. fMRI gauges brain activity by measuring blood oxidation levels in the brain of a subject. These data are then often displayed as an "activation map," a three-dimensional representation of a brain with the increased blood oxidation levels (Devlin et al. 2005). An increase in brain activity in a particular area results in an increased need for oxygen. By measuring this increase in oxygen, fMRI technology is able to measure which parts of the brain are more active at any given time. For example, let us pretend that I am currently receiving an fMRI scan. While in the scanner, a researcher jabs my foot with a sharp instrument, which in turn produces a significant amount of pain. The part of my brain associated with pain reception would see a spike in activity, and as such an increased level of blood oxidation. The scanner would measure this increase, which a researcher could then use to create a map of my brain with the pain receptors "lit up" so to speak. This technology has a number of potential uses, ranging from medical diagnosis to increasing our understanding of memory formation. However, there is one potential use of fMRI technology which will drive our investigation. This is the attempt to use fMRI technology as a lie detector.

It is widely accepted that polygraph lie detecting technology, currently used by a variety of public and private groups, is not a precise way to measure the veracity of any given statement (National Research Council 2003, 61).[2] Some enterprising companies have seized on this fact and offer what they claim to be a far more accurate alternative.[3] By utilizing fMRI technology, these companies claim that they are able to determine when an individual is lying by observing blood oxidation levels in the brain. If you ask a subject a question to which they lie in response, the "lying centers" of the brain lights up and those conducting a scan can act accordingly. These scans bypass the problems of a polygraph by going straight to the brain, so the companies claim. Marketing efforts target a number of sectors. Of particular interest, are the campaign efforts focusing on government agencies and departments, such as the U.S. Department of Defense (DoD) and the U.S. Department of Homeland Security (DHS) (No Lie MRI Inc. 2012). These two departments, among others, already utilize polygraph tests. The sales pitch claims that this more accurate tool would greatly benefit our military and serve as a more effective tool for ensuring national security. After all, when gathering information which is critical to national security, one would want this information to be as accurate as possible. I will now apply this claim to an imaginary test case in an effort to demonstrate the national security application of fMRI technology.

Let us say that during the course of U.S. operations in Afghanistan, a special forces team captures a high-level insurgent named X. X is designated as an enemy combatant and transported to Guantanamo Bay for interrogation and eventual prosecution. While X is under detention, intelligence reports arise which indicate that X has detailed knowledge of an impending attack on U.S. soil. If the U.S. government could acquire this knowledge, the attack would be thwarted and lives would be saved. Throughout the course of X's interrogation, agents of the U.S. government utilize fMRI-based lie detecting technology. At first X gives false answers to the questions regarding time and location of the attack. These answers result in measurable increases in the blood oxygen level (BOLD) signed in "lying centers" of X's brain, and the interrogators continue their efforts. Eventually, X supplies information which does not trigger lying centers of his brain, and this presumably true information is forwarded on to the Department of Homeland Security in an effort to thwart the attack.

This imaginary example clearly demonstrates the use of fMRI lie detecting technology in a national security context, and it allows us to narrow our focus onto one of the State's most likely applications of neuroscientific advancement. The "high-profile" detainees at Guantanamo Bay, such as Khalid Sheikh Mohammed, presumably have information which countries such as the United States consider vital to improving national security. Guantanamo is largely out of the public eye, making it an ideal testing ground for new interrogation techniques. Couple this with the compromised rights status of the detainees, and you have the perfect test case for neuroscientific technology applied in the name of national security. Two questions arise. First, has the United States used fMRI (or similar) technology at Guantanamo? This is a difficult question to answer given the secretive nature of this prison. An easier question to answer is whether the United States is interested in using fMRI (and or similar, if not more advanced) technology in a national security context. It seems that such interest exists. The United States has already demonstrated that it will resort to extreme (even internationally illegal) measures in an effort to gain access to "high-profile" information. Polygraphs have been a standard tool for the DoD and DHS, in spite of the fact they produce unreliable results. New and potentially more effective techniques for obtaining information would reasonably attract the eye of the DoD and DHS. The National Research Council (2009, 89), an organization which serves as an advising body for the federal government, lists fMRI as one of several "high-priority opportunities for army investment in neuroscience technologies." Given the supposed promise of fMRI lie detectors, this is profoundly unsurprising. In short, interest likely exists in the DoD and DHS for the use of brain scanning technologies as advanced lie detectors. Now it is time to briefly discuss the underlying ethical issues regarding this usage.

AN ILLEGITIMATE NONCLINICAL USAGE OF fMRI?

Depending on how we answer the question regarding the current effectiveness or potential future effectiveness of fMRI technology as a lie detector, we run into a set of varying ethical issues. If the technology is currently ineffective or unreliable, then our focus falls on those who advocate for fMRI toward such ends and how they are manipulating the powers-that-be with pseudoscientific nonsense. Let me be clear. Given the current research climate, it seems that fMRI does not live up to the ideal of an undefeatable lie detector (Klein 2010a).[4] They are, in short, unreliable. Given this fact, there is no upside that can serve as a justification for the use of this technology. Let us leave aside the issue of utilizing medical technology on vulnerable populations for nonmedical reasons. If the technology is ineffective, then those who attempt to market fMRI for use as a lie detector are charlatans. Either they knowingly suppress evidence in order to further their own interests, or they are making truth claims regarding the effectiveness of this technology without sufficient evidence. Both are unethical, though likely to different extents. In addition, given the ramifications of these scans on high-profile prisoners, and the actions a government such as the United States might take based on this false information, we can see that the potential consequences of unreliable tests may be dire. Innocent individuals may be arrested and supposed combatant targets may be struck. People could lose their livelihood or their very lives based upon inaccurate readings, especially if those using this technology have a large amount of faith in the accuracy of the scans. Compare these two significant ethical issues with the lack of any real upside, and the use of fMRI technology as an advanced lie detector is flatly unjustified.

However, if we set aside the issue of accuracy, we encounter an interesting set of ethical and political issues. I will focus on two of these problems in the pages that follow. The first concerns the use of medical technology for a nonmedical purpose. fMRI technology was created as a tool for use in the medical environment. It is a tool that can potentially diagnose neurological disorders, aid in psychological in research, and so on. What has concerned some individuals, such as Mandavilli and others, is the prospect of this tool being pulled from the hospitals and put into prisons (Mandavilli 2006). Something designed for diagnostic use in a healing context is being used for a different purpose. Jonathan Marks sums this point up quite nicely regarding the use of fMRI in national security situations:

> ...the time has come for a broader public debate about the *legitimate nonclinical applications of neuroscience* (emphasis mine)-... that takes into account the concerns addressed here and seeks to learn from the abuse of medicine, behavioral psychology, and polygraphy in the national security context. (Marks 2010, 10; see also Chapter 11)

I will return to this claim further along, specifically regarding how we should understand the use of fMRI in Guantanamo as illegitimate and nonclinical. For now it is sufficient to point out that the nonmedical use of medical technology raises some potential problems in the field of health care ethics, in general, and neuroethics, in particular.

The second related problem concerns the use of this technology on vulnerable populations. Vulnerable groups, in general, already live under conditions of compromised protection and rights recognition by the State. Prisoners at Guantanamo Bay certainly fall into this category. Many have been shuffled around in secret Central Intelligence Agency (CIA) prisons before they ended up in this "enemy combatant" prison camp, where they are held indefinitely without formal charges. Widespread allegations of prisoner abuse and torture surround the camp. The human dignity of these individuals has been utterly compromised. To be sure, this camp is a stain on the fabric of human rights. What issues arise when we utilize fMRI technology as an interrogation tool in these environments? How are these prisoners, whose rights have been compromised, affected in this process?

We need to understand what would drive the State to utilize this sort of medical technology and why the subjects are members of a vulnerable population whose status as rights-bearers is compromised. Fortunately, this is far from uncharted territory. While the work of Michel Foucault is not widely characterized as relevant to neuroethics or ethics at all for that matter (by exception and in neuroethical focus, see Anderson 2011; Anderson et al. 2012; Giordano 2013), his late discussions regarding power structures in society can provide a unique insight into why the State would take on a quasi-clinical character. Foucault's descriptive account of power circulation, specifically regarding what he refers to as "biopower," leads to some disturbing revelations. One could read individuals such as Marks and Mandavilli as defenders of the distinction between the clinical and the political, as astute commentators regarding the danger of blurring the lines between the two. Following my analysis, I hope to show that these lines are already blurred, if not destroyed. The fox is already in the henhouse, so to speak, and if we are to combat this problem, we must view it as an ongoing and deeply entrenched issue in our political system. Marks claims toward the end of his own analysis that we must ask ourselves some difficult questions if we are to make any significant headway on the issue of brain imaging in the prison camp

(Marks 2010, 11; see also Chapter 11). I agree. My point is that before we ask "how can we stop this," we must answer the question "why is the State driven to do this"? It is the latter question that motivates me in this work, and Foucault's descriptive (as opposed to normative) account of State power provides a unique insight into the use of fMRI technology as it is used on vulnerable populations.[5]

BIOPOWER AND THE MEDICAL CHARACTER OF THE STATE

Put briefly, Foucault's account is as follows. During the nineteenth century, the State began to exercise power in a new way, which Foucault refers to as "biopower" or "biopolitics." Biopolitical systems arose, Foucault claims, in response to the Hobbesian way in which autocrats ruled by the sword. The works of the eighteenth-century jurists make this transformation clearer. After all, the jurists claimed, did we not enter into contract with the sovereign in order to protect our lives? "Mustn't life remain outside the contract to the extent that it was the first, initial, and foundational reason for the contract itself?" (Foucault 1997, 241). It is this problematizing of life and death that led to the rise of biopower. Whereas the already existing disciplinary power focused on the individual body in a number of ways (surveillance, exercise, drill, etc.), the newly emerging power focused on "man-as-species" (Foucault 1997, 242).[6] The former reduced the population into numerous individual bodies as objects of control, while the latter surveyed a global mass affected by processes of biology (life, birth, health, illness, reproduction, death). As Foucault (1997, 243) states, "we have, at the end of that century, the emergence of something that is no longer anatomo-politics of the human body, but what I would call a 'biopolitics' of the human race."

Let us pause momentarily and take stock of this claim. The problem of how the State and its subjects related to life and death in the eighteenth century highlights a shift in how the State came to utilize life and death in order to control the population. Disciplinary power, which focuses on the individual, is now complemented by biopower, which focuses on the population as a whole. While I will clarify this distinction in the coming pages, we can already see a foundational quality of biopower emerging. Biopower is, as stated above, focused primarily on the health of the population as a whole. This new function of State power focuses all of its efforts on tracking and controlling the health of a population. It is sufficient for now to highlight the change that is taking place; the relation between life and death as utilized by the State shifts in a way that favors the former.

Let us continue with a brief outline of several areas of biopolitical concern. The first of these are reproduction and endemics. Birth rate, mortality rate, and longevity became the first objects of biopolitical concern in the later half of the eighteenth century (Foucault 1997, 243). Next, the State engaged in the creation of public hygiene campaigns and attempts to centralize medical knowledge in an effort to stave off illnesses prevalent in the population. State agencies compiled this knowledge and disseminated it through a network of clinics, hospitals, and public service programs (Foucault 1997, 244). Biopolitical intervention continued into the realms of health insurance, correction of medical problems caused by industrialization, and alteration of the environment in order to create sanitary living spaces.[7]

Three distinct aspects of biopower emerge in conjunction with these methods of intervention. The first of these is biopower's object. While previous power structures were concerned with the individual subject, biopower views the population itself as a political, scientific, and biological problem that must be addressed (Foucault 1997, 245). The second concerns the nature of the phenomena taken into consideration by biopower. Issues of health in general are quite unpredictable on an individual basis, but when biopower examines them on a large scale, predictable trends begin to emerge. These are "serial phenomena," or phenomena that occur over a certain period of time and must be studied over that period of time (Foucault 1997, 246). The third and final aspect of biopower addresses the mechanisms used in biopolitical intervention and the goal to which they aim. Examples of these mechanisms include the use of forecasts and statistical estimates regarding the overall health of the public. But for what purpose is biopower utilizing these mechanisms? The answer is simple, to establish an equilibrium, a statistical norm, in the various areas of a population's health (normalized birth rates, mortality rates, etc.). As Foucault (1997, 246) states, "in a word, security mechanisms have to be installed around the random element inherent in a population of living beings so as to optimize a state of life." When something deviates from the established statistical norm, biopower utilizes its various mechanisms to return to the established norm.[8] Shortly, we shall see this normalizing aspect assert itself in a heretofore unmentioned way.

STATE RACISM AND NEUROETHICAL APPLICATIONS

Foucault's account is challenged by a pressing question regarding this life-centered system. If biopower is focused on the preservation of life, then how do we explain State-sponsored killings, such as executions and military action? Foucault accounts for this by referring to racism's intervention in the State. While Foucault (1997, 254) is not claiming that biopower invented racism, he does claim that a new type of racism emerges with biopower.[9] But how does racism (in this instance) function? It functions by dividing society into two parts: those that must live and those that must die. The State performs this division under the principle that in order for society to exist, the inferior must die out (Foucault 1997, 255).[10] It is helpful to think of this in terms of an "us vs. them" mentality. If the others do not die, they will destroy society. In the biopolitical system, "killing or the imperative to kill is acceptable only if it results not in a victory over political adversaries, but in the elimination of the biological threat to the improvement of the species or race" (Foucault 1997, 256). A biopolitical system kills in order to wipe out "degenerate" elements. Killing a "biological threat" becomes no different than excising a tumor.

Let us continue with this theme and turn to two key components of Foucault's 1976 lecture, insofar as they relate to neuroethics and national security. The discussion above indicates the first theme, the broad concept of health care. One cannot describe biopower without some reference to the maintenance of a population's health. Since I have already discussed this theme, I will leave it aside and move forward. The second theme in Foucault's lecture is a thoroughly political one. While biopolitical intervention frequently takes the form of health care, it is always a tool utilized by the State. At bottom, biopower is a political rationality whose aim is to control all that pertains to the sphere of life. The means to do this are the tracking of societal health norms and the management of said norms.[11] Foucault's account always places such events as the formation of centralized bodies of medical knowledge, and the rise of public sanitation, in the realm of a state's power structures. It is telling that the majority of the scholarship surrounding Foucault's account, in particular the works of Giorgio Agamben, focuses most intensely on the political applications of biopolitical intervention (Agamben 1998, 2005).

As I have hinted above, these two themes are not isolated from one another. On the contrary, the concerns of health care and the concerns of the State become deeply intertwined. On Foucault's (1997, p. 246) account, the State concerns itself primarily with optimizing a "state of life." In order to accomplish this goal, the State establishes a base line and utilizes the numerous instantiations of health care to maintain the base line. Biopower provides us with an example of the intersection between modern political power and health care application. One does not operate without the other. Even when biopolitical intervention leads to war and genocide, Foucault explains the phenomena in terms of optimizing a "state of life." Biopolitical violence aims at cleansing society of what the State views as degenerate elements. The previously used example of removing a tumor accurately captures this concept. It is no mistake that Foucault makes use of the term "biopolitics."

The general neuroethical application of Foucault's work is as follows. Foucault's account of biopower demonstrates a close relationship between health care (care for life in general) and political power structures. The power structures utilize health care systems in order to track and maintain norms, while health care provides the support needed for biopolitical intervention to function. Given this intimate connection between politics and the health care system, Foucault's account of biopower demonstrates that the object of neuroethical inquiry in a medical context has a profoundly political character. Medical neuroethics concerns itself with optimizing ethical conditions in health care. Biopolitical intervention demonstrates that in order to fully accomplish this task, a portion of the efforts of neuroethical inquiry must aim themselves at the intersection between health care and political power structures. If one ignores the political implications and machinations that are enmeshed in the care for life, then neuroethical inquiry will suffer from a stunted perspective. This is what Foucault's account of biopower appears to provide the neuroethicist. It shows a shared interest between the health care system and the political realm, insofar as the focus on the maintenance of life is inscribed in the functions of State power.

METHODOLOGICAL CONCERNS

There are two methodological concerns regarding Foucault's descriptive account of biopower. The first is that Foucault is attempting to provide a value-neutral account while utilizing value-laden language. Nancy Fraser (1989, p. 28), for example, claims "it is clear that Foucault's account of power in modern societies is anything but neutral and unengaged." Words and phrases such as "domination," "resistance," and "State-sponsored racism" are all value laden. Domination is something that, in the political arena, we hope to avoid. It is freedom-denying and flies in the face of respect for individual autonomy. Resistance against domination is something to be encouraged. We look at those who fight against authoritarian regimes as heroes, as those who are willing to sacrifice all they have in the name of liberation. The Arab Spring is an excellent example of this trend. Last but not least, State-sponsored racism conjures up images of genocide which carry a substantial amount of normative weight. In short, it may be that Foucault is cheating with his account. He is sneaking value-laden content through the back door.

In response, one could claim that it is impossible for us to completely separate words such as "domination" and "resistance" from some sort of moral valuation. However, this methodological concern does not seriously undermine Foucault's account. We are free to make whatever ethical claims we like based upon the various ways in which biopower manifests itself. Foucault cannot, and did not, attempt to block such action in any significant way. The following is evidence of this claim:

> The problem is not of trying to dissolve them (relations of power) in the utopia of perfectly transparent communication, but to give one's self the rules of law, the techniques of management, and also the ethics, the *ethos*, the practice of self, which would allow these games of power to be played with a minimum of domination. (Foucault 1988, 18)[12]

It is simply the case that Foucault himself did not aim to label biopolitical power structures as good or evil. It was not his concern.

The second concern is a result of how Foucault characterizes power structures in society. According to Foucault, power flows throughout the community in a remarkably decentralized fashion. That is to say, Foucault does not discuss power as it is wielded by one party against another. Biopower (and its complimentary structure disciplinary power) circulate through the community in a way that resists zeroing in on those who hold the reigns. To put it another way, discussing power in relation to those who control government structures is unimportant for Foucault. This approach is somewhat problematic. It seems obvious to most that power is not something which spontaneously circulates throughout society. It is something which is held by one group and exercised against another. To use the current case, one would not say that biopolitical intervention enters Guantanamo Bay on its own. It is a function of the State which we can trace back to specific leaders and specific decrees. Examples range from Nazi leadership in the Holocaust to the Public Health Service in the Tuskegee experiments. As a result, Foucault's refusal to acknowledge this centralized characteristic of power seems to complicate his overall account.

There is a response to this concern. It is not the case that Foucault's account succeeds or fails based upon the acceptance of a wholly decentralized model of power. The important point is that these power structures outlast those who control the State. They are characteristics of the modern State, not characteristics tied to particular leaders. In this light, Foucault's reluctance to discuss the "who" of biopower makes more sense. Focusing on the "who" implies that individual leaders or groups of leaders are the font of biopolitical intervention, and as such when power shifts from one leadership group to another the preexisting power structures shift as well. This is a claim which Foucault would deny, and one which we can reasonably deny as well. While it is true that societies have changed the way in which power is utilized, these changes are quite slow and require large political/cultural shifts in order to gain traction. These shifts transcend individual leaders or leadership groups. Again, Foucault's focus on decentralized power structures is a methodological approach which can coexist with the factual claim that power is wielded by individuals. Now that I have addressed these concerns, I shall move on to the importance of biopower in understanding the use of fMRI technology in nonclinical environments.

GUANTANAMO BAY: NATIONAL SECURITY CLINIC

As you may recall, there are three lingering questions that I must answer. These are why would the State be interested in utilizing medical technology for political ends; why would the State focus this medical technology on vulnerable populations such as detained enemy combatants; and how is this use both nonmedical and illegitimate? Regarding the first question, critics have often (and at times justifiably) called Foucault to question regarding the accuracy of his historical claims. As this discussion of lie detecting brain scans indicates, however, Foucault's discussion of biopower does have a real-world correlate. Even if the science behind fMRI-based lie detection is questionable at best, the potential use of such scans on detainees is a picture perfect example of medical technology utilized for political ends. In an attempt to promote national security, the U.S. government brings the fMRI out of the hospital and into the interrogation room. I will not belabor this point, given the fact that I have already discussed it in some detail. What I would like to add is a comment regarding the connection between the political and medical realms, which in turn will lead to an answer to the second question.

By continuing the analysis, one can make the following claim regarding this connection between politics and the medical realm. If the State as a whole has, as one of its political ends, the tracking and maintaining of established statistical norms regarding the health of the population, then the State cannot surround itself solely with safety mechanisms aimed at what might be considered "standard" medical threats. The State must do more than remind us to get our flu vaccines, tell us to regularly wash our hands, and provide functional waste disposal infrastructure in order to prevent the spread of disease. A State characterized by biopower must expand its focus in order to maintain a nation's overall health and prevent spikes in morbidity. The use of fMRI in national security contexts is simply another example of this expansion. The DoD and DHS are attempting to maintain the health of the overall population by creating a body of national security-related intelligence, much in the same way that early biopolitical States began to collect and centralize medical knowledge. Given this fact, the utilization of medical technology makes more sense. As odd as it may sound, we can group organizations like the Center for Disease Control (CDC), the Department of Health and Human Services, and the Department of Homeland Security under the same biopolitical umbrella. The only real difference, according to a biopolitical understanding, is the "foe" each is attempting to combat. For the CDC, it may take the form of a new communicable disease or next year's flu virus. For the U.S. Department of Health and Human Services (USDHHS), the goal is often to provide practical information for living a healthier lifestyle. Examples include nutritional advice and insurance information. For the DHS, the "foe" does not resemble a typical biological threat. This leads to the second question. Why would the U.S. government focus on the use of medical technology on vulnerable populations?

Recall that one of Foucault's reasons for discussing biopower was his interest in describing State racism. The word "racism" may be somewhat misleading, given that the subjects of State racism are not always members of a particular race. They are individuals who the State deems to be a threat to the health of the overall population. As such, they are subject to repression or extermination in an effort to maintain the health of the nation. As I claimed before, this marginalization or genocide can be metaphorically compared to a surgeon removing a tumor. The tragedies of Nazi-era Germany certainly are clear examples of this sort of State racism, but those of us in the United States should not be deluded into thinking that the nation is above such atrocities. The Tuskegee experiments, overseen by the U.S. Public Health Service, are but one clear example where the State used medical means to inflict harm on a marginalized population (Brandt 1978). Returning to the Nazis, we can look to the Nuremburg trial and find another example of the U.S. State racism in the testimony of concentration camp physicians. One of the defenses that these physicians made was that U.S. researchers had deliberately infected hundreds of prison inmates with deadly diseases in order to increase the U.S. body of medical knowledge. This includes over 800 inmates who were infected with malaria plasmodia in the 1920s (Agamben 1998, p. 90).[13] According to this argument, the doctors at Auschwitz and the University of Chicago were not so different. We can find another clear example of this State racism at Guantanamo Bay.

This discussion of biopower allows one to see Guantanamo in a new light and helps to show why the State would want to utilize medical technology, such as fMRI, on detainees. According to this analysis, Guantanamo is not a prison so much as it is a quarantine zone. The prisoners can be regarded as biological threats that the State must isolate in an effort to guard the general population. The prisoner becomes a "disease" to be contained or eradicated. This dehumanization is evident from frequent human rights abuses, allegations of torture, denial of due process, and so on. They truly are considered by the government to be "those who were missed by the bombs" (Žižek 2004). Given that the State characterizes these individuals as a biological threat, it should come as no surprise that the State would use a medical tool to assist in mitigating or removing the threat to the greatest possible extent. In this context, fMRI is more than a form of lie detector. It is a tool used to diagnose a threat to the State in the same way that a physician utilizes a blood test to diagnose a threat to the body. It seems that we now have answers to the first two questions. The DHS is drawn to medical technology due to the fact that a biopolitical State takes on a profoundly clinical character. The neat line between the political and the medical disappears, and the State uses medical technology for political ends. This is particularly apparent with regard to the use of medical technology on vulnerable populations. These dehumanized groups are treated as biological threats, as a sort of anthropomorphic disease, which the State must isolate, study, and eradicate.

This leads to a final question. Marks claims that we ought to search for "legitimate nonclinical applications" of fMRI technology. The tone of his argument implies that the current national security landscape and the inaccuracy of fMRI lie detection technology render this nonclinical application of medical technology illegitimate.

It seems reasonable to assume that for Marks the term "legitimate" carries ethical force. If this is the case, then claims could be made to the effect that use of inaccurate functional neuroimaging would violate some right of the detainee or some duty we have to protect such a vulnerable population. Some potential ethical problems that render this technology illegitimate could include the compromised rights status of the subjects in general, the violation of some right to privacy, the likelihood of this technology being used in conjunction with torture, or the fact that this medical technology is being used in such a way that it offers no benefit to the subject. This is not an exhaustive list, but it helps to frame some of the potential ways in which performing fMRI scans on Guantanamo detainees are illegitimate in a normative sense. As I will show, this discussion of a biopolitical system and the way in which fMRI scanning maps onto this system complicates such a normative move.

A biopolitical system reconfigures the way by which one can view fMRI scanning technology as illegitimate. Is it the case that the application is still illegitimate in this particular case? Most likely. However, it is illegitimate only in the sense that the application of neuroscientific techniques does not produce accurate results. Remember, a biopolitical State is not particularly concerned with human rights issues, especially regarding the rights of those who it designates as threats to the overall health of the population. Guantanamo Bay provides a real-life example of this. To put it bluntly, fMRI-based lie detection is not illegitimate in the ethical sense. The use of fMRI technology is only illegitimate because it provides unhelpful data, but if it did, then the State could legitimately utilize this tool for political ends (for deeper discussion of the importance of "helpful"-ness of issues of legitimacy in the legal sense, see Chapter 9). Comprehension of biopower complicates any understanding of the "legitimate" part of "legitimate nonclinical applications of medical technology." It strikes me that most neuroethicists might argue that these fMRI scans are illegitimate in more than a purely scientific sense. Unfortunately, it is not easy to demonstrate that the establishment of a legitimate application of neuroscience carries ethical force. A closer look at the use of the term "nonmedical" will further complicate ethical undertakings of fMRI technology in a variety of contexts.

The claim that fMRI-based lie detection is nonmedical is fairly straightforward. As discussed above, fMRI was initially designed for a variety of medical purposes. It belongs in the hospital or in a research lab, with the ultimate aim of curing or understanding ailments. While it is not necessarily the case that all nonmedical uses of medical technology are unethical, it is most likely the case that the bioethical community as a whole would cast a long and questioning gaze at any use that does not fall in line with the goal of healing the body. At Guantanamo Bay, this gaze appears to be quite justifiable. Here you have technology which is used in a harmful environment, with no benefit to the subject of said technology. Whether or not we go the extra mile and tack on a normative claim regarding this nonmedical use, most would agree that such utilization is nonmedical in a factual sense.

Yet, there are two immediate problems with this claim given our biopolitical analysis. The first is that the State itself has, as I have said, taken on a clinical character. The boundaries between the medical community and the political sphere begin to disappear. Medical aims coincide with political aims in such a way that clinical care has leaked out into previously nonmedical realms. This is not to say that the hospital is exactly like the prison camp. Of course, there are differences between the two. However, given the clinical character of the State, one cannot simply point to Guantanamo as a nonclinical space without further analysis. Members of the neuro-ethical community might balk at the notion that a space associated with such potential harm could have anything in common with a house of healing. But question we must, if we are to truly confront the problem of conducting brain scans in the prison camp.

In addition, one could make a claim that is stronger than a simple caution regarding the difficulty in clearly identifying spaces as medical or nonmedical. Given our analysis, one could go so far as to claim that Guantanamo Bay is a clear example of a biopolitical clinical space. Remember, fMRI diagnostic technology is being used against a marginalized group in order to protect the overall health of the nonmarginalized population. This sort of clinic may trade its sterile examination rooms for dank prison cells, but the ultimate goal remains the same. Protect the health of the public and eliminate threats to this maintenance of health. Guantanamo Bay is more a "prison clinic" than "prison camp," as defined by the aims of State racism. In summation, to claim that the use of fMRI in lie detection in Guantanamo Bay is illegitimate and nonclinical is severely complicated by our biopolitical reading. At best, we could claim that the use of this technology in its current form is an illegitimate clinical use of fMRI. Again, it is illegitimate only if it does not produce proper results. In spite of this, the scans and the space in which they occur are quite clinical.

I have tried to show that, according to a biopolitical analysis, the State is interested in utilizing medical technology in a national security setting due to the medical character of the State. The care for life that characterizes the application of biopower is changed by State racism, and those who are deemed to be enemies of the State become no more than a biological threat for a new sort of clinical group to address. In this case, enemy combatants imprisoned in Guantanamo Bay represent the threat, and the U.S. DoD and DHS represent the new clinical group. Extracting information from high-profile detainees is akin to medical research. The knowledge helps the State better understand the threat it faces and allows the State to install increased security measures to protect the statistical health norms of its population. To claim that the use of functional neuroimaging is illegitimate is only true in a scientific sense, according to this analysis. To claim that these scans are nonclinical in the prison camp is flatly mistaken. Marks, Mandavilli, and others who seek to protect the neuromedical realm from the encroachments of national security are quite frankly too late.

WHERE DO WE GO FROM HERE?

At this point, I have provided a merely descriptive account regarding complications in defining the national security context, in general, and Guantanamo, in particular, as nonclinical. I would like to end by providing something in the way of positive next-steps to go along with my critical application of biopower. As I previously stated, the criticisms of thinkers such as Marks are admirable contributions to the discourse addressing neuroethics and national security. It is simply the case that the national security landscape has a more medical character than previously recognized. The avenues for future research that I provide may seem disjointed, but I believe they could serve as fruitful complement to the current discussion.

I will begin with the issue of the danger of hypothetically effective brain scans. Let us assume once more, for the sake of argument, that the use of fMRI technology for lie detection is sufficiently effective to produce consistently accurate results. The use of fMRI technology on vulnerable populations does not necessarily result in the creation of a new ethical dilemma, at least at first glance. That is to say, an fMRI lie detector test may not add new problems to the extensive list of human rights abuses perpetrated in place such as Guantanamo Bay. In fact, if it is an extremely effective tool, it may help to right some of the wrongs associated with this prison camp. Take, for example, Naqib Ullah. Naqib is a Pakistani national who was captured in a Taliban camp in Afghanistan and transferred to Guantanamo Bay in January 2003 (United States Department of Defense 2011). This 14-year-old boy was armed at the time of his capture, but did not fire his weapon. It took the Joint Task Force of Guantanamo until August 11 of the same year to determine that Naqib was an extremely low-level threat. Naqib had not voluntarily joined the Taliban. He had been kidnapped and served as a forced conscript when captured. This oversight on the part of the United States is but one example of many where either wholly innocent or low-level threats are mistakenly treated as high-level enemy combatants (Bell 2011).[14] While an effective use of fMRI could not wholly solve the human rights violations at Guantanamo, it is possible that it could lead to the expedited release of parties whom no one could reasonably claim are enemy combatants.

Be that as it may, the sinister side of biopower still hangs over the clinical character of the prison camp. An extremely effective diagnostic tool, such as this hypothetical use of brain scanning technology, could actually lead to increased rights violations, rather than a decrease. How might this be possible? By providing the DoD and DHS with an extremely effective tool to measure veracity, they could overcome one of the most significant barriers to the use of torture in interrogation settings. They would have a tool that could parse which information is accurate and which information is simply supplied in order to "make the pain stop." Allow me to clarify: One of the leading arguments against the effectiveness of torture is that the victim will tell the torturer anything to get them to stop (Horton 2009). As such, any information gathered through torture is potentially unreliable, making the practice questionable on factual as well as moral grounds. An extremely accurate lie detector, such as hypothetically precise neuroimaging technology, would remove the factual barrier due to an ability to cross-check information. If the victim is lying, then the process of torture would begin again. The likelihood of this occurring in a place such as Guantanamo Bay seems quite high given our biopolitical analysis. The State is clearly not concerned with the rights of those subjected to State racism. As previously asserted, these individuals are dehumanized and viewed as little more than biological threats. For the biopolitical State there are no moral grounds for refraining from torture, only factual grounds. Torture is, as we have seen, illegitimate only in that it does not produce accurate results. If neuroimaging were to provide an effective way around this factual barrier, then it seems likely that the State could resort to more rather than less torture. This leaves us in a rather murky place from a neuroethical standpoint, but it seems reasonable to address this sort of issue before the technology outpaces our analysis.

The other avenue for continuing exploration arises given that a biopolitical analysis of neuroethics and national security expands inquiry outside of the field of neuroscience. There are two interconnected ways with which to engage this expansion. The first is to acknowledge that even if we were able to halt the use of fMRI at the gates of Guantanamo and elsewhere, it will still be the case that medical technology will creep out of its "proper place" in an effort to support the national security agenda. The DoD and DHS are just as likely to use surgical procedures, nonneurological screening, and chemical treatments as they are to utilize functional neuroimaging (Sydney Morning Herald 2007).[15] Remember, the requirement for use of medical technology in the prison camp is effectiveness. If other technologies show more promise, then the State will most likely abandon imaging for more fruitful medical procedures. Those neuroethicists who are interested in what they call illegitimate uses of medical technology (in a normative sense) will need to expand their focus in an effort to capture this fact. This point is hardly problematic, although it will likely require a larger coalition of ethicists and experts due to the expanded medical scope.

The second aspect of expansion pushes the traditional neuroethicist a bit further outside the usual area and focus of expertise and into the field of political philosophy (McDermott 2009).[16] If my analysis is correct in revealing the cause of DoD and DHS interest in neuroscience, then we have moved beyond the field of neuroethics alone and into the much broader field of political studies. I have proposed that the issue of fMRI-based lie detection in Guantanamo Bay is a symptom of a much larger problem, namely that of biopower and State racism. Efforts to tackle the problems posed by Marks and others would require substantial efforts to understand and overcome what any reasonable person would consider to be the negative aspects of biopolitical intervention. This is not to say that we must tear the whole system down and start again, even though some Foucauldians may argue that biopower is so deeply entrenched in modern democracy that there is no other option. We can challenge the government when we see the abuse of vulnerable populations, always with the mind-set that they face a danger so extreme it threatens to crush their status as human beings. This problem seems to be much bigger than illegitimate uses of neuroimaging. It cuts to the core of institutionalized State racism, and any future efforts to address this issue should take this fact into consideration. In other words, it is not enough for the neuroscientific and neuroethical community to simply say "keep your hands off our technology." These communities must conjoin their expertise to other disciplines' in an effort to solve what appears to be a systemic issue in the modern national security setting.

It is easy to describe Foucault's work as overly pessimistic, as a series of connected descriptive accounts that offer no escape from existing power structures. With no clear "up side" to his writings, one may be tempted to dismiss his work as mere cynical musings; or if one takes these writings seriously to abandon any project aimed at transcending the injustices inscribed in the state apparatus. It is my contention that neither response is appropriate. This biopolitical analysis has shown how we can avoid preemptive dismissal and deep despair. Descriptive accounts such as Foucault's help us to understand why the United States would attempt to bring neuromedical technology into Guantanamo Bay. It shows us why the prison camp takes on the character of a prison clinic. With this understanding, we can seek to correct the injustices that plague the vulnerable individuals housed in these camps and eventually address a more general problem of dividing the population into those who must be protected and those who must be destroyed. If one is to hold that this use of brain scanning technology is truly nonmedical and illegitimate, it is simply the case that the road ahead is more complicated than previously imagined.

NOTES

1. See my previous commentary on the subject of fMRI lie detectors at Guantanamo, based upon the work of Jonathan Marks.
2. The National Research Council stated the following regarding the accuracy of polygraph testing. "There is no direct scientific evidence assessing the value of the polygraph as a deterrent, as a way to elicit admissions or confessions, or as a means of supporting public confidence."
3. Two examples are No Lie MRI Inc. (2012) and Cephos Corporation (2012).
4. Some claim that neuroimaging provides no evidence that any particular region of the brain plays a causal role during the performance of a specific task, thus undercutting the possibility of the fMRI-as-lie detector.
5. The distinction between descriptive and normative accounts is a result of Foucault's general project. Throughout his various works, Foucault claims that he is merely attempting to describe the way power structures operate in society. The accounts are meant to be value-neutral. This is opposed to accounts which describe societal power structures, and then label these structures according to some sort of moral valuation. Foucault resists the later, though he does not deny that we can coherently assign such valuations. It simply wasn't his project.
6. It seems wise to isolate some of the key aspects of disciplinary power in order to better understand the distinction between disciplinary power and biopower. A brief definition of the key aspects of disciplinary power can be found in the work by Foucault (1995, 136–137). In short, disciplinary power functions as a constant corrective force which breaks the individual down into component parts, which are then conditioned through subtle coercion to be utilized in a particular fashion. Take, for example, military drills in a training camp. The individual soldier is under constant surveillance. Every mechanical movement, every swing of the arm and step, is an object of power's coercion. Should a soldier step out of line, a corresponding corrective action is utilized in order to fix the proverbial misstep. This is an excellent example of disciplinary power in action, which Foucault argues is in practice throughout society. Schools and prisons are the most prevalent examples.
7. One should note that the industrial shift is not some sort of realization of a moral ideal to protect the worker. The sole concern is creating a healthy and productive workforce. After all, it is difficult to work while injured.
8. It should be noted that the statistical norms that are tracked through biopolitical intervention are not completely static. That is to say, it is not the case that a statistical norm is established at some point in history and then security measures protect the norm forever. The biological trends of a population shift from time to time, and it is biopower's function to track these trends and secure them. One can look to the example of Western infant mortality rates, and the shift in this trend from the nineteenth to the twenty-first century, as an example of this.
9. Foucault states here that while racism did exist before the advent of biopower, it was not inscribed in the mechanisms of the State. That is to say, according to Foucault old racism was not crucial to the functionality of the State. However, the newer form of racism is crucial to a biopolitical regime.

10. Foucault is quick to point out that the notion of killing to prevent death was originally used in war (kill the enemy to survive), which predates the rise of biopower. However, when this principle of war is combined with racist ideology, Foucault claims that it morphs into a biological-type relationship. "In order for my species to survive, the inferior must die out." This helps to shed more light on what Foucault means by a new State racism.

11. I am indebted to Dr. Andrew Cutrofello for this distinction between "aim" and "means."

12. I am indebted to Dr. David Ingram's (2005) essay "Foucault and Habermas," contained within the second edition of the *Cambridge Companion to Foucault*, for the assistance in locating this quotation.

13. The Nazi physicians also submitted as evidence a University of Chicago medical release form for death row inmates.

14. Another example of this trend involves Abdul Badr Mannan, an anti-extremist Pakistani author who was sent to Guantanamo after Pakistani Intelligence services framed him.

15. One chemical treatment already used is thiopental sodium, a general anesthetic which is supposed to induce relaxation and compliance in individuals who are subject to interrogation.

16. This is not the first time an author has expressed interest in this sort of dialogue between neuroscience and political studies.

REFERENCES

Agamben, G. 1998. *Homo Sacer.* Stanford, CA: Stanford University Press.

Agamben, G. 2005. *State of Exception.* Chicago: Chicago University Press.

Anderson, M.A. 2011. "Ethical considerations in international biomedical research." *Synesis: A Journal of Science, Technology, Ethics, and Policy* 2:G56–61.

Anderson, M.A., N. Fitz, and D. Howlader. 2012. Neurotechnology research and the world stage: Ethics, biopower, and policy. In *Neurotechnology: Premises, Potential, and Problems*, ed. J. Giordano. Boca Raton, FL: CRC Press, pp. 287–300.

Bell, J. 2011. Guantanamo bay files: Anti-extremist author framed and whisked to Cuba. *The Guardian*, April 24. http://www.theguardian.com/world/2011/apr/25/guantanamo-files-framed-author-mannan.

Brandt, A. 1978. "Racism and research: The case of the Tuskegee syphilis study." *The Hastings Center Report* 8(6):21–29.

Cephos Corporation. 2012. The fMRI testing process. Accessed on April 28, 2012, http://www.cephoscorp.com/lie-detection/index.php#testing.

Devlin, H., I. Tracey, H. Johansen-Berg, and S. Clare. 2005. Introduction to fMRI. Nuffield Department of Clinical Neurosciences. Accessed on April 27, 2012. http://www.fmrib.ox.ac.uk/education/fmri/introduction-to-fmri.

Fenton, A., L. Meynell, and F. Baylis. 2009. "Ethical challenges and interpretive difficulties with non-clinical applications of pediatric fMRI." *American Journal of Bioethics: Neuroscience*, 9(1):3–13.

Foucault, M. 1988. The ethic of care for the self as a practice of freedom. In *The Final Foucault*, eds. J. Bernauer and D. Rasmussen, Cambridge, MA: Massachusetts Institute of Technology Press, pp. 1–20.

Foucault, M. 1995. *Discipline and Punish: The Birth of the Prison.* New York: Vintage Books.

Foucault, M. 1997. *Society Must be Defended.* New York: Picador.

Fraser, N. 1989. *Unruly Practices: Power, Discourse, and Gender in Contemporary Social Theory.* Minneapolis, MN: University of Minneapolis Press.

Giordano, J. 2013. "Ethical considerations in the globalization of medicine—An interview with James Giordano." *BioMed Central Medicine* 11:69.

Horton, S. 2009. Torture doesn't work, neurobiologist says. Harper's, September 22. http://harpers.org/blog/2009/09/torture-doesnt-work-neurobiologist-says/.

Ingram, D. 2005. Foucault and Habermas. In *The Cambridge Companion to Foucault*, ed. G. Gutting. New York: Cambridge University Press, pp. 240–283.

Klein, C. 2010a. "Images are not the evidence in neuroimaging." *The British Journal for the Philosophy of Science* 61(2):265–278.

Klein, C. 2010b. "Philosophical issues in neuroimaging." *Philosophy Compass* 5(2):186–198.

Littlefield, M. 2009. "Constructing the organ of deceit: The rhetoric of fMRI and brain fingerprinting in post-9/11 America." *Science, Technology, and Human Values* 34(3):365–392.

Mandavilli, A. 2006. "Actions speak louder than images." *Nature* 444(7120):664–665.

Marks, J.H. 2010. "A neuroskeptic's guide to neuroethics and national security." *American Journal of Bioethics: Neuroscience* 1(2):4–12.

McDermott, R. 2009. "Mutual Interests: The case for increasing dialogue between political science and neuroscience." *Political Research Quarterly* 62(3):571–583.

National Research Council. 2003. *The Polygraph and Lie Detection*. Washington, DC: National Academies Press.

National Research Council. 2009. *Opportunities in Neuroscience for Future Army Applications*. Washington, DC: National Academies Press.

No Lie MRI Inc. 2012. Customers-Government. Accessed on April 28, 2012. http://www.noliemri.com/customers/Government.htm.

Sydney Morning Herald. 2007. Truth serum used on child killers. September 12. http://www.smh.com.au/news/world/truth-serum-used-on-serial-child-killers/2007/01/12/1168105166282.html.

Thomsen, K. 2010. "A Foucauldian analysis of 'A neuroskeptic's guide to neuroethics and national security.'" *American Journal of Bioethics: Neuroscience* 1(2):29–30.

Tovino, S. 2007. "Functional neuroimaging and the law: Trends and directions for future scholarship." *American Journal of Bioethics: Neuroscience* 7(9):44–56.

United States Department of Defense. 2011. Guantanamo Detainee File: Naqib Ullah US9AF-000913DP. *The Guardian*, April 24. http://www.theguardian.com/world/guantanamo-files/US9AF-000913DP.

Žižek, S. 2004. What Rumsfeld doesn't know he knows about Abu Ghraib. *In These Times*, May 21.

13 Between Neuroskepticism and Neurogullibility

The Key Role of Neuroethics in the Regulation and Mitigation of Neurotechnology in National Security and Defense

Paolo Benanti

CONTENTS

INTRODUCTION: A CONTEMPORARY FRAMEWORK FOR NEUROSCIENCE AND NEUROTECHNOLOGY IN NATIONAL DEFENSE AND SECURITY

Improvement in the neurosciences, not least through the development of increasingly sophisticated neurotechnologies, is being viewed for its potential in national security and defense contexts. Surely, this is not new, as many have noted (Achterhuis 2001; Ihde 2009; Singer 2009; Benanti 2010; Giordano et al. 2010; Marks 2010; Canton 2012; see also Chapter 11). But what is new is the extent to which multiple scientific subdisciplines are being incorporated into the realm of the neurosciences and the relative pace at which concepts are being translated into operationalizable tools and techniques.

Neuroscientific experiments, often characterized by low invasiveness, are intended and implemented for affording better understanding of brain structures and functions and the relationship of neural activity to thought and behavior. These trials have engendered significant contribution to cognitive sciences, stimulated philosophical debate about free will, responsibility, and autonomy, and attracted interest from both the academic community as well as the public. With this surge—as yet relatively nascent—in neuroscientific capability and information, some see the potential for militarized neuroscience as being far too incipient to be of any serious concern; conversely, other scholars posit that we are on the cusp of a massive shift in military technology that will have profound effects both at the front lines, as well as within the spheres of national and international politics (Achterhuis 2001; Ihde 2009; Singer 2009; Benanti 2010; Giordano et al. 2010; Marks 2010). To be sure, these points may be debatable. However, recent funding investments and allocations in brain research by the U.S. and other nations' defense agencies appear to mitigate the former stance in favor of the latter. On some level, this may be viewed as a reality check. In this chapter, I argue that neuroscience and neurotechnology and their use in national security and defense should be well understood and evaluated—not only to establish a more rational view of when and how neurotechnology can be used in national defense and security agenda, but also to address—if not challenge—strategic and political questions that foster serious (neuro)ethical implications about the use of neuroscience in such enterprise.

THE CORE QUESTION: NEUROSKEPTICISM OR NEUROGULLIBILITY?

As these new possibilities arise, a hard debate has emerged that its scope and implications are too broad to fully explicate here. In short, there are two recognizably fundamental positions in the differing perspectives that characterize a view of neurosciences and the applicability of its tools and techniques. One side is characterized by a thesis that is based on a so-called *neuroskepticism*: a perspective informed by scientific studies that entails considerable scrutiny when viewing the practical implications and real-world applications of recent developments in neuroscience (Marks 2010). This perspective asserts that the use—and possible misuse—of neuroscience in contexts of national security science demand urgent evaluation in light of actual utility, assumed viability, and applications in practice (Giordano et al. 2010; Marks 2010; see also Chapters 11 and 17). For example, Marks expresses concerns about the naming and utilization of neuroscientifically based outcomes and products in national security contexts and focuses upon the practical and ethical hazards that arise from the deployment of this opaque terminology. Marks' considerations rest upon the observation that neuroscience offers unparalleled opportunities to transform our lives on the one hand, and on the other hand simultaneously fosters new ethical questions, issues, and problems (a point Giordano [2012] has emphasized in characterizing the "demiurgical" potential of neuroscience and neurotechnology—and one that is exceedingly relevant to national security and defense agendas and operational employments; see also, Singer 2009). Marks' reflection unequivocally addresses core examples of psychoactive drugs and neuroimaging, with the potential for using neuroscientific techniques on detainees who may represent a most vulnerable population in the "war on terror." His assumptions produce a blanket vision of neuroscience as a homogeneous field. Therefore, Marks suggests creation of a national advisory committee on neurosecurity, staffed by professionals who possess the relevant scientific, ethical, and legal expertise (Marks 2010; see also Chapters 10, 11, 15, and 17).

We can also recognize a different perspective, herein called *neurogullibility*. This perspective describes the opportunity created by unifying neuroscience and integrating neurotechnology, and recommends transforming the ideas, outcomes, and products of these endeavors to advance the relative flourishing of individuals and society. According to these arguments, the early decades of the twenty-first century will evidence concentrated efforts to bring together nanotechnology, biotechnology, information technology, and new, humane neurotechnology focused upon augmenting cognitive science—and the capabilities it confers upon human users (Canton 2012; Giordano 2012). The core assertion is that in a world where the very nature of warfare is changing rapidly, national defense requires the uptake and leveraging of innovative technology (inclusive of neurotechnology) that projects power so convincingly that any threats to the current Western superpowers (e.g., the United States and its North Atlantic Treaty Organization [NATO] allies) are deterred, minimized, or eliminated and that danger to Western warfighters from hostile or friendly fire can be mitigated, and training costs reduced by more than an order-of-magnitude through applications of neurotechnology (Bainbridge et al. 2006; Giordano 2011). What communicates a feeling of gullibility is that the inherent and derived ethical questions generated by this perspective are seemingly bypassed by the premise that any defense application of neuroscience in the highly competitive environments of deterrence, intelligence gathering, and lethal combat dictate technological advancement so as to remain as far ahead of potential opponents' efforts as possible (Roco and Bainbridge 2003; Bainbridge et al. 2006). Both perspectives and their relative forms of analysis—whether *neuroskepticism* and/or *neurogullibility*—underscore the need for a more critical evaluation of any use of neurotechnology in national security and defense.

NEUROSCIENCE, NEUROTECHNOLOGY, AND NEUROETHICS: THREE FACETS OF A SINGLE LENS

To avoid reductionism or partial views in any ethical address of the use of neuroscience in national security, I believe that we should first analyze technology and its development relative to scientific knowledge and culture and then (and perhaps only then) analyze the use of neurotechnology in the national security milieu. Essential to such reshaped ethical consideration is the use of a more complete terminology. Neuroscience, neurotechnology, neuroethics, neuroskepticism, and neurogullibility (and arguably any term bearing the "neuro" prefix) should be clarified terms of unique meanings in and for authentic ethical reflections (Schein 2010; Giordano and Benedikter 2012).

The literature provides an extended discussion and debate about the relationship between science, culture, and technology. We can recognize that one of the most important tenets established within the philosophy of technology over the last century has been the absolute novelty of the technological approach to reality. The new, more empirically oriented philosophy of technology as developed both in America and post–World War II technocentric nations (many of which are now aligned as U.S. allies) during the last 30 years speaks to the coevolution of technology and society; this does not view technology as autonomous, but instead seeks to explicate the numerous social forces that give rise to—and act upon—technology (as both construct and activity) (Ihde 2009). The differences between the classical view of technology and the understanding offered by contemporary philosophy of technology can be summarized in three ways.

First is to note how classical philosophy of technology tended to be concerned with technology overall and not *specific* technology (Achterhuis 2001). The classical philosophers of technology were more occupied with the historical and transcendental conditions that made modern technology possible and tended to be less concerned about the real changes accompanying the development of a technological culture (Giordano 2012). Therefore, if and when analyzing *neuro*technology, we must face the real changes that its development for national defense and security produce an understanding of what the said technologies can and cannot really do.

Second is a view (Achterhuis 2001; Ihde 2009) that avoids dystopian interpretations of technology: In effect, we must understand this new wave of neurotechnology, rather than to reject it nostalgically in demanding a return to some prior, seemingly more harmonious and less problematic relations with technological artifacts for national defense and security.

Third, a somewhat new philosophy of technology assumes a more empirical—or concrete—turn. This enables an understanding of technological development not as an independent force that externally impinges upon society, related only to scientific knowledge (as neuroskepticism and neurogullibility tend to assume), but rather, views technology as a social activity in and of itself, which reflects the particulars of setting in time and place, and arises from the dreams, purposes, and relationships of people (Achterhuis 2001).

Thus, it is more neurotechnology than neuroscience that raises moral, ethical, and legal questions and problems. If viewed in this way, neurotechnology can be seen to be not so much a question, but an answer: technological artifacts (everything that is human-made) summarize in themselves the answers that humankind gave to a provocative reality in a certain time and place (Schein 2010). According to Schein (2010), neurotechnology can be analyzed at several different levels, with the "term" level referring to the degree to which the neurotechnologic phenomenon is visible to the observer. Some of the confusion surrounding the definition of what neurotechnology really is results from not differentiating the levels at which it is manifest. These levels range from very tangible, overt manifestations that one can see and feel, to more deeply embedded, unconscious, basic assumptions that can be defined as the "essence" of neurotechnology. Between these layers are various espoused beliefs, values, norms, and rules of behavior that members of a culture (as users of neurotechnology) employ as ways of depicting neurotechnology to selves and others

(Schein 2010). At the surface is the level of neuro-artifacts. Neuro-artifacts include the visible products of neuroscience. The most important point to be made about this level of neurotechnology is that it is both easy to observe and very difficult to decipher. In other words, observers can describe what they see and feel, but cannot reconstruct from that alone what those things mean in the given group, or whether they even reflect important underlying assumptions. It may be especially dangerous to try to infer the deeper assumptions from neuro-artifacts alone, because one's interpretations will inevitably be projections of one's own feelings and reactions (Schein 2010).

Analyzing a deeper level of neurotechnology, all artifacts that humanity produces ultimately reflect someone's original beliefs and values, their sense of what ought to be, as distinct from what actually is. When a group is first formed or first faces a new task, issue, or problem, the first solutions proposed characteristically reflect some individual's (or particular group of individuals') assumptions about what is right or wrong, and what will work or not work. Beliefs and values that emerge at this conscious level will predict much of the behavior that can be observed at the artifacts' level. If the espoused beliefs and values are reasonably congruent with the underlying assumptions, then the articulation of those values into an operating philosophy can indeed be helpful in bringing the group together, serving as a source of identity, and solidifying a core mission. But in analyzing beliefs and values, one must discriminate carefully between those that are congruent with underlying assumptions and those that are, in effect, either rationalizations or mere aspirations for the future. Often, such beliefs and values are so abstract that they can be mutually contradictory, as when a company claims to be equally concerned about stockholders, employees, and customers, or when it claims both highest quality and lowest cost. Espoused beliefs and values often leave large areas of behavior unexplained, generating a feeling that we understand a piece of the culture, but still do not have the culture as such in hand. To access deeper levels of understanding, to decipher the pattern, and to predict future behavior correctly, it becomes important to more fully understand the category of basic underlying assumptions (Schein 2010).

Basic assumptions, in the sense in which I want to define that concept, have become so taken for granted that one finds little variation within a social unit. This degree of consensus results from repeated success in implementing certain beliefs and values, as previously described. In fact, if a basic assumption comes to be strongly held in a group, members will find behavior based on any other premise to be almost inconceivable (Schein 2010). This type of multileveled analysis of neurotechnology—operating at the level of its artifacts, the level of its espoused beliefs and values, and the level of its basic underlying assumptions—illustrates the potency of implicit, unconscious assumptions and shows that such assumptions often deal with fundamental aspects of life: the nature of time and space, human nature and human activities, the nature of truth and how it is discovered and revealed, ways for individuals and the group to relate to each other, and the relative (if not changing) roles and importance of work, family, and self-development (Ihde 2009; Schein 2010; Benanti 2012b). Neurotechnology must be studied at each and all of these three. If one does not examine and intuit the pattern of basic assumptions that may be operating, one will not know how to correctly interpret the artifacts or recognize how much credence to give to the articulated values. In other words, the "essence" of neurotechnology lies in the pattern of basic underlying assumptions, and once these are understood, one can easily comprehend other, more superficial levels and appropriately deal with them (Ihde 2009; Schein 2010; Benanti 2012a, 2012b).

We can find traces of this phenomenon in language. Latin roots of the English word *provocative* are made by elements *pro* and *vocatio*: something that calls forth or advances (*pro*) the human to draw up ideas, tasks, or reflections (*vocatio*). Certainly, uses of neurotechnology—especially as related to national security and defense agendas—are provocative. Simultaneously, the human act of response (*respondeo*) is contained in roots of the English word *responsibility*. Morality and ethics are forged by responsibility. Evidently, any such use of neurotechnology—inclusive of its operationalization in national security and defense—incur, if not demand, responsibility in intent, planning, and action. Therefore, neurotechnology is intrinsically related to ethics: Ethical questions do not arise around practical use of neuroscience; instead, they are born and live in the essence of each neurotechnological artifact (Ihde 2009; Benanti 2012a, 2012b). To remove opaque terminology from neuroscience, we must distinguish between neuroscience and neurotechnology. Morality and ethics are elements built from a declaration that neurotechnological artifacts have a nonneutral moral constitution. The use of such artifacts is intrinsically involved in the process that brings neurotechnology to the market. Cultural needs are infused into moral choices, and these indirectly offer the supposed promise(s) of national security via neuroscience. So, we must ask why we are developing these tools instead of others—why do we need some kind of neurotechnology, and/or what kind of human relationships will—or should—these artifacts forge? (Ihde 2009).

CONTEMPORARY ETHICAL PARADIGMS FOR EVALUATING NEUROTECHNOLOGY

To develop perspectives in neuroethics, it cannot be ignored that some forms and extent of ethical evaluation for the use of various neurotechnologies are already in place and being applied. I would like to summarize these ethical evaluations before I offer my own perspective. Looking at ethical arguments to evaluate the use or misuse of neurotechnology, I found three recurrent paradigms that I have called: (1) fear of the uncertain; (2) pursuit of equality and happiness; and (3) emphasis on policies (Benanti 2012a, 2012b).

The first paradigm, fear of the uncertain, is used to regulate or mitigate use of neurotechnology in a double sense. Some ethicists argue that we should use only those neurotechnologies that can be previewed and controlled. In this way, neurotechnology use will be safe and protected by unwanted effects. In another way, some ethicists assert that the future of national defense and security is what really remains uncertain, and thus, only the concerted use of neurotechnology can transform uncertainty to any realistic form of national safety. Both are focused on fear: fear of what can happen in the future if neurotechnology is either allowed or disallowed in national security scenarios. I believe that we cannot allow fear to play such a prominent, if not preemptive role in neuroethical assessment and adjudication of neurotechnological applications in national defense agendas (Benanti 2012b).

The second paradigm, pursuit of equality and happiness, is focused upon the relationship between people within a given nation. From this view, fundamental rights should be granted in every situation: in no one circumstance, inclusive of national defense and security, it is allowable that neurotechnology be employed to violate human rights as granted by foundational documents and tenets (such as the U.S. Constitution and World Health Organization Declaration of Human Rights). This perspective offers a defensible point that human integrity within societies is supported and sustained by such rights and law. However, I believe that national defense and security are issues that are larger than rights and laws defined by national borders (Benanti 2012a, 2012b).

The third paradigm, emphasis upon policies, establishes that regulatory language and doctrine be formulated and enacted to guide and govern the use of neurotechnology in national defense agendas in ways that are independent from legal prescription of any one particular nation, per se. This paradigm is reflective of the views of scholars working in conjunction with, under the auspices of, and/or who are supportive of the precepts of large international institutions such as the United Nations (UNESCO). Within this view, only supranational and independent institutions can prescind to the extent necessary to develop and implement policy language and effect(s) capable of achieving an equitable and sound use of neurotechnology in national defense and security applications (Benanti 2012a).

TOWARD GOVERNANCE

Schein's analysis shows that neurotechnology, like every technology, must be dissected to its levels of artifacts, espoused beliefs and values, and basic underlying assumptions (Schein 2010). To ignore this renders any analytic—and guiding—approach vulnerable to a form of techno-neuro-reductionism, as neuroskepticism and/or neurogullibility arguably reveal. To avoid this, we should strive to create and sustain neuroscience as a reflexive practice that responds to social and cultural challenges posed both to the field of science and to world society, as consequential to recent advances in brain sciences. To achieve this goal, it will be important to recognize the complexity of culture and to develop effective tools to foster meaningful practice. A clarified terminology of neuroscience—and neuroethics—will be an important first step (Giordano 2011; Giordano and Benedikter 2012). Establishing pragmatic distinctions between neuroscience and neurotechnology may foster improved understanding, and in so doing, may lead to a form of neuroskepticism that does not offer such a pessimistic view of neuroscience's future, but instead, prompts a deep and urgent request for moral commitment to developing and using neurotechnology in national defense and security agenda, as well as more broadly, in health care and the conduct of daily life (Giordano 2012; Giordano and Benedikter 2012; see also Chapter 17).

A clear and transparent terminology illuminates differences between neuroscience and neurotechnology, allows better explication of the cultural forces that undergird and direct technological development, and provides instruments to challenge and address urgent problems of technological use, nonuse or misuse, such as those that are likely to occur in national security contexts. To realize a truly analytic, critical, and therefore valuable neuroethics, it will be crucial to develop moral argumentation about the nature of neurotechnology and neuroscience in such contexts. Yet, this critical approach can only find voice and truly effective power through governance. Governance has been defined as the rules of a political system used to address and resolve conflicts between actors or agents and to adopt decisions that prescribe or proscribe agents' actions (i.e., the process of legality and laws). Governance also has been defined to refer to the proper functioning of institutions and their acceptance by the public (i.e., the process and effect of legitimacy). As well, it has been defined to describe the efficiency of government and the achievement of consensus by democratic means (i.e., the process of participation).

Because the processes of governing involve a variety of private and public actors/agents, the governing oversight of neurotechnology must be regarded as a complex issue; in effect, it can only be obtained with new forms of participatory governance focusing upon deepening democratic engagement through the participation of citizens in concert with the state (see Chapter 17). In this light, I argue that citizens should play more direct roles in effecting public decision making or at very least more fully engage political issues focal to national defense and security use(s) of neurotechnology. But such public participation cannot fall upon deaf ears; government officials must be informed about neuroscience and neurotechnology, so as to be responsive to this kind of engagement (Jeannotte et al. 2010). In practice, participatory governance can supplement the roles of citizens as voters or as watchdogs through more direct forms of involvement. I believe that this form of critical neuroethics—and reflective neuroscience—can be actualized only through creation of national and international advisory committees on neurosecurity (Ihde 2009; Benanti 2012a, 2012b). Only an incentivized direction of participatory governance, mediated and represented by such advisory committees, can realize and make effective the key role of neuroethics in the regulation of neurotechnology in national defense and security agendas, contexts, and scenarios (Ihde 2009; Benanti 2012a, 2012b).

REFERENCES

Achterhuis, H. 2001. *American Philosophy of Technology: The Empirical Turn.* Bloomington, IN: Indiana University Press.
Bainbridge, W.S., M.C. Roco, National Science Foundation (U.S.), and World Technology Evaluation Center. 2006. *Managing Nano-Bio-Info-Cogno Innovations: Converging Technologies in Society.* Dordrecht, the Netherlands: Springer.
Benanti, P. 2010. "Neuroskepticism to neuroethics: Role of morality in neuroscience that becomes neurotechnology." *American Journal of Bioethics: Neuroscience* 1(2):39–40.
Benanti, P. 2012a. Cyborgization. In *Neurotechnology: Premises, Potential, and Problems,* ed. J. Giordano. Boca Raton, FL: CRC Press, pp. 191–198.
Benanti, P. 2012b. *The Cyborg: Corpo e Corporeità Nell'epoca del Post-Umano.* Assisi, Italy: Cittadella.
Canton, J. 2012. Toward our neurofuture: Challenges, risks and opportunities. In *Neurotechnology: Premises, Potential, and Problems,* ed. J. Giordano. Boca Raton, FL: CRC Press, pp. 233–241.
Giordano, J. 2011. "Neuroethics: Traditions, tasks and values." *Human Prospect* 1(1):2–8.

Giordano, J. 2012. Neurotechnology as demiurgical force: Avoiding Icarus' folly. In *Neurotechnology: Premises, Potential and Problems*, ed. J. Giordano. Boca Raton, FL: CRC Press, pp. 1–14.

Giordano, J. and R. Benedikter. 2012. "An early-and necessary-flight of the Owl of Minerva: Neuroscience, neurotechnology, human socio-cultural boundaries, and the importance of neuroethics." *Journal of Evolution and Technology* 22(1):14–25.

Giordano, J., C. Forsythe, and J. Olds. 2010. "Neuroscience, neurotechnology, and national security: The need for preparedness and an ethics of responsible action." *American Journal of Bioethics: Neuroscience* 1(2):35–36.

Ihde, D. 2009. *Postphenomenology and Technoscience: The Peking University Lectures.* Albany, NY: SUNY Press.

Jeannotte, A.M., K.N. Schiller, L.M. Reeves, E.G. DeRenzo, and D.K. McBride. 2010. Neurotechnology as a public good. In *Scientific and Philosophical Perspectives in Neuroethics*, eds. J. Giordano and B. Gordijin. Cambridge, MA: Cambridge University Press, pp. 302–320.

Marks, J.H. 2010. "A neuroskeptic's guide to neuroethics and national security." *American Journal of Bioethics: Neuroscience* 1(2):4–12.

Roco, M.C. and W.S. Bainbridge. 2003. *Converging Technologies for Improving Human Performance: Nanotechnology, Biotechnology, Information Technology and Cognitive Science.* Boston, MA: Kluwer Academic Publishers.

Schein, E.H. 2010. *Organizational Culture and Leadership.* 4th edn. San Francisco, CA: Jossey-Bass.

Singer, P.W. 2009. *Wired for War: The Robotics Revolution and Conflict in the Twenty-First Century.* New York: Penguin Press.

14 Why Neuroscientists Should Take the Pledge
A Collective Approach to the Misuse of Neuroscience

Curtis Bell

CONTENTS

The fact that knowledge is power and power can be used for good or ill is as true for neuroscience as for any other branch of knowledge. For many of us who are neuroscientists, the interest and excitement that we feel when learning about important discoveries in neuroscience are shadowed by fear of the harm that can flow from the same discoveries. These concerns have been made acute by the rapid growth of neuroscience and its associated neurotechnologies over the last few decades. The current condition of our conflict-laden world in which state and nonstate actors are tempted to use whatever power or technology might advance their causes adds to the concern.

This chapter describes a pledge as a course of action for neuroscientists who share these concerns. Signers of the pledge commit to (1) making themselves aware of the potential applications of their work and that of others to applications that violate basic human rights or international law such as torture and aggressive war; and (2) refusing to participate knowingly in the application of neuroscience to violations of basic human rights or international law (Bell 2010). The pledge began circulating internationally in 2010 and has been signed by neuroscientists in 17 different countries. The pledge can be read and signed online (http://www.tinyurl.com/neuroscientistpledge).

Why such a pledge? Because neuroscientists' identity as ethical and compassionate human beings must take precedence over their identity as scientists, and because the danger of using neuroscience in violation of human dignity, human rights, and international law is real.

The pledge calls on neuroscientists to follow the basic ethical principles of recognizing the consequences of their actions, taking responsibility for those consequences, and obeying the law. The pledge fits well within the broader framework of "Critical Neuroscience" as introduced by Choudhury et al. (2009) and as presented more fully by several authors in a recent book (Choudhury and Slaby 2012). "Critical Neuroscience" calls for the examination of the historical, political, social, ethical, and economic contexts of neuroscience and for neuroscientists to maintain awareness of these larger contexts (see also Chapter 17).

The pledge proscribes work on all applications of neuroscience that violate basic human rights and international law but it focuses on two egregious examples: torture and aggressive war. Both torture and aggressive war are not only immoral but are also illegal under international and U.S. national law. Torture is illegal under the U.N. Convention against Torture and other Cruel, Inhuman or Degrading Treatment or Punishment, and the Geneva Conventions Relative to the Treatment of Prisoners of War. The U.N. Convention against Torture defines torture as "... any act by which severe pain or suffering, whether physical or mental, is intentionally inflicted on a person for such purposes as obtaining from him or a third person information or a confession ... when such pain or suffering is inflicted by, or at the instigation of, or with the consent or acquiescence of a public official acting in an official capacity" (United Nations 1984).

Aggressive war is illegal under Article 39 of the Charter of the United Nations where it is defined under international law as a war that is neither in self-defense nor sanctioned by the United Nations Security Council. Aggressive war was considered to be the supreme international crime at the Nuremberg trials and preventing such wars was the fundamental reason for founding the United Nations. Torture and aggressive war are also illegal under U.S. law because the international treaties that the United States signs are binding within the nation as well.

At present, each neuroscientist who signs the pledge must decide for themselves whether particular actions of their government constitute "torture" or "aggressive war." These are, however, legal terms and their application to particular cases will ultimately depend on tribunals and courts with appropriate jurisdiction. For example, the Nuremberg tribunals convicted officials of Nazi Germany of the crime of "aggressive war." Similarly, a 2010 gathering of representatives from 100 different countries in Kampala Uganda considered adding "aggressive war" to the list of crimes to be judged by the International Criminal Court (Simons 2010).

Someone who knowingly assists in the violation of a law is an accessory to the crime and can be legally sanctioned. Current enforcement of the laws against torture or aggressive war varies from minimal to nonexistent, so indictment of perpetrators of these crimes is unlikely, and indictment of scientists as accessories is even more remote. But the obligation to obey a law—even if that law is not enforced—remains.

It is important to note that the pledge does not proscribe working in an area of neuroscience that has the *potential* for application to torture or aggressive war. Such a broad proscription is manifestly impossible. Every single area of neuroscience, from the most molecular to the most clinical, has such potential. What the pledge proscribes is *knowingly* working on *applications* of neuroscientific knowledge to torture or aggressive war. Nor does the pledge proscribe working for a country's military or taking money from its military. The pledge only proscribes working for a military that is engaged in torture or aggressive war.

National and regional professional neuroscience societies can incorporate the substance of the pledge into their ethical statements. The neuroscience society of one country, Uruguay, has already done so. Opposition to work on applications that violate basic human rights and international law is consistent with the ethical positions that many professional neuroscience societies have already taken. Basic documents of the Society for Neuroscience, for example, affirm such goals as ethical treatment of humans and animals in research; human health and well-being; and continuing discussions on ethical issues relating to the conduct and outcomes of neuroscience research (Society for Neuroscience 2012).

EXAMPLES OF ETHICAL STANDS TAKEN BY SCIENTISTS AND HEALTH CARE WORKERS

The pledge is one of many instances of scientists and health care workers taking responsibility for the larger social and political effects of their actions. One well-known example is that of the petition signed in 1945 by 155 atomic scientists. The petition asked the U.S. government to consider the demonstration of the bomb on a remote island rather than using it on population centers. Unfortunately, their petition was too late and was ignored. The bomb had already been developed and the decision had been made to drop it on Japanese cities, an act that would probably now be judged a war crime (Gerson 2007).

Another example of scientists taking responsibility for the uses of their knowledge is a pledge issued by the Network of Concerned Anthropologists in 2005 in relation to the U.S. "war on terror." The pledge declares that "anthropologists should refrain from directly assisting the US military in combat, be it through torture, interrogation or tactical advice" (Network of Concerned Anthropologists 2009; Anthropologists' pledge 2012). Over 1000 anthropologists have signed the pledge and the American Association of Anthropology has issued a statement in accord with the pledge (American Anthropological Association 2012). The anthropologists believe that assistance to the U.S. military in counterinsurgency is contrary to the ethics of their profession that call for support of the tribes they work with, rather than control or domination.

Governing bodies of the World Medical Association, the International Council of Nurses, the American Medical Association, and the American Psychiatric Association have all issued statements against their members participation in torture (Miles 2009).

Members of the American Psychological Association (APA) have also acted to oppose torture. Fifty-eight percent of APA members signed a petition in 2008 declaring that "... psychologists may not work in settings where persons are held outside of, or in violation of, either International Law (the U.N. Convention against Torture and the Geneva Conventions) or the U.S. Constitution" (American Psychological Association 2008). The petition was a grassroots effort by the membership in response to acceptance by the leadership of the APA of participation by American psychologists in acts of "coercive interrogation" by the U.S. military at Guantanamo and elsewhere.

The accepted acts of "coercive interrogation" included waterboarding, sleep deprivation, stress positions, humiliation, and even slamming people against a wall (Miles 2009). Disagreement between the APA leadership and many APA members on the ethics of psychologist participation in torture continues (Kaye 2011).

APPLICATIONS OF NEUROSCIENCE TO "NATIONAL SECURITY"

In much of the discussion about "National Security" and neuroscience, the term "National Security" is used in the rather narrow sense of a country's military and intelligence agencies. In this article, I place the term in quotes when it is used in this sense in order to remind us that real national security is more than that. Real national security will certainly include freedom from fear of foreign invasion and terrorism which "National Security" purports to provide, but will also include freedom from other fears such as hunger, arbitrary arrest, ill health, old age, loss of income, and no future for one's children. Because of the interconnected world in which we live, real national security will require freedom from such fears for all peoples, not just the people of one country. Our concern must be for security that is both real and international.

Current and potential applications of neuroscience to a country's military and intelligence agencies, that is, "National Security," have been extensively described throughout this volume and elsewhere (Rose 2005; Moreno 2006; National Research Council 2008, 2009; Giordano and Wurzman 2011; Giordano 2012; Neurdon 2012; The Royal Society 2012; Tennison and Moreno 2012). Only a brief overview can be given here.

Giordano (2012) distinguishes two major categories of applications of neuroscience to a country's military and intelligence agencies, *assessment* and *intervention*. *Assessment* is the relatively passive use of neural indicators such as electroencephalogram (EEG), evoked potentials, and functional magnetic resonance imaging (fMRI) for purposes such as monitoring alertness and other psychological states in soldiers while they are watching videos or radar screens; determining suitability of individuals for different tasks; training; and determining whether someone is lying during an interrogation. This last use, determining if someone is lying, has received a great deal of attention (National Research Council 2008; Marks 2010; Tennison and Moreno 2012). The consensus among neuroscientists seems to be that lie detection by monitoring brain function is not possible at present, but commercial companies have claimed otherwise and are marketing the use of fMRI (No Lie fMRI Inc. 2012) and EEG (Government Works Inc. 2012) for this purpose. One of these companies even claims that it is possible to identify terrorists or the intention to commit terrorist acts by recording brain activity, a process they refer to as "brain fingerprinting" (Government Works Inc. 2012). Such "mind reading" capacities may be largely fanciful at present, but this may not always be the case.

The possible use of such technology for lie detection or "mind reading" raises such ethical and legal issues as the right to privacy and protection from self-incrimination, rights granted to U.S. citizens under the Fourth and Fifth Amendments to the U.S. Constitution. If it is against the law to search someone's house without a warrant, it should also be against the law to search someone's brain without a warrant.

Intervention is the active external control of neural processes by such means as drugs, brain stimulation, and brain inactivation. Drugs that enhance the performance of their own soldiers can be of obvious value to a military. Drugs that reduce the need for sleep are already in use in the military (Tennison and Moreno 2012), and one can imagine drugs that make soldiers more aggressive, less fearful, or less sensitive to pain. Drugs that affect memory have been discussed to be of possible use in preventing the spread of information or the after-effects of traumatic experiences (National Research Council 2009; Tennison and Moreno 2012).

Drugs and chemicals that impair the function of enemy soldiers are also of obvious military value. "Nonlethal" or "less lethal" agents such as calmatives have received much attention in this regard (Wheelis and Dando 2005; British Medical Association 2007). With regard to intelligence gathering, coercive interrogation might make use of drugs that increase anxiety, fear, or pain. The possible use of the hormone oxytocin to elicit unwarranted trust in interrogation has been widely discussed (National Research Council 2008; Zak 2011; Tennison and Moreno 2012).

Brain stimulation by electrical or other forms of energy is another method of intervention that could be used for some of the same military and interrogation purposes as drugs. Such stimulation would not necessarily require the disturbing prospect of placing electrodes in the human brain but could also be done from outside the cranium using transcranial magnetic stimulation (TMS), transcranial direct current stimulation (tDCS), focused acoustic energy, or optical activation (Tennison and Moreno 2012). Such technologies might be used to target brain centers responsible for recall, executive function, or reward in prisoners being interrogated. They might also find a use in enhancing the effectiveness of a military's own soldiers or damaging the effectiveness of the enemy's soldiers. The latter would require some type of focused energy beam that could affect enemy soldiers at a distance, something comparable perhaps to Raytheon's heat ray, the "active denial system," that has been used in Afghanistan and is being marketed for crowd control domestically (The Royal Society 2012). Intervention includes the possibility of temporary or permanent inactivation of parts of the central nervous system, something which could also be done from outside the cranium. Bypassing a person's agency or will without their consent or altering their personality is manifestly unethical, whether these techniques are applied to a military's own soldiers or to prisoners under interrogation.

Giordano includes brain–machine interfaces (BMIs) within his intervention category. BMIs involve both recording and stimulation of neural activity through electrodes. BMIs can therefore bypass the relatively slow sensory and motor pathways, that is, the normal input and output pathways of the nervous system. Information transmission over these pathways can take hundreds of milliseconds. BMIs could therefore increase the speed of human–machine or human–human communication. A rather fanciful article in Wired Magazine describes squads of soldiers equipped with these devices communicating silently with each other, with distant commanders, and with devices (Piore 2011).

A third category of applications of neuroscience to "National Security," besides those of assessment and intervention, is *biomimesis*—understanding how biological systems work and implementing that understanding in artificial systems. Military forces of many countries are beginning to rely very heavily on robots (Singer 2009).

Animals and humans have evolved remarkably successful means of moving through complex environments, perceiving objects, and acting in accord with their perceptions and goals. Neuroscientists investigate these mechanisms. Many scientists use robots to test their hypotheses about these complex processes, so the application of neuroscience to robotics can be quite direct. Accordingly, research on animal locomotion and other functions of the nervous system is supported by military agencies such as the U.S. military's Defense Advanced Research Projects Agency (DARPA) (DARPA-BAA-11-65 2012; DARPA-Our Work 2012).

Animals move, perceive, make decisions, and act as autonomous agents. Much discussion surrounds the possibility of similarly autonomous robots. Currently, robots such as drones are used for surveillance and for killing, but the decision to kill is made by a human being. The longtime delays involved in keeping a human being in the loop as a decision maker and the vulnerability of the communication links to interference have led to the potential use of autonomous robots. The U.S. military has argued, for example, that its drones have the same right to defend themselves from enemy radar that human pilots have (Singer 2012).

Robots and especially autonomous robots raise the legal and ethical issues of accountability (Singer 2009, 2012). Who bears responsibility for what robots do? They also raise the ethical issues of making war both easier to start and harder to stop because they remove the possibility of human casualties for one side, perhaps the most important traditional impediment to starting and continuing a war (Howlader and Giordano 2013). Reliance on robots is part of a larger mind-set of overconfidence in the superiority of one's technology that can also make war more likely.

Animals and humans have formidable perceptual and cognitive abilities that cannot be easily matched at present by machines but are of critical importance for the military and for intelligence gathering. Surveillance drones, for example, provide massive amounts of video images that require hundreds of human analysts to monitor for useful information (Benjamin 2012). Other forms of surveillance such as monitoring of phone calls or email messages require the same human skills and employ thousands of analysts (Bamford 2012). Understanding the mechanisms of animal and human cognition, and implementing that understanding in machines, is of clear utility for "National Security." Hybrid systems have been developed in which brain responses are recorded from soldiers as they watch successive images (Bardin 2012). Images that evoke brain responses associated with detection of "objects of interest" can be selected for further analysis.

The "dual-use" dilemma is always present in any discussion of neuroscience and the military. Almost every application of neuroscience can be used for benign peaceful uses as well as for military purposes. As pointed out by Nagel (2010), the potential for beneficial uses is often used as a means to silence those who raise fears of misuse. Such critics pose the question, "Surely you are not against helping quadriplegics with BMI devices or finding earthquake victims with robots?" The pledge does not oppose such uses of neurotechnology, rather it relates to the dual-use issue by asking neuroscientists to stay aware of the potential for misuse of neuroscience and by asking that they refuse to participate knowingly in such misuse. It is not sufficient to view BMIs only from the perspective of helping quadriplegics, or autonomous robots only from the perspective of finding earthquake victims.

Marchant and Gulley (2010) have pointed to a "reverse dual-use" dilemma that adds to concerns about the use of neurotechnology for war and intelligence gathering (see also Chapter 10). Such technology can be and is being brought back into the civilian sector where its uses can threaten civil liberties and commonly held values (Arike 2010). Surveillance drones are now being used by police forces in the United States, and there may soon be pressure to arm such drones. As with the initial arming of military surveillance drones, logic suggests that you "neutralize" an enemy or malefactor once you have them in view. A French company has, for example, proposed arming drones with tasers to capture criminals (Singer 2012). "Less lethal" technology for use against enemy soldiers is being brought home for domestic law enforcement and crowd control (Arike 2010). The recent strengthening of the relationship between military and police facilitates this transfer of war and intelligence gathering technology into the civilian sector (Baker 2011).

The brief description in the preceding paragraphs and more extensive discussion elsewhere in this volume and in the literature make clear that neuroscience can be applied to "National Security" in many ways and that these applications have serious social, ethical, and legal consequences (Wheelis and Dando 2005; British Medical Association 2007; Marks 2010; Giordano and Wurzman 2011; DARPA-BAA-11-65 2012; Giordano 2012; The Royal Society 2012; Tennison and Moreno 2012; see also Chapter 7). Concern over these issues is reflected in some of the titles of recent reports and articles such as "Biologists napping while work militarized" (Dando 2009), "A Faustian bargain" (Rose 2011), and "Neurobiology, a case study of the imminent militarization of biology" (Wheelis and Dando 2005). The problems are heightened by the likelihood that much of the current work on applications of neuroscience to "National Security" is classified and the current state of such applications is not fully known.

THE PLEDGE AND NEUROSCIENTISTS

The pledge is one approach that is open to neuroscientists concerned about neurotechnology applications to "National Security." Other approaches include the following:

1. Development of awareness through education and discussion is a necessary first step, but only a first step. The danger is in relying on this approach and burying the issues in academic courses, conferences, and journal articles.
2. Development of committees or working groups to examine the issue and consider ethical parameters that might guide work on "National Security" applications or legislation. A committee of the British Royal Society has, for example, recently issued a report that includes recommendations for national and international oversight of applications of neuroscience to military and law enforcement agencies (The Royal Society 2012). Comparable reports by committees of the U.S. National Research Council have also been published, although these reports are mainly about the ways in which

neuroscience can serve "National Security" and give only minimal attention to ethical issues (National Research Council 2008, 2009).

3. Strengthening and bringing up-to-date existing international law as embodied in the Geneva Conventions, the Chemical Weapons Convention, the U.N. Convention against Torture, and other Cruel, Inhuman or Degrading Treatment or Punishment, and the Biological and Toxin Weapons Convention. Malcolm Dando and colleagues have written extensively on the need for such strengthening and how it might be accomplished (Wheelis and Dando 2005; Atlas and Dando 2006; British Medical Association 2007; Dando 2009, 2012). Attention should also be given to national or supranational laws and resolutions. For example, in 1999, a European Parliament committee called for "a global ban on all research and development, whether military or civilian, which seeks to apply knowledge of the chemical, electrical, sound vibration, or other functioning of the human brain to the development of weapons which might enable any form of manipulation of human beings, including a ban on any actual or possible deployment of such systems" (European Parliament Committee 1999).

4. Embedding concern over applications of neuroscience that violate fundamental human rights or international law into ethical statements by neuroscience societies. Neuroscience is one of the "health sciences," and the Hippocratic oath to do no harm should apply to it as well as to medical practitioners.

None of these approaches will prevent the misuse of neuroscience but together they can help move us toward a scientific culture that is focused on enhancing human well-being, a culture that neither ignores nor minimizes the possibility of applications that damage such well-being.

The pledge is an important part of the overall effort. It is a powerful means of education and raising awareness because it asks for more than passive reception of information or ideas. It moves beyond discussion and provides a way for individual scientists to act in accord with their conscience. It is a simple act, requiring only a signature, in contrast to the long-term work of changing international treaties and ethical statements of professional organizations.

The pledge has been signed by many well-established neuroscientists but has also been signed by many young scientists in graduate school or in their early post-doctoral years. Young scientists are especially sensitive to ethical issues in neuroscience. The pledge allows young scientists to stay in touch with the ethical and compassionate side of themselves while still doing their science. Young people have fears of being only super technicians providing tools to distant and uncertain power. A young student from Brazil commented about the pledge as follows: "As a student I've always been afraid about the day neuroscience would be employed for private and army interests ... I'm glad others like me think this way and worry about the future. Besides it's important that researchers' worries show up to students like me, because we need models to reflect ourselves. Thanks a lot to have made this pledge and thanks a lot to remember that scientists can make politics too."

The pledge includes but goes beyond ethical and legal concerns such as invasion of privacy with assessment techniques or denial of agency and personhood with intervention techniques. Such concerns would apply whether a military is involved in a "good" war such as a war of self-defense or an illegal aggressive war. But the pledge goes further and proscribes knowingly working on applications for a military engaged in aggressive war, even if the application is not in itself of ethical concern. For example, a neuroscientist might be involved in a study of reaction times for personnel engaged in emergency responses on board navy ships. Such a study is itself completely benign, but the pledge would proscribe carrying out such a study for a military that is engaged in aggressive war. The use of a ship in an aggressive war is not benign.

Most neuroscientists will readily agree to the first part of the pledge calling for awareness of how neuroscience can be used for applications that violate basic human rights or international law. But many neuroscientists will not agree to the second part of the pledge that calls for a refusal to participate knowingly in such applications. Some may feel that taking ethical or political stands in relation to their science conflicts with the professional goal of "objectivity" (Schmidt 2000). Others may feel that they should not be "gate keepers" and that they should rely on democratically elected governments to determine how their knowledge is actually used. Those who wish to rely exclusively on democratic processes should ask themselves whether the evidence supports the effectiveness of such reliance in preventing abuses of basic human rights and international law. The evidence is in fact to the contrary. Governments, including democratic ones, do violate basic human rights and international law.

Some neuroscientists will be reluctant to sign the pledge because of fear of alienating their colleagues. Decisions about hiring, promotions, and funding can be based on opinions that go beyond a candidate's scientific abilities and include their ethical or political views. The fear is legitimate, but does not negate the need to act in a moral and lawful manner.

One neuroscientist objected to the pledge as weird and "kind of creepy." He found it odd to pledge not to do things he had no intention of doing. Similar objections could be made against the Hippocratic oath by medical doctors who do not intend to do harm or against a professional society's proscription of participation in acts of torture by those professionals who do not intend to participate it such acts. But the Hippocratic oath, ethical statements by professional societies, and the pledge serve an important purpose. They constrain undesirable behavior and help create a culture of ethics, responsibility, and accountability within the field.

CONCLUSION

Some readers may consider themselves "realists" and view the pledge as naïve. They may believe that nation states will generally do whatever is necessary to maintain power and that little can be done about it. Some will go even further and declare that a nation may have to "go over to the dark side" when dealing with unscrupulous enemies or a serious loss of national power. Torture as well as wars that could be judged to be aggressive will then be necessary. Such readers may recognize the need for appearing to support human rights and international law, but will also understand that this is largely lip service.

Acquiescing in or actively supporting violations of human rights and international law means taking sides in a continuing struggle between two different cultures that are at play in the world. The first culture views the world in terms of power conflicts between nation states and between nation states and nonstate actors. The major framework of this culture is one of coercion, whether through war or other means. The second culture views the world in terms of the great mass of ordinary people. The major framework of this second culture is one of respect for human rights, including real democracy and international law. This second culture holds the hope of a life of dignity for all. The struggle between the two cultures is made more acute by the increasing power and sophistication of technology, including neurotechnology.

Which of these cultures will be dominant is in flux and will be determined by all manner of actors and actions. The neuroscientist pledge is one way in which neuroscientists can join with other professional groups and civil society groups in moving the world toward a culture of peace, human dignity, and respect for international law.

ACKNOWLEDGMENTS

The author thanks Karl Hufbauer, Sandra Oster, and Michael Arbib for their comments on earlier drafts of this article.

REFERENCES

American Anthropological Association. 2012. Statement by executive board of American Anthropological Association. http://www.aaanet.org/pdf/eb_resolution_110807.pdf (Accessed May 11, 2012).

American Psychological Association. 2008. APA petition resolution ballot. http://www.apa .org/news/press/statements/work-settings.aspx (Accessed May 11, 2012).

Arike, A. 2010. The soft kill solution: New frontiers in pain compliance. *Harper's Magazine*, March 2010, pp 38–47.

Anthropologists' Pledge. 2012. Network of concerned anthropologists pledge of non-participation in counterinsurgency: Text of the Anthropologists' Pledge. http://www .ncanthros.org/internationalpledge.

Atlas, R.M. and M. Dando. 2006. "The dual-use dilemma for the life sciences." *Biosecurity and Bioterrorism* 4(3):276–286.

Baker, A. 2011. When the police go military. *New York Times Sunday Review*, December 3. http://www.wired.com/2012/03/ff_nsadatacenter/.

Bamford, J. 2012. The NSA is building the country's biggest spy center (Watch what you say). *Wired Magazine*. http://www.wired.com/2012/03/ff_nsadatacenter/.

Bardin, J. 2012. From bench to bunker: How a 1960s discovery in neuroscience spawned a military project. *Chronicle of Higher Education,* July 9. http://chronicle.com/article/ From-Bench-to-Bunker/132743/ (Accessed June 11, 2012).

Bell, C. 2010. Neurons for peace: Take the pledge, brain scientists. *New Scientist,* 2746:1–2.

Benjamin, M. 2012. *Drone Warfare: Killing by Remote Control.* New York: OR Books.

British Medical Association. 2007. *The Use of Drugs as Weapons: The Concerns and Responsibilities of Healthcare Professionals.* London: BMA Science and Education Department.

Choudhury, S., S.K. Nagel, and J. Slaby. 2009. "Critical neuroscience: Linking neuroscience and society through critical practice." *BioSocieties* 4:61–77.

Choudhury, S. and J. Slaby. eds. 2012. *Critical Neuroscience: A Handbook of the Social and Cultural Contexts of Neuroscience.* West Sussex: Wiley-Blackwell.

Dando, M. 2009. "Biologists napping while work militarized." *Nature* 460(20):950–951.

Dando, M. 2012. Blog at Bulletin of the Atomic Scientists. http://www.thebulletin.org/web-edition/columnists/malcolm-dando (Accessed June 11, 2012).

DARPA website for some of its neuroscience programs: http://www.darpa.mil/Our_Work/DSO/Focus_Areas/Neuroscience.aspx (Accessed June 11, 2012).

DARPA-BAA-11-65. 2012. Defense Sciences Research and Technology. Solicitation number for brief listing of projects of interest to DARPA in basic neuroscience research, operational neuroscience, and behavioral neuroscience. http://www.darpa.mil/Opportunities/Solicitations/DARPA_Solicitations.aspx (Accessed May 11, 2012).

European Parliament Committee. 1999. European Parliament Committee on Foreign Affairs, Security and Defence. Policy report on the environment, security and foreign policy, January 14. www.envirosecurity.org/ges/TheorinReport14Jan1999.pdf (Accessed June 11, 2012).

Gerson, J. 2007. *Empire and the Bomb: How the U.S. Uses Nuclear Weapons to Dominate the World.* London: Pluto Press.

Giordano, J. 2012. "Neurotechnology as demiurgical force: Avoiding Icarus' folly." In *Neurotechnology: Promises, Potential and Problems,* ed. J Giordano. Boca Raton, FL: CRC Press, pp. 1–13.

Giordano, J. and R. Wurzman. 2011. "Neurotechnologies as weapons in national intelligence and defense—An Overview." *Synesis: A Journal of Science, Technology, Ethics, and Policy* 2(1):T55–T71.

Government Works Inc. 2012. "Brain Wave Science." http://www.governmentworks.com/bws/ (Accessed May 11, 2012).

Howlader, D. and J. Giordano. 2013. "Advanced Robotics: Changing the nature of war and thresholds and tolerance for conflict–implications for research and policy." *The Journal of Philosophy, Science, and Law* 13:1–9.

Kaye, J. 2011. "APA 'Casebook' on psychologist ethics, interrogations fails to convince." The Public Record, August 25. http://pubrecord.org/torture/9667/casebook-psychologist-ethics/ (Accessed May 11, 2012).

Marchant, G. and L. Gulley. 2010. "National security, neuroscience and the reverse dual-use dilemma." *American Journal of Bioethics: Neuroscience* 1(2):20–22.

Marks, J.H. 2010. "A neuroskeptic's guide to neuroethics and national security." *American Journal of Bioethics: Neuroscience* 1:4–12.

Miles, S.H. 2009. *Oath Betrayed: America's Torture Doctors.* Berkley, CA: University of California Press.

Moreno, J.D. 2006. *Mind Wars: Brain Research and National Defense.* New York: Dana Press.

Nagel, S.K. 2010. "Critical perspective on dual-use technologies and a plea for responsibility in science." *American Journal of Bioethics: Neuroscience* 1:27–28.

National Research Council. 2008. *Emerging Cognitive Neuroscience and Related Technologies.* Washington, DC: National Academies Press.

National Research Council. 2009. *Opportunities in Neuroscience for Future Army Applications.* Washington, DC: National Academies Press.

Network of Concerned Anthropologists. 2009. *The Counter-Counterinsurgency Manual: Or Notes on Demilitarizing American Society.* Chicago, IL: Prickly Paradigm Press.

Neurdon 2012. DARPA SyNAPSE, http://www.neurdon.com/tag/darpa-synapse/ (Accessed May 11, 2012).

The Neuroscientists' Pledge-may be read and signed online. http://www.tinyurl.com/neuroscientistpledge.

No Lie fMRI Inc. 2012. http://www.noliemri.com/ (Accessed May 11, 2012).

Piore, A. 2011. The army's bold plan to turn soldiers into telepaths. *Discover Magazine*, April 2011. http://discovermagazine.com/2011/apr/15-armys-bold-plan-turn-soldiers-into-telepaths (Accessed May 11, 2012).

Rose, S. 2005. *The Future of the Brain: The Promise and Perils of Tomorrow's Neuroscience.* Oxford: Oxford University Press.

Rose, S. 2011. "A Faustian bargain." *EMBO Reports* 12(11):1086.

The Royal Society. 2012. *Brain Waves Module 3: Neuroscience, Conflict and Society.* London: The Royal Society Science Policy Center.

Schmidt, J. 2000. *Disciplined Minds: A Critical Look at Salaried Professionals and the Soul-Battering System That Shapes Their Lives.* Lanham, MD: Rowman and Littlefield.

Simons, M. 2010. "International court may define aggression as a crime." *New York Times*, May 30. http://www.nytimes.com/2010/05/31/world/31icc.html?r=0 (Accessed November 19, 2012).

Singer, P.W. 2009. *Wired for War: The Robotics Revolution and Conflict in the 21st Century.* New York: Penguin Press.

Singer, P.W. 2012. "A world of killer apps." *Nature* 477:399–400.

Tennison, M.N. and J.D. Moreno. 2012. "Neuroscience, ethics, and national security: The state of the art." *PLoS Biology* 10(3):E1001289.

Society for Neuroscience. 2012. Mission Statement (http://www.sfn.org/about/mission-and-strategic-plan) and Treatment of animals and humans (http://www.sfn.org/Advocacy/Policy-Positions/Policies-on-the-Use-of-Animals-and-Humans-in-Research) (Accessed May 11, 2012).

United Nations. 1984. Convention against Torture and other Cruel, Inhuman or Degrading Treatment or Punishment. http://www.un.org/documents/ga/res/39/a39r046.htm.

Wheelis, M. and M. Dando. 2005. "Neurobiology: A case study of the imminent militarization of biology." *International Review of the Red Cross* 87:859.

Zak, P.J. 2011. "The physiology of moral sentiments." *Journal of Economic Behavior and Organization* 77(1):53–65.

15 Military Neuroenhancement and Risk Assessment

Keith Abney, Patrick Lin, and Maxwell Mehlman

CONTENTS

Much of the debate on human enhancement technologies starts from the standpoint of traditional bioethics. The usual ethical principles applied are familiar to medicine, such as nonmaleficence, the physician's injunction to do no harm. But emerging technologies blur the line between what is medicine and what is engineering. In circumstances such as the human-enhancement debate, it is appropriate to use conceptual tools from engineering ethics as well, such as risk–benefit analysis (RBA).

This extra perspective helps to fill gaps in bioethical analysis, which is made more complicated by enhancements used in a military context as well as those affecting the mind. Military research is a major driver of scientific and technological innovations, from basic science and energy research to robotics and human enhancements; so we cannot ignore military applications, especially since they involve ethically difficult issues related to life and death (Lin 2010). Neuroenhancements, further, deal with perhaps the least understood and most complex biological system—the human brain—with implications for moral and personal identity, and so pose both medical and metaphysical risks.

A LOOK BEYOND TRADITIONAL BIOETHICS

To better explain the role of an RBA here, much of bioethics commonly uses some version of Principlism, from the Nuremberg Code (1948) through Beauchamp and Childress' influential textbook (1977), to the official Belmont Report (1979). Typical statements of Principlism assert that medical professionals must uphold nonmaleficence, beneficence, autonomy, justice, and other relevant principles in their work, while following sometimes-complicated recipes to resolve conflicts among those principles in difficult cases. This complexity partly exists because new cases—especially involving novel technologies—challenge common inter-pretations of how to apply the principles, as well as the usefulness or even the validity of the principles themselves.

In particular, standard applications of these principles are often rooted in certain presuppositions about the limitations and features of the human brain and mind, and these presuppositions are upended by the emerging human enhancement technolo-gies. For example, to force prisoners of war to stay awake for 48 consecutive hours would seem to be unethical and illegal: for normal humans with normal brains, such actions are torturous. But it may not be objectionable to employ a drug that safely enables a soldier to stay awake and alert for that duration, for example, for stand-ing guard or in actual combat. Moreover, where the traditional focus of bioethics is on the welfare of the individual, in a military setting, the welfare of the individual legitimately may be subordinated to the interests of the unit, the mission, or the state; and so, we need something else to reconcile any discrepancies between the two. In the following, we propose that a risk-assessment approach can serve as a useful instrument in the larger ethical toolbox.

THE RISK-ASSESSMENT MODEL

Bioethical dilemmas, then, are exacerbated when core principles come into conflict, or when exact consequences or circumstances of application are uncertain. Under these conditions, it is reasonable to turn to an RBA, sometimes understood as a form of cost–benefit analysis, as a way to assess the permissibility of possible actions. The Belmont Report (1979) had such concerns listed as desiderata under the principle of beneficence (and nonmaleficence):

Assessment of Risks and Benefits

1. The nature and scope of risks and benefits.
2. The systematic assessment of risks and benefits.

But the vagueness of these terms is a recurrent problem in bioethics. While more rigorous RBA is widely used in policy making, such as evaluating the impact of engineering projects, it may be an unfamiliar territory for bioethicists and thus worth explicating here.

"Risk" is an unavoidable concept in the ethics and policy of military neuroen-hancement, yet the term is often used much too loosely. Without a clear under-standing of the range, quality, quantity, diversity, or other aspects of the risks at hand, it would be difficult to arrive at practical guidance for future action. So let us examine the concept more closely: The risks we address herein are primarily related to harmful but *unintended* behavior that may arise from human enhancement in the military. We will describe a fuller range of other risks and issues involving *intentional* harm in the course of the conduct of war later. Also, while much of the literature's discussion of risk deals predominantly with harms to the individual warfighter (e.g., Lin et al. 2008; Wang 2011), we expand their range here to include possible harms to others.

Following the discussion in Lin et al. (2008), let us first define risk simply in terms of its opposite: safety. Risk is the probability of harm; and safety is the degree of freedom from risk. Safety in practice is merely relative, not absolute, freedom from harm, because no activity is ever completely risk-free. For instance, even a training run raises the risk of heat stroke or heart attack; taking aspirin raises the risk of blood not clotting properly or stomach bleeding, even if it lowers the risk of heart attack. Many risks are uncontroversially worth taking, but how can we determine that?

It may help to recognize that risk can be understood in at least four distinct ways. Following on the work of Sven Ove Hansson (2004) and Fritz Allhoff (2009), we can first understand a "risk" as a chance of some unwanted event, or lack of a wanted event, which one is uncertain will occur. If instead an enhancement definitely had some specific impact, such as causing all such patients to die within a year, then it would be more appropriate to term it a "consequence" of that enhancement, rather than a risk: uncertainty is one of the features of risk.

Second, we can understand a risk as the cause of an uncertain but unwelcome event or of the possible nonoccurrence of a desired event. A human enhancement may cause an inability to sleep, or sexual dysfunction, or decreased inhibitions and resultant inappropriate behavior, or other side effects in a way not perfectly predictable. We sometimes call such statistical causal claims a "risk" of such side effects.

The third conception holds that risk is the numerical probability of an unwanted event or lack of a wanted event, expressible as a percentage outcome. Imagine that we ask about the risk of an enhancement to have a certain health impact. For example, how likely is it that taking a particular antisleep medication, which enhances alertness, will result in paranoia or seizures? The appropriate answer is stated as a probability, for example, that the risk is 20% according to clinical studies.

Fourth, risk can be understood as a measure of the expected outcome of unwanted, or lack of wanted, events; this is best understood for groups of events, rather than for a single instance. So, imagine that there are 1000 soldiers who will be given a new mind-altering biotechnology designed to increase their ability to process information and decrease their response time during stressful situations, such as battle. Further, imagine that some of the soldiers will have adverse reactions to the neuroenhancements and be paralyzed as a result. We do not know which soldiers will be paralyzed, but given previous studies or clinical trials, we estimate a rate of 15%. The risk, then, is 150 out of the 1000 soldiers, in the sense that we expect that number of soldiers to become paralyzed due to the biotechnological intervention.

These last two ways of understanding risk are more quantitative, as opposed to qualitative. The third sense of risk gives us the likelihood that something will happen, usually expressed as a percentage; whereas the fourth sense gives us an expected outcome, usually in terms of some number of valued entities lost, or some number of valued entities that we fail to gain, or some number of disvalued entities gained. This fourth sense of risk is the most common sense of "risk" in professional risk analysis. In particular, this concept of "risk" can be defined as "a numerical representation of severity, that is obtained by multiplying the probability of an unwanted event, or lack of wanted event, with a measure of its disvalue" (Allhoff 2009).

In RBA, it is this fourth conception of expected value that is often of most interest to decision makers (see, e.g., Sen 1987). That is, what people usually most want to know is the expected value of the result, sometimes conflated with the "expected utility." This allows a quantitative assessment of both risk and benefit in a way that gives a clear numerical answer for a course of action—a "decision algorithm" of sorts. For example, we could decide that causing paralysis to 150 soldiers is unacceptable and demand changes to the bioenhancements to make them safer before they are used. But if the expected loss can be reduced to, say, 0.5%—that is, we expect five soldiers out of 1000 to be paralyzed as a result—we may deem the enhancement "safe enough" to use. Such judgments are routinely made for vaccines and other public health interventions that bear some risk for the individual while enhancing the whole. Such judgments are also routine for commanders of troops in wartime, assessing whether particular tactics in battle are too risky or not.

But of course, while this sense of risk as expected value may be desirable for policy makers, it often greatly oversimplifies the intractable problem of ascribing mathematically exact probabilities to all the undesired outcomes of our policies. It often suggests an aura of false precision in ethical theorizing. It also ignores a common issue concerning risk assessment in bioethics: the distinction between "statistical victims" and "identifiable victims." RBA might well assert a statistical certainty that we would save more lives (or quality-adjusted life years or whatever the unit of assessment) by diverting money we would spend on "last-chance treatments" to instead campaigns to, say, prevent smoking. But the "rule of rescue" (Jonsen 1986) and related ethical rules of thumb rely on the idea that we actually value saving identifiable lives more than statistical lives. That is, we tend to care more about using every last measure to save grandma from her stage IV cancer than to save many more lives of future strangers. Or, in the military, I may unquestioningly risk the future well-being of myself and even my entire unit in the mad dash to rescue a wounded brother-in-arms, in a way that RBA would consider irrational but in fact may result in a medal of valor, even if posthumously awarded. As long as the difference in our moral attitudes toward statistical victims and identifiable victims is defensible, attempts to use RBA are problematic at best.

What then can we say for certain about risk, especially with respect to military neuroenhancement? How can we answer the question of determining acceptable risk? We can begin by seeing that risk and safety are two sides of the normal human attempt to reduce the probability of harm to oneself and others, even as we are often unsure of the exact probabilities involved. To make things even more difficult, war is a strange human activity, not least because it reverses this tendency: in war, one ordinarily wishes to increase the probability of harm to one's enemies. But the laws of armed conflict and the typical rules of engagement make clear that not all ways of increasing risk for one's enemy are morally legitimate, and some ways of increasing risk for one's own side may be morally legitimate and even morally required. These facts considerably complicate the ethics of risk assessment for military human neuroenhancement.

In the absence of precise probabilities, can we say anything useful about how to determine whether or not particular neuroenhancements pose an acceptable risk or not? Perhaps some further conceptual clarification will help.

ACCEPTABLE-RISK FACTOR

To begin, the major issues in determining "acceptable risk" include, but are not limited to, the following five factors (Lin et al. 2008).

Consent/Voluntariness

Is the risk voluntarily endured, or not? For instance, secondhand smoke is usually considered more objectionable than firsthand, because the passive smoker did not consent to the risk, even if the objective risk is smaller. Will those who are at risk from military human enhancements reasonably give consent? When, if ever, would it be appropriate to engage in enhancement without consent of those affected?

Morality ordinarily requires the possibility of consent: to be autonomous is, at minimum, to have the capacity to either give or withhold consent to some action. But warfighters often have no choice about substantial parts of their roles and duties; once an individual has volunteered for service, military ethics accepts that many choices open to civilians are no longer options to military personnel. But which choices exactly—that is, under what circumstances could a human enhancement be required for warfighters?

Informed Consent

Another possible problem is the uncertainty or unpredictability arising from enhancements: Will they actually work as promised? Even if they do, will the enhancements have unintended consequences or side effects? This leads to a second aspect of consent, familiar from the bioethics literature.

The worry here begins with the usual requirement in civilian bioethics to inform patients of the details about his or her diagnosis, prognosis, alternative treatment options, and side effects of each alternative, before treatment is morally permitted. For enhancement ethics, this is already problematic: a "diagnosis" is commonly understood as a physician's theory of what ails a patient, but nothing ails the soldier undergoing enhancement; enhancement is typically understood to stand in contrast to therapy (Allhoff et al. 2010). Instead, the "diagnosis" refers to whatever ability the enhancement is intended to improve or optimize—possibly regardless of its effect on the rest of the warfighter's life. The "prognosis" then refers to the expected future with respect to that ability given the enhancement, versus the expected future without it; only if alternative enhancement treatments are offered would further alternatives be relevant to discuss. And if the enhancements are given prior to the completion of clinical trials, the side effects may be merely speculative, or even completely unknown. Are warfighters entitled to all of this information before they consent to enhancement, if their consent is indeed required?

There is yet a further risk factor that falls under "informed consent," though not to the warfighter but to other people. Is the risk—of enhancement malfunction, increased probability of disproportionate violence or even war crime, or other harm—by neuroenhanced warfighters to enemy combatants required to be disclosed? Under usual interpretations of the laws of armed conflict, there is no general "duty to disclose" the nature of one's attack upon one's intended target, as long as it adheres to principles of discrimination and proportionality; surprise is well understood as a legitimate tactic in war.

But neuroenhancements may pose novel difficulties if they increase risk to unintended targets—the noncombatants, specifically, the civilian population of the enemy, or even of neutrals or one's own population while housing and training enhanced warfighters. Is it morally permissible to have neuroenhanced warfighters who pose a risk to civilian populations without informing the populations of the risk? For example, suppose warfighters take drugs or other psychological enhancers that reduce inhibitions and fear in order to enhance battlefield performance, but in a civilian setting, these drugs cause more traffic accidents. This is reported to be exactly the risk with toxoplasmosis, a parasitic infection of interest to the military (Sapolsky and Vyas 2010).

The Affected Population

This leads us to consider that even if consent or informed consent for the warfighter is not morally required with respect to human enhancements, we may need to focus on the affected population as another factor in determining acceptable risk.

Who else is at risk, besides the enhanced soldiers themselves—does it include groups that are particularly susceptible or innocent, such as the elderly or young children, or merely those who broadly understand that their role with respect to enhancements is risky, even if they do not know the particulars of the risk? In military terms, civilians and other noncombatants are usually seen as not morally required to endure the same sorts of risks as military personnel, especially when the risk is nonvoluntary or involuntary. Will the use of military neuroenhancements pose the risk of any new special, unacceptable harms to noncombatants?

An immediate issue pertains to the reliability of military neuroenhancements: Will they degrade over time or have side effects that only slowly come to light? Will they be easily reversible upon reentry into civilian life, or will their effects be permanent? Will they have vast and/or unpredictable differences between different human subjects? Will they exacerbate underlying physical or psychological problems, and potentially cause physical or psychological difficulties for the loved ones, friends, family, and communities of enhanced soldiers?

For instance, any neuroenhancements that increase aggression may then cause warfighters to attack indiscriminately or disproportionately, similar in effect to landmines as well as nuclear, biological, and chemical weapons, and likewise would be immoral to deploy. Even worse is when enhancements foreseeably may cause damage outside a combat zone, for example, in ordinary interactions with shopkeepers, friends, or family.

Seriousness and Probability

We thereby come to the two most basic facets of risk assessment: seriousness and probability, that is, how bad would the harm be and how likely is it to happen?

Seriousness

A risk of death or serious physical or psychological harm is understandably seen differently than the risk of a scratch or a temporary power failure or slight monetary costs. But the attempt to make serious risks nonexistent may turn out to be prohibitively expensive. What, if any, serious risks from military neuroenhancements are acceptable—and to whom: soldiers, noncombatants, one's family, the rest of one's environment, or anything else?

Probability

This is often conflated with seriousness but is conceptually quite distinct. The seriousness of the risk of a 15-km asteroid hitting Earth is quite high (possible human extinction), but the probability is reassuringly low (though not zero, as perhaps the dinosaurs discovered). What is the probability of harm from military neuroenhancements? How much certainty can we have in estimating this probability? What probability of serious harm is acceptable? What probability of moderate harm is acceptable? What probability of mild harm is acceptable?

Who Determines Acceptable Risk?

In all social theorizing, the understanding of concepts retains a certain degree of fluidity, dependent in part upon how those in power or epistemic authority determine their meaning. The concept of risk, which includes psychological, legal, and economic considerations as well as ethical ones, is certainly no different. Hence, the concept of an acceptable risk—or an unacceptable one—is at least in part socially constructed. In various other social contexts, all of the following have been defended as proper methods for determining that a risk is unacceptable (Lin et al. 2008).

Good-faith subjective standard: Under this standard, it would be left up to each individual to determine whether an unacceptable risk exists. That would involve questions such as the following: Can soldiers in the battlefield be trusted to make wise choices about acceptable risk? The problem of nonvoluntary risk borne by civilian noncombatants makes this standard impossible to defend, in addition to the problems raised by the idiosyncrasies of human risk-aversion and the requirements of the chain of command and the reasonable expectation that orders will be carried out.

The reasonable-person standard: An unacceptable risk might be simply what a fair, informed member of a relevant community believes to be an unacceptable risk. Can we substitute military regulations or some other basis for what a "reasonable person" would think for the difficult-to-foresee vagaries of conditions in the field and the subjective judgment of soldiers? Or what kind of judgment would we expect an enhanced warfighter to have: would we trust them to accurately determine and act upon the assessed risk? Would they be better—or worse—than an "ordinary" soldier in risk assessment? Would their enhanced powers distort their judgment?

Objective standard: An unacceptable risk requires evidence and/or expert testimony as to the reality of, and unacceptability of, the risk. But there is the "first-generation problem" to consider: how do we understand that something is an unacceptable risk unless some first generation has already endured and suffered from it? How else could we obtain convincing objective evidence? (Lin et al. 2008).

With regard to the military use of neuroenhancements, the last standard seems most plausible, given that it is most often defended in law and practice, despite the first-generation problem. One solution could be to assert an ethical obligation for extended testing of enhanced warfighters in a wide range of environments before risking dangerous interaction between the enhanced and unenhanced. This testing must be thorough, extensive, realistic, variegated, and come in stages, so that full deployment with possible or actual civilian contact comes only at the end of a long training regimen and safety inspection. From the risk–reward perspective of RBA, it may very well be acceptable to deploy enhanced warfighters as soon as such extensive testing indicated their mistakes, and other risks were, on average, no worse than that of the typical human soldier.

PRECAUTIONARY PRINCIPLES

It should be noted that, unlike an RBA, an alternate view of risk is one that ignores benefits entirely: a precautionary principle. Although it has been variously formulated (Allhoff 2009), here is a representative statement of a strong version of the precautionary principle: "When an activity raises threats of harm to the environment or human health, precautionary measures should be taken even if some cause and effect relationships are not fully established scientifically" (Wingspread 1998).

Neuroenhancements that had not been adequately researched may violate this version of the precautionary principle. Such a principle takes the uncertainty inherent in RBA and in effect endorses a kind of "maximin" (or maximizing the minimally acceptable results [Rawls 1971]) mode of assessing acceptable risk: unless we can be sure the worst-case scenario will be acceptable, we ought not to undergo the risk.

But taking this precautionary principle as a blueprint for risk assessment is vastly at odds with standard procedures, not only in the military but also in civilian bioethics. For instance, we do not in fact require that a new vaccine or other medical treatment be guaranteed to produce no deaths or other negative side effects, and the point of clinical trials is to attempt to establish cause–effect relationships, not to completely prohibit use; there are well-understood circumstances in which an experimental treatment may be made available before it is fully causally understood and has full regulatory approval. And military necessity is one of those circumstances.

OTHER RISKS

A perpetual risk remains with respect to security issues for enhanced warfighters, although the issues here are common to many aspects of technological culture. For example, how susceptible would a neuroenhanced warfighter be to "hacking," for example, after capture? That is, especially given "arms race" concerns, are there enhancements that would create a security risk if they fell into enemy hands? For example, suppose we deploy warfighters with enhanced immunity to brain damage or to biological or chemical pathogens that normally disable the brain, such as neurotoxins; if they are captured and thereby have that technology discovered and replicated by a rogue state or terror group, would it unduly risk a biological or chemical attack on our citizens, at no risk to the warfighters themselves?

Besides policy risks, there are also specific legal risks to monitor. Some experts have pointed out that neuropharmacological agents—either enhancing or incapacitating—may violate the Biological Weapons Convention and the Chemical Weapons Convention (The Royal Society 2012) as usually understood. We might also suggest that the Biological Weapons Convention could be breached in an unusual but nevertheless possible way: a bioenhanced person or animal could count as a "biological weapon" or "biological agent," since these terms not clearly defined in the agreement (Lin 2012).

Some commentators have raised risks of a more abstract sort. For instance, is there a risk of, perhaps fatally, affronting human dignity or cherished traditions (religious, cultural, or otherwise) in allowing the existence of enhanced warfighters or "Supermen"? Do we "cross a threshold" in creating such superhuman warfighters, possibly in a way that will inevitably lead to some catastrophic outcome? Is this "playing God" with human life? (Evans 2002).

What seems certain is that the rise of neuroenhanced warfighters, if mishandled, will cause popular shock and cultural upheaval, especially if they are introduced suddenly and/or have some disastrous safety failures early on. That is all the more reason that a lengthy period of rigorous testing and gradual rollout (a "crawl-walk-run" approach) appears a moral minimum for the ethical deployment of enhanced warfighters. Further, this points to the early, prior need to identify a full range of possible ethical, technological, and societal issues of military enhancements in order to better account for risk.

CONCLUSION

Of course, much more can be said of the above risk-assessment model and its risk factors: this chapter is meant to sketch out its frame in a military context. Nor is it suggested that other forms of bioethics analysis should be replaced here (see Chapter 17 for additional discussion). Again, bioethics is a natural and appropriate starting point in discussing human enhancements. But in a novel, hybrid field as human enhancement—often blending medicine with engineering—we need all the conceptual tools we have to perform a full ethical analysis.

ACKNOWLEDGMENTS

This chapter is adapted from a research report, in progress as of this writing, funded by The Greenwall Foundation, California Polytechnic State University's College of Liberal Arts, Philosophy Department, and Research and Graduate Programs (San Luis Obispo). Any opinions and findings expressed in this chapter do not necessarily reflect those of the supporting organizations.

REFERENCES

Allhoff, F. 2009. "The coming era of nanomedicine." *The American Journal of Bioethics* 9(10):3–11.

Allhoff, F., P. Lin, J. Moor, and J. Weckert. 2010. "Ethics of human enhancement: 25 questions & answers." *Studies in Ethics, Law, and Technology* 4(1):4.

Ashford, N., K. Barett, A. Bernstein, R. Costanza, P. Costner, C. Cranor, and P. deFur, et al. Wingspread Statement on the Precautionary Principle (1998). In *Protecting Public Health and the Environment: Implementing the Precautionary Principle*, eds. Raffensperger, C. and J. Tickner. Washington, DC: Island Press, pp. 353–355.

Beauchamp, T.L. and J.F. Childress. 1977. *Principles of Biomedical Ethics*. New York: Oxford University Press.

The Belmont Report. 1979. "The Belmont report ethical principles and guidelines for the protection of human subjects of research." Report from the national commission for the protection of human subjects of biomedical and behavioral research. Department of Health, Education, and Welfare, Washington, DC. http://www.hhs.gov/ohrp/humansubjects/guidance/belmont.html.

Evans, J.H. 2002. *Playing God?: Human Genetic Engineering and the Rationalization of Public Bioethical Debate*. Chicago, IL: University of Chicago Press.

Hansson, S.O. 2004. "Fallacies of risk." *Journal of Risk Research* 7(36):353–360.

Jonsen, A.R. 1986. "Bentham in a box: Technology assessment and health care allocation." *The Journal of Law, Medicine & Ethics* 14(3/4):172–174.

Lin, P. 2010. "Ethical blowback from emerging technologies." *Journal of Military Ethics* 9(4):313–331.

Lin, P. 2012. "More than human? The ethics of biologically enhancing soldiers." *The Atlantic*, February 16. http://www.theatlantic.com/technology/archive/2012/02/more-than-human-the-ethics-of-biologically-enhancing-soldiers/253217/.

Lin, P., G. Bekey, and K. Abney. 2008. "Autonomous military robots: Risk, ethics, and design." Report for the U.S. Department of Defense/Office of Naval Research. California Polytechnic State University, San Luis Obispo, CA. http://ethics.calpoly.edu/ONR_report.pdf.

Rawls, J. 1971. *A Theory of Justice*. Cambridge, MA: Belknap Press.

The Royal Society. 2012. *Brain Waves Module 3: Neuroscience, Conflict and Security*. London: Royal Society.

Sapolsky, R. and A. Vyas. 2010. "Manipulation of host behaviour by *Toxoplasma Gondii*: What is the minimum a proposed proximate mechanism should explain?" *Folia Parasitologica* 57(2):88–94.

Sen, A. 1987. "Rational behavior." In *The New Palgrave: Utility and Probability*, eds. Eatwell, J., M. Milgate, and P. Newman. New York: The Macmillan Press Ltd, pp. 198–216.

Wang, J. 2011. "The impact of individual-, unit-, and enterprise-level factors on psychological health outcomes: A system dynamics study of the U.S. military." Massachusetts Institute of Technology, Cambridge, MA. http://hdl.handle.net/1721.1/68447.

16 Can (and Should) We Regulate Neurosecurity?

Lessons from History

James Tabery

CONTENTS

INTRODUCTION

From Leonardo da Vinci's tank and Galileo's military compass to the Manhattan Project and human terrain teams, the relationship between science and the military has enjoyed a long history. We should expect the neurosciences to be co-opted for and into national security and defense use as well, and there are already efforts underway toward such ends (National Research Council 2008). As elucidated in the chapters of this volume, neuroimaging technologies are being utilized for intelligence, neuro-pharmacological mechanisms are being investigated to enhance warfighter performance, and brains–machine integration is being developed to facilitate training and engage remote cognitive augmentation (see also Huang and Kosal 2008). Jonathan Moreno has conveniently referred to the intersection of neuroscience and the military as "neurosecurity" (Moreno 2006).

In this light, the task of this chapter is to ask and answer the following question: Can (and should) we regulate neurosecurity? By "regulate" I mean to establish some organization/committee charged with the task of overseeing military-purposed neuroscience. I will address this question in two steps: first, by reviewing the history of the science–military relationship in order to draw out generalizable lessons from that history; and second, by reviewing existing regulatory frameworks that oversee scientific research. The lessons from history will show that some of these regulatory frameworks are better than others at capturing the unique features of the science–military relationship.

THE LESSONS FROM HISTORY

LESSON ONE: CONTROVERSIAL ... NO MORE, NO LESS

One episode in particular dominates modern discussions of the relationship between science and the military, the Manhattan Project (Hughes 2002). The atomic bombings of Hiroshima and Nagasaki continue to be controversial. Were these necessary so as to save countless lives by quickly ending the war? Or, were the bombings, which killed hundreds of thousands of Japanese, an act of terrorism that intentionally targeted civilians? The ongoing nature of this controversy was reinforced in August 2010, when the U.S. Ambassador to Japan, John Roos, attended the annual memorial of the bombings, the first time a U.S. representative did so. Was Roos' attendance an inappropriate act of contrition, or an overdue act of empathy? (Fackler 2010). Scientists are representative of the broader communities from which they come, and so debate exists among scientists about particular episodes (like the Manhattan Project) and about the more general relationship between science and the military. Some scientists foreswear any defense industry funding (a perspective strongly advocated by Bell in Chapter 14); others acknowledge the importance if not need to participate in national defense (Giordano et al. 2010).

Lesson One, then, is that the current relationship between science and the military is *controversial ... but no more, no less*. The controversy is a reality, and it must be acknowledged that there are legitimate criticisms of intentionally targeting civilians with a weapon of mass destruction. But that controversy surrounding an act in a time of war should remain distinct from something like the Tuskegee syphilis study, for which no controversy exists. Everyone now agrees that intentionally misleading human research subjects about the nature of a disease and actively preventing them from receiving treatment is a gross violation of human research ethics (Reverby 2009).

LESSON TWO: SUCCESSES ... AND FAILURES

Indeed, the Manhattan Project dominates current discussions of the relationship between science and the military. And that is a problem. The problem is that the Manhattan Project is a rather unique episode in the history of that relationship in that it was a complete "success." Theoretical physicists such as J. Robert Oppenheimer and Edward Teller set out and succeeded in creating a containable, controllable, deliverable fission-based nuclear device (York 1976). The problem with using the Manhattan Project as exemplar of the relationship between science and the military is that the ethical considerations focus only on the dangers of scientific *successes*. In fact, the history of the relationship is replete with failures.

On the one hand, there have been failures in design; these are failures that result from bad science. For instance, a freedom of information act request by the (now defunct) *Sunshine Project* in 2005 revealed that the U.S. Air Force considered spending several million dollars in an attempt to develop a gay bomb. As noted on the Project's website (offline since February 2013), the use of such a weapon could adversely alter discipline and morale in enemy units; a distasteful but nonlethal example would be the use of strong aphrodisiacs to induce rampant (homo)sexual behavior. With the gay bomb deployed, or so the thought went, enemy soldiers would be too preoccupied with sexual lust for their fellow soldiers to take up arms and engage in military actions. Needless to say, a gay bomb was never developed; it was never developed because it was a failure from the very beginning: a failure in design, a failure in being based upon flawed science and logic. Homosexuals have served openly in the military of countries such as Australia, Canada, Denmark, Ireland, Israel, and the Netherlands (McLean and Singer 2010). In 2011, the United States joined this list with its abandonment of "don't ask, don't tell." Soldiers from these countries were/are just as capable as their heterosexual counterparts in fulfilling their military duties, despite the "gay bomb" effect(s) of their nature and nurturance.

There are also failures in implementation: these are based on sound science, but fail at the point at which sound science is implemented for military ends. A tragic example of this unfolded in 2002 in Moscow. Chechen militants took nearly 900 civilians hostage at the Dubrovka Theater. After a two-day siege, Russian forces pumped aerosolized fentanyl (an opiate analgesic commonly used during medical procedures) into the theater. Russian forces weaponized a medical intervention as a means to disable terrorists. The problem was that the use of fentanyl did not discriminate between terrorist and hostage. When Russian forces subsequently assaulted the theater, it became obvious that terrorists and hostages alike had not just been sedated but were killed by the drug. The Russian government has closely guarded the details and cause of death of those hostages involved, but best estimates posit roughly 150–200 dead from fentanyl overdose, and several hundred more suffer long-term disability from fentanyl poisoning (Pilch 2003).

Lesson Two from history is that the relationship between science and the military presents a mixed bag of *successes ... and failures.* What is more, neurosecurity seems to be particularly vulnerable to failures, in part because neuroscience itself is a complex—and somewhat nascent—science. So let us not simply look to the Manhattan Project as representative of science "getting it right." There are plenty of gay bombs and fentanyls that "get it wrong," which raises significant concerns in the ethical discourse.

Lesson Three: Neurosecurity Is Not Totally Unique

This volume is devoted to neurotechnology in national security and defense. It is focused specifically on the intersection of neuroscience and the military because this raises a unique set of ethical concerns. But it is important to realize that (1) the military engaged science and technology long before neuroscience was a discipline, and (2) other sciences also can raise ethical issues.

The U.S. military has long been in the business of neuropharmacologically altering soldiers, albeit with caffeine and nicotine. Soldiers' rations included stimulants such as coffee, chocolate, and cigarettes long before neuroscience was institutionalized in the 1960s (Fisher and Fisher 2011). Moreover, the military did not need neuroscience to affect soldiers' empathy. After World War II, research revealed that U.S. soldiers were often firing their guns but intentionally missing enemy soldiers because of unease with killing. In response, the U.S. military switched from using round targets to using human-shaped targets during military training. The switch to a human-shaped target was designed to reduce soldiers' empathy by making them more comfortable aiming and firing their guns at other humans (Grossman 2009). The military did not rely upon neuroscience to alter soldiers' cognitive and emotional function.

What's more, neuroscience is not alone in efforts to integrate mind and machine. Bioengineers at Raytheon have created a bionic exoskeleton that individuals can wear and that robotically enhances the wearer's normal muscular movements and capacities (Marrapodi and Lawrence 2010). Although the robotic suit now only functions to increase a soldier's carrying capability, it is viable that the suit is to be developed with both defensive and offensive technologies incorporated. How will soldiers conceive of their robot-mediated actions? Is robot-mediated killing the same as unmediated killing? Is robot-mediated heroism the same as unmediated heroism? The bionic exoskeleton does not rely upon neuroscience to alter a soldier's sense of self.

Lesson Three is that *neurosecurity is not totally unique*. We should be genuinely concerned about altering a soldier's capacity for empathy or a soldier's sense of self. But such consequences are not solely the products of technology. The danger in focusing only on neurotechnology is that we may mistakenly confine analysis to that particular intersection, when issues are broader and older.

Lesson Four: Legitimate Concerns

Lesson One was that the relationship between neuroscience and the military is controversial, but no more and no less. Still, we must recognize that there are genuine, legitimate concerns associated with military-purposed neuroscience. Intentionally killing, maiming, or permanently disabling another person is, *prima facie*, wrong. Of course, killing, maiming, and permanently disabling are only *prima facie* wrong because there are circumstances such as self-defense and national defense in which the acts may be justified. But the scientist who contributes to making the act of killing, maiming, or permanently disabling more efficient should (at the very least) be wary because he or she will rarely have any say in whether or not the implementation of his or her science is justified or not.

Turning to human subjects who participate in research, critics have noted that research designed to enhance a soldier in one context (such as a time of war) may in fact so specialize him/her that the result is a general disability outside of that context (Moreno 2006). For instance, neuroscientific research designed to reduce empathy may make a soldier more efficient on the battlefield, while simultaneously making that soldier a less efficient spouse, parent, or citizen upon returning home. The defense industry has a somewhat checkered past when it comes to protecting human research subjects. Projects such as MK-ULTRA, which sought to use lysergic acid diethylamide (LSD) as a drug for "mind control" in the 1950s and 1960s, never achieved this goal, but did in fact permanently disable and even kill human research subjects in the process (Streatfeild 2007).

Finally, scientists must be wary of the rush to apply basic science in operational settings before it is ready (Giordano 2012). This legitimate concern was fully displayed during the fentanyl debacle in Moscow. It can also be seen in attempts to utilize neuroscientific techniques and technologies to detect liars. The neuroscience of lie detection is yet infant, but that has not prevented some from deploying the technology in such pursuits (Greely and Illes 2007; see also Chapters 9 and 12).

Hence, Lesson Four from history is that there are *legitimate concerns* pertaining to the relationship between neuroscience and the military. This is not to say or imply that neuroscientists who assist the military are unethical, abusers of research subjects, or promoters of bad science. But it is to say that history justifies legitimate concerns associated with military-purposed science: concerns about killing, concerns about conduct, concerns about research subjects, and concerns about poorly implemented and operationally translated research.

OPTIONS FOR REGULATING NEUROSECURITY

Can and should we regulate neurosecurity? There are two trends in the literature that attempt to answer this question. First, a number of authors have simply called for the discourse, more *questions*, and more *awareness*. After asking a series of hypothetical questions about government-supported neural monitoring, Nita Farahany (2008) concluded, "These are just some of the questions we must ask as we balance scientific advances and the promise of enhanced safety against a loss of liberty." After reviewing Moreno's *Mind Wars*, Hugh Gusterson (2007) concluded, "Time to start talking!", and Charles Jennings (2006) added, "[*Mind Wars*] should help bring these questions into the open." Such calls are a valuable starting point, but we must inevitably ask, "Then what?" Calls for questions and talking don't actually provide solutions.

Second, and at the opposite extreme, others have suggested international accords, or engagement of nongovernmental regimes to take responsibility for regulation (Moreno 2006; Huang and Kosal 2008). A problem with this, as noted by Green (2008), is that there is a diversity of perspectives on science, the military, and governance/regulation across different nations and scientists from different countries, cultures, and politics. As a result, it is hard to imagine a consensus statement that would bring such diverse parties into agreement on something directly related to national security.

Here, I focus on options available for regulation at the intersection of academic neuroscience and the military in the United States. There are several advantages for focusing such energies here. First, by keeping discussion and emphasis confined to U.S. applications, we may avoid problems associated with diversity at the international level. And second, there already are a number of existing models for regulating academic science in the United States; thus, these models can be evaluated in terms of their appropriate application to military-purposed neuroscience rather than attempting to reinvent an entirely new approach.

But before I turn to the options I propose, one clarification is in order: From this point, I will be discussing options for regulating military-purposed science *generally*. This follows from Lesson Three that neurotechnology is not unique. Thus, the question, "can (and should) we regulate neurosecurity?" must be situated within the broader question, "can (and should) we regulate military-purposed science and technology?" If military-purposed science generally can and should be regulated, then any considerations of neuroscience and technology go with it; if on the other hand, military-purposed science cannot or should not be regulated, then neuroscience and neurotechnology are not exceptions.

Option One: Ban

One regulatory option would be to simply ban military-purposed science altogether. Existing models for such an extreme move include temporary pauses by scientists themselves (as was the case with recombinant DNA research following the Asilomar conference in the 1970s; in the contexts of neuroscience, refer to the proposal of Bell in Chapter 14) or federal bans on funding by the government (as was the case with research on all but a limited number of human embryonic stem cell lines under the George W. Bush administration) (Fredrickson 2001; Seelye 2001). The basic idea would be that scientists themselves or the government would decide to avoid practicing or funding military-purposed science.

The problem with this option is that it ignores Lesson One from history: that such science is controversial but no more and no less. A collective ban is much too extreme a response to something that is only controversial in nature. That is, individual scientists may choose not to participate in military-purposed science, but this is unlikely to occur collectively. Moreover, the federal government is obviously not going to deprive itself of the resources of science where national security and defense are concerned.

Option Two: Strong Regulation

A second regulatory option follows the model of an institutional review board (IRB), which oversees all human research at an institution, or an institutional animal care and use committee (IACUC), which oversees all animal research at an institution. These models may be regarded as *strong* regulation. These committees have powers of approving or disapproving research, of impeding funding, of investigating allegations of abuse, and of reviewing research on a regular basis (Shamoo and Resnik 2009).

The problem with responding to military-purposed science with a strong regulatory model is that it again ignores Lesson One from history. Committees with the power of strong regulation have historically arisen in response to overt and extreme violations of ethical norms, such as was the case with IRBs following the revelation of the Tuskegee syphilis study. The Tuskegee syphilis study revealed and fortified that informed consent was absolutely necessary for the conduct of human research and that researchers could not be left to themselves to handle the task of informed consent. Moreover, if the concern is about protecting human subjects participating in military-purposed research, then this worry has already been addressed through the existence and activities of IRBs. Thus, strong regulation also is a poor fit for further regulation.

OPTION THREE: WEAK REGULATION

An institution's conflict of interest committee provides another model of regulation. In contrast to IRBs or IACUCs, conflict of interest committees do not have strong powers of approving/disapproving research or removing funding. Instead, conflict of interest committees are notified by researchers of research involving a conflict of interest by answering a few quick questions, such as whether or not a researcher stands to significantly benefit financially from the research (on the order of $5000/year or more) and/or whether or not a researcher stands to benefit financially from purchasing decisions associated with the research. If a researcher does not have a significant conflict of interest, the lack of a conflict is noted, and the researcher can continue his/her project. Conversely, if a researcher does have a significant conflict of interest, the conflict of interest committee works with the researcher to eliminate, reduce, or manage the conflict so that research can continue without the danger of it being corrupted (Shamoo and Resnik 2009). This model may be regarded as *weak* regulation. This is not to say that conflict of interest committees are weak; rather, it is a relative term to contrast the design of conflict of interest committees with IRBs and IACUCs. If conflict of interest committees find violations, the matter is reported to academic administration.

A weak regulatory model for military-purposed science would be as follows: When academic researchers report their proposed research to the institution's office of sponsored projects, the researcher answers a few quick questions just as s/he would for conflicts of interest.

1. Is the research funded by the defense industry?
2. Is the intended purpose of the research to
   a. kill, maim, or permanently disable persons?
   b. covertly monitor persons? or
   c. permanently alter persons?

If the researcher answers "no" to both questions, then the researcher continues with his/her project. If the researcher answers "yes" to the first question (about defense industry funding) but "no" to the second question (about purpose), then the military-purposed research project is noted, but no action or management is necessary. If the researcher answers "yes" to the first question and "yes" to the second about any one of the purposes, then a committee (let us call it the "military-purposed science committee") works with the researcher to proceed appropriately. For instance, if the research is designed to covertly monitor persons (via, say, some type of lie detection), then the committee could put the researcher in touch with legal counsel to insure that privacy laws are not (or could not be) violated through the use of technology. If the research is designed to permanently alter persons (with, say, neuropharmacology), then the committee could suggest adding a physician (psychiatrist) to the research team to focus on the impact of such alterations. If the research is designed to kill, maim, or permanently disable, then the committee could notify academic administration that genuinely controversial research is being proposed, so that the institution can decide whether or not this type of research is advocated and/or supported by the university governance.

The case for a weak regulatory model of military-purposed science is derived from Lesson Four: legitimate concerns. The idea is to identify when legitimate concerns are in play, so that steps can be taken to address such issues. In the same way that conflicts of interest are not inherently wrong but inherently must be managed, so too is military-purposed science not inherently wrong, but inherently susceptible to legitimate concerns that would necessitate, and benefit from oversight.

OPTION FOUR: ACTIVE EDUCATION

A less-invasive model that still retains some element of regulation can be found in mandated responsible conduct of research (RCR) training. The National Institutes of Health (NIH) and the National Science Foundation (NSF) both require RCR training for all students (graduate and undergraduate) who receive financial support for research. The frightening history of irresponsibly conducted research (be it data falsification, plagiarism, improperly managed conflicts of interest, or abuse of human research participants) motivated these federal organizations to mandate training, so that future scientists would recognize research ethics issues when they arise and know how to reason through them when they are encountered.

With regard to military-purposed science, an option is to list military-purposed science (or something more general about the relationship between the scientist and society) within the standardized core areas that must be covered as part of RCR training. At present, there are two separate lists of RCR core areas, one from the NIH and one from the Office of Research Integrity (ORI). Both include human subjects research, non-human animal research, conflicts of interest, data management, authorship/publication, peer review, collaboration, mentor/mentee responsibilities, research misconduct, and conflicts of interest. The difference is that the NIH list also includes "the scientist as a responsible member of society, contemporary ethical issues in biomedical research, and the environmental and societal impacts of scientific research," while the ORI list does not (Steneck 2007; National Institutes of Health 2009). The NIH and ORI lists of RCR core areas should be coordinated so as not to breed confusion about what is and is not required in mandated training. More specifically, the ORI list should add a core area pertaining to the scientist as a responsible member of society. Listing "military-purposed science" as a standalone domain of RCR training might be asking a lot, but treating that topic under the more general heading of "scientist as a responsible member of society" is both feasible and arguably necessary, particularly in light of ongoing efforts to advance brain research through societally innovative applications of neurotechnologies.

OPTION FIVE: KEEP TALKING

The last model is the existing discursive model—that is, just keep talking. The idea here is motivated by the fact that any new regulatory system at the ORI/NIH/NSF level, which would have to be enacted by individual institutions, would be incredibly difficult to implement. What's more, advocates for the existing model could point to Lesson One about military-purposed science being "just controversial." So, the argument goes, the best option at present is to just keep talking for now and watch how things play out.

The problem with this option is that it ignores Lessons Two and Four. Lesson Two reminds us of the costly failures that have followed from poorly designed or implemented science. Lesson Four, about legitimate concerns, suggests that we should not wait for military-purposed science to incur some profound ethical issue or problem before acting, lest the response will likely be in the form of strong regulation. We should not interpret regulation as suggesting that military-purposed science is inherently wrong. Rather, formalizing the relationship assists the military in "getting it right" by helping to prevent failures that come from bad science or inapt implementation.

CONCLUSION

When regulating military-purposed science at the academic/institutional level in the United States, there are a number of existing models that may be considered. Of course, there is always the option of looking beyond existing models, but that would require an *ab initio* approach. So, the focus of this chapter affords a survey of the various ways that scientific research is currently regulated in the United States in order to address if and how any extant models fit the needs and parameters required to guide ethical issues associated with military-purposed neuroscience.

The lessons from history concerning the relationship between science and the military are as follows: (1) it is controversial, but no more and no less, (2) it is an amalgam of successes and failures, (3) there is nothing totally unique about neuroscience and technology, and (4) there are legitimate concerns. Based on these lessons, I recommend at least Option Four (Active Education); however, this requires first standardizing the RCR core areas to specifically address "the scientist as a responsible member of society." Option Three (Weak Regulation) should also be considered as a viable approach to directing the focus and conduct of military-purposed bioscientific research. Of course, this would incur costs associated with establishing and operating another regulatory body at academic institutions across the United States. However, in exchange this offers a feasible mechanism of oversight for those legitimate concerns that are currently unregulated. (For further discussion of how Options Three and Four could be formalized and operationalized, refer to Chapter 17.) Options One (Ban), Two (Strong Regulation), and Five (Keep Talking) are all poor fits for contemporary military-purposed science generally, and neuroscience and neurotechnology, more specifically, as these approaches fail to acknowledge the momentum in the field, and the pace and extent of effect(s) that such momentum can—and likely will—incur via the (neuro)ethical, legal, and social domains.

REFERENCES

Fackler, M. 2010. "At Hiroshima ceremony, a first for a U.S. envoy." *The New York Times*, August 6. http://www.nytimes.com/2010/08/07/world/asia/07japan.html?_r=0.

Farahany, N. 2008. "The government is trying to wrap its mind around yours." *The Washington Post*, April 13. http://www.washingtonpost.com/wp-dyn/content/article/2008/04/11/AR2008041103296.html.

Fisher, J.C. and C. Fisher. 2011. *Food in the American Military: A History*. Jefferson, NC: McFarland & Co.

Fredrickson, D.S. 2001. *The Recombinant DNA Controversy, A Memoir: Science, Politics, and the Public Interest 1974–1981*. Washington, DC: ASM Press.

Giordano, J. 2012. "Neurotechnology as demiurgical force: Avoiding Icares' folly." In *Neurotechnology: Premises, Potential and Problems*. ed. J. Giordano. Boca Raton, FL: CRC press, pp. 1–15.

Giordano, J., C. Forsythe, and J. Olds. 2010. "Neuroscience, neurotechnology, and national security: The need for preparedness and ethics of responsible action." *American Journal of Bioethics Neuroscience* 1(2):35–36.

Greely, H.T. and J. Illes. 2007. "Neuroscience-based lie detection: The urgent need for regulation." *American Journal of Law and Medicine* 33:377–431.

Green, C. 2008. "The limits of traditional arms control models." *Bulletin of the Atomic Scientists*, September 11. http://thebulletin.org/military-application-neuroscience-research/limits-traditional-arms-control-models.

Grossman, D. 2009. *On Killing: The Psychological Costs of Learning to Kill in War and Society*. New York: E-Rights/E-Reads, Ltd.

Gusterson, H. 2007. "The militarization of neuroscience." *Bulletin of the Atomic Scientists*, April 9. http://thebulletin.org/militarization-neuroscience.

Huang, J.Y. and M.E. Kosal. 2008. "The security impact of the neurosciences." *Bulletin of the Atomic Scientists*, June 20. http://thebulletin.org/security-impact-neurosciences.

Hughes, J. 2002. *The Manhattan Project: Big Science and the Atomic Bomb*. New York: Columbia University Press.

Jennings, C. 2006. "Battlefield between the ears." *Nature* 443:911.

Marrapodi, E. and C. Lawrence. 2010. "Future soldiers may be wearing 'Iron Man' suits." CNN, November 12. http://www.cnn.com/2010/TECH/innovation/11/11/iron.man.suit/index.html.

McLean, C. and P.W. Singer. 2010. "What our military allies can tell us about the end of 'don't ask, don't tell.'" The Brookings Institution, June 7. http://www.brookings.edu/research/opinions/2010/06/07-dont-ask-dont-tell-singer.

Moreno, J.D. 2006. *Mind Wars: Brain Research and National Defense*. New York: Dana Press.

National Institutes of Health. 2009. NOT-OD-10-019: Update on the requirements for instruction in the responsible conduct of research. http://grants.nih.gov/grants/guide/notice-files/NOT-OD-10-019.html.

National Research Council. 2008. *Emerging Cognitive Neuroscience and Related Technologies*. Washington, DC: National Academies Press. http://www.nap.edu/catalog.php?record_id=12177.

Pilch, R. 2003. "The Moscow theater hostage crisis: The perpetrators, their tactics, and the Russian response." *International Negotiation* 8:577–611.

Reverby, S.M. 2009. *Examining Tuskegee: The Infamous Syphilis Study and Its Legacy*. Chapel Hill, NC: The University of North Carolina Press.

Seelye, K.Q. 2001. "The president's decision: The overview; Bush gives his backing for limited research on existing stem cells." *The New York Times*, August 10. http://www.nytimes.com/2001/08/10/us/president-s-decision-overview-bush-gives-his-backing-for-limited-research.html.

Shamoo, A.E. and D.B. Resnik. 2009. *Responsible Conduct of Research*. 2nd Edition. Oxford: Oxford University Press.

Steneck, N.H. 2007. *ORI: Introduction to the Responsible Conduct of Research*. Revised Edition. Washington, DC: U.S. Government Printing Office.

Streatfeild, D. 2007. "*Brainwash: The Secret History of Mind Control*." New York: St. Martin's Press.

Sunshine Project. http://www.sunshine-project.org/incapacitants/jnlwdpdf/wpafbchem.pdf.

York, H.F. 1976. *The Advisors: Oppenheimer, Teller, and the Superbomb*. Stanford, CA: Stanford University Press.

17 Engaging Neuroethical Issues Generated by the Use of Neurotechnology in National Security and Defense

Toward Process, Methods, and Paradigm

Rochelle E. Tractenberg, Kevin T. FitzGerald, and James Giordano

CONTENTS

THE REALITIES OF NEUROTECHNOLOGICAL USE—AND NEUROETHICAL ISSUES—IN NATIONAL SECURITY AND DEFENSE

As Aristotle noted over 2000 years ago, every human enterprise and tool can be construed as being purposed to achieve some definable "good" (Aristotle 1966). As methods to sustain and optimize survival and promote the flourishing of individuals, groups, and communities, science and technology (S/T) can most certainly be seen in this light. In the ideal, S/T would be employed only in benevolent ways; yet, definitions of the "good" may vary, and what is deemed to be good for some may in fact be burdensome if not harmful to others. New knowledge and technical capability confer considerable power, and as history illustrates, new developments in S/T continue to possess particular appeal in agendas of national security and defense.

The axiomatic goal of any country's efforts in national security, intelligence and defense (NSID) is the protection of the population. Toward this end, knowledge— and engagement—of real and potential threats is vital to preventing or mitigating events before they escalate into scenarios of large-scale harm. As illustrated throughout this volume, the potential for neuroscience and neurotechnology (i.e., neuro S/T) to assess and affect the cognitive, emotional, and behavioral dimensions of individuals and groups, on scales that range from the synaptic to the social, renders these approaches specifically enticing. Yet, these very same capabilities generate somewhat distinctive issues, questions, and dilemmas that yoke any analysis of neuro S/T to ethical examination (Roskies 2002; Giordano 2010, 2011, 2012a, 2014; Giordano and Olds 2010; Levy 2010; Racine 2010). The goal is not to be merely proscriptive or dismissive, but rather to be critically perceptive to the potential for innovation and viable ways that neuro S/T could be developed, used and/or misused, to a variety of ends, and by a number of nations, groups, and individual actors (Benedikter et al. 2009; Benedikter and Giordano 2012; Giordano and Benedikter 2012a, 2012b; Giordano 2014; Shook and Giordano 2014).

PREMISES, CONDITIONS, AND NECESSITIES

At this point, it is important to begin with what we believe to be three crucial premises: First, it is likely that neuro S/T will be ever more widely incorporated into approaches to intelligence gathering, implementation of counterintelligence, and other enterprises of national security and defense. Second, like any scientific technique or technology, neuro S/T has potential for misuse and harm; and third, international investment in neuro S/T is rapidly growing (Lynch and McCann 2009). Thus, it is probable that many countries (as well as corporations and subsidized actors) are developing neuroscientific capabilities in NSID (Giordano et al. 2010). From these premises, we posit three necessities that undergird the ethical address and guidance of neuro S/T in NSID. First is the need for realistic evaluation of (1) the actual capabilities and limitations of the types and extent of neuro S/T being developed and employed and (2) the ethico-legal issues generated by apt or inapt use and/or blatant abuse in specific NSID contexts. Second is the need to recognize how the pace and extent of neuro S/T advancement warrant a stance of preparedness for the realities conferred by the use of neuro S/T in NSID agendas on the twenty-first century world-stage; and third is the need to avoid the fallacy of two wrongs and not cavalierly commit to neuro S/T research and translation in NSID agendas in reaction to global activities in these areas without explicit dedication to the ethico-legal probity of any such work.

But it is also important to ask what constitutes "ethico-legal probity"? In this volume, and elsewhere, William Casebeer asserts that any neuroethical evaluation of neuro S/T (in NSID) should consider the "three Cs: consequences, character, and consent" (see also Casebeer 2013). That is, what outcomes and ends will be effected by the use of a given neuroscientific technique and/or technology? How might such uses modify the fundamental characterological bases (i.e., thoughts, feelings, values, motives, and expressions) of those affected? and are such interventions rendered with the permission and affirmation of the recipients? We concur with Casebeer, adding an additional "C"—context—which we believe would be important to establish those specified situations, environments, exigencies, contingencies, and parameters that are relevant to the ways that ethical analysis and guidance can and should be articulated (Giordano and Benedikter 2012a, 2012b; Shook and Giordano 2014; Lanzilao et al. 2014). Note that the context of NSID might necessitate consideration of additional "Cs": Cosmopolitanism (i.e., regard for the ways that neuro S/T might be used to effect international economic, medical, social, and military needs and relations), Cooperation (with domestic and international governmental and civilian groups and organizations), Communication with those that can and will be affected by various trajectories of neuro S/T utilization, and some extent of Concealment, given that neuro S/T in NSID will likely be employed in ways that contend against groups of others that have been identified as possessing viable threat. This latter consideration can be exceedingly provocative, if not contentious, given the specter of secrecy, and real or imagined conceptions of nefarious use (*vide infra*, and see also Chapter 7; Giordano et al. 2010).

FUNDAMENTAL QUESTIONS AND DIRECTIONS

To wit, we propose the following questions as important to defining and shaping the conduct of neuro S/T research and use within NSID:

- Is there some "inviolability of mind" that negates the use of such approaches, irrespective of circumstance?
- Or, are there particular circumstances under which certain neuro S/T approaches may be employed to mitigate real and present danger to the populace?
- Does the use of neuro S/T incur greater or lesser risks and harms than other methods of intelligence, security, and deterrence?
- Are there limits to the ways that neuro S/T should be used in NSID situations, and if so, how should such criteria be developed and enforced?

There are assertions that neuro S/T research and its outcomes should not be employed in NSID agendas because of the potential for escalation and misuse and the view that using neuro S/T in such ways would incur violation(s) of broadly accepted, fundamental human rights (for further discussion see Chapters 11 through 14). We respect the validity and value of these claims and in this light offer three possible options:

1. Abstaining from implementing neuro S/T research, development, or translation in any/all NSID agendas and situations
2. Utilizing neuro S/T only in specific situations that would dictate—and ethico-legally justify—the need for this type and level of assessment and/or intervention

3. Making appropriate neuro S/T approaches available and employable in all national security endeavors, but only in accordance with strictly defined and implemented ethico-legal parameters (that would then need to be surveyed and enforced on a variety of scales and levels)

While perhaps noble in intent, we hold the first option to be unrealistic and thus untenable given (1) that almost any published neuro S/T research can be used for NSID purposes; (2) the potential for dual-use neuro S/T research; and (3) the directly subsidized neuro S/T within various militaries and defense silos (worldwide). This leaves options 2 and/or 3. When considering these options, it must be borne to mind that the appointed goal of intelligence and security efforts (at least of the United States and its allies) is not to cause harm without purpose, but rather to uphold and protect the rights of the polis. But, to reiterate, neuro S/T research and use in NSID is not limited to the United States, the North Atlantic Treaty Organization (NATO), and allied nations; and as history has shown, law enforcement and military authority can be misappropriated and abused (in any country), and these possibilities must be taken into account and abuses should be prevented or mitigated as best possible.

Therefore, we hold that it will be necessary to discern whether neuroethical deliberation and decisions relevant to NSID applications should be based upon the following approaches:

1. A particular philosophical approach (e.g., utilitarianism? deontology? standpoint?) to ground any ethical posture toward the use/non-use of neuro S/T in NSID
2. The spirit of the law, which might allow uses of neuro S/T with particular legal constraints (that could be amended and/or modified in accordance with legal procedure) as usually afforded to circumstances of public safety (Ferguson 1999; Giordano et al. 2014; see also Chapter 9)
3. Philosophical and applied constructs of military ethics (e.g., *jus in bello*) that could be applied to the use of new and emerging neuroscientific techniques and tools (*vide supra*; see Simon 1999; Gross 2013; Chapter 15)
4. Some extant or new combination of these approaches that might afford precepts and principles that are more reflective of and responsive to the rapidly shifting capabilities of neuro S/T in the social and political spheres that define arenas of national security concerns on local and global levels (Giordano and Benedikter 2012a, 2012b; Lanzilao et al. 2014; Shook and Giordano 2014)

Irrespective of the neuroethical approach and/or system embraced, we advocate sensitivity to what we have referred to as "footfall effects," namely, that it is not the intention to impede the pace or momentum of forward progress of neuro S/T (given the aforementioned difficulties, if not impossibility of doing so), but rather to scrutinize where each forward step may fall, "... so as to tread wisely with appropriate lightness or force, and remain upright and balanced both in the course of usual events, and if pushed or stricken" (Giordano 2013a).

Footfall sensitivity and responsivity necessitates three interacting considerations: First is whether (a given) neuro S/T is sufficiently mature to be used in ways proposed within NSID applications. In some cases, it appears that actual capabilities are lacking, thereby limiting or contradicting such use (e.g., deception detection, predictive capacity, covert use of indwelling neuromodulatory devices; see elsewhere in this volume for overviews and literature examining the current scope of neuro S/T in NSID applications). However, today's limitations often foster ongoing research, which can lead to delimiting techniques and technologies that confer ever-greater sophistication and competence.

Despite these possible trajectories, it is still vital to fully and realistically comprehend what neuro S/T actually can—and cannot—do, so as to accurately deliberate upon and decide those ways that neuro S/T should or should not be utilized. From this, the second consideration must focus upon the risk of misperception, miscommunication, and misappropriation of neuroscientific information and capabilities to wage arguments that fortify fallacious (sociopolitical) positions that can be used to support particular beliefs and actions. Third, is that when taken together, these two considerations prompt concerns of whether ethical methods and systems are in-place and viable—and therefore possess merit—in addressing, analyzing, guiding, directing, and governing the possible use of new and emerging neuro S/T in various NSID scenarios. In the main, we contend that they are not; at least not to the extent that we believe necessary and sufficient to account for the contingencies spawned by the rapid advancement of neuro S/T and the shifting social, economic, political, and military architectonics that shape and define the milieu of twenty-first century NSID operations.

THE TASK(S) OF ADDRESSING NEUROETHICS IN NATIONAL SECURITY AND DEFENSE

To be sure, the changing environment(s) and exigencies of NSID establish the need for dexterous exercise of knowledge, receptivity, and responsible decision making and action. The tasks(s) at hand will be to attend to (1) the technical rectitude of any and all neuro S/T to be used in NSID; (2) situational variables germane to NSID engagement; (3) ethico-legal implications and manifestations of such use (or nonuse) in these situations; (4) the evaluation, revision, and/or generation of ethical concepts and systems that may be used to assess and guide decision making and action; and (5) frameworks for establishing and executing a systematic approach to ethical engagement.

We argue that any such approach must possess what we denote as *TASKER* properties, that is, it should be *t*emporally and *t*ask-*a*gile, *s*cientifically and *s*ituationally *k*nowledgeable, and *e*xperientially and *e*thically *r*eceptive, *r*esponsive, and *r*esponsible. The overarching goals are to develop a stance of competence (in the various domains that affect and are affected by neuro S/T relative to the ethical issues fostered by NSID), and preparedness (for the advancement of neuro S/T at-large, and the interactive effects that neuro S/T will manifest in the sociopolitical dimensions critical to NSID). As Jonathan Moreno (2006, 2012), Malcolm Dando (2007), S.E. White (2008), William Casebeer (2013), and others have noted, this would necessitate specifically dedicated groups that are keenly aware and focused upon the science, and the issues and implications of brain research and its applications (see also Chapters 10, 11, 14, and 16). We agree and recognize that any such groups should consist of multidisciplinary professionals (e.g., neuroscientists, neuroengineers, historians, anthropologists, social and political scientists, ethicists, lawyers, security and military operations' specialists) from both government and civilian sectors who possess considerable experience, knowledge, and expertise in several areas that are instrumental to understanding and engaging ethical decisions affecting the employment, restraint, and outcomes of neuro S/T in specific contexts and considerations of NSID (DiEuliis and Giordano 2014).

As well, it will be necessary to develop a process through which such groups could identify, address, analyze, and resolve neuroethical, legal, and social issues (NELSI) arising in and from neuro S/T research and use in NSID. We posit the need for a number of groups (both nationally and internationally) that would be detailed/ assigned a portfolio of neuro S/T-related projects, in which key NELSI would be assessed, analyzed, and engaged. These groups would be interactive and also communicate with key stake- and shareholders—including the public (Giordano et al. 2010; Giordano and Benedikter 2012a; DiEuliis and Giordano 2014)—in order to balance consensus and dissensus in a dialectical approach to decisions affecting potential use of neuro S/T in NSID circumstances. This process, schematically depicted in Figure 17.1, should be iterative, reflexive, and enable both analysis

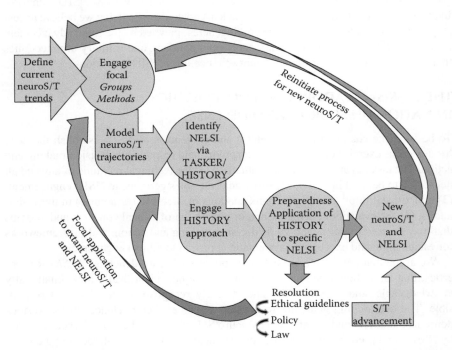

FIGURE 17.1 Schematic depiction of elements, processes, and methods involved in addressing and analyzing NELSI fostered by neuro S/T. Preparedness entails (a) prediction of emergent NELSI and their effects generated in specific situations by neuro S/T and (b) reasoned approaches to preventing, mitigating, or actively dealing with NELSI through the formulation of guidelines, which can then inform policy and national and/or international law(s) to direct and govern neuro S/T research, and use. This allows for subsequent advancement of neuro S/T, which in turn can foster new NELSI, thereby reengaging the process. NELSI = NeuroEthical, Legal, and Social Issues; TASKER = Task and Temporal Agility, Scientific and Situational Knowledge, Experiential and Ethical Reflexivity, Responsivity and Responsibility; HISTORY = Historicity and Implications of Science and Technology, Ombudsmanship and Responsible Yeomanry. (Figure © S. Loveless.)

of prior ELSI/NELSI that may be applicable to present circumstances, as well as establish bases for casuistically approaching future NELSI spawned by proposed or envisioned use of neuro S/T in NSID.

A METHODOLOGICAL APPROACH

Rapid advancement of neuro S/T and shifts in global sociopolitical dynamics may yield novel situations and contingencies (Benedikter and Giordano 2012; Giordano and Benedikter 2012a, 2012b). Still, an analytical approach to historicity can be useful for developing heuristics to (1) examine current uses of neuro S/T in light of previous successes, failures, and NELSI; (2) inform neuroethical focus, deliberation, and decision-making to guide present and future applications of neuro S/T; and in this way (3) avoid both repetition of previous mistakes, as well as current misdirection and/or misuses of neuro S/T (Giordano 2013a, 2013b, 2014; see also Chapter 16). This methodological approach, which we refer to by the acronym *HISTORY*, addresses *H*istoricity and *I*mplications of *S/T* in those prior instances and situations that offer relevant information to address current circumstances, contingencies, issues, questions and problems; and engages *O*mbudsmanship (the realistic evaluation of capabilities, limitations, benefits, burdens, risks, and harms of neuro S/T), toward *R*esponsible *Y*eomanry in the pragmatic elucidation and address of NELSI and problems generated by neuro S/T in specific contexts (Giordano 2013b; Giordano and Benedikter 2013).

Yet, TASKER-in-practice and HISTORY-as-process are reliant upon, reflective of, and presume a level of expertise in those professionals who are involved. Much has been written about the validity of ethical expertise, and a complete discussion of that literature is beyond the scope of this chapter (for overviews, see Rasmussen 2005; Selinger and Crease 2006). We seek to side-step the debate and directly define what ethical expertise should entail and obtain. The multidisciplinarity of the focus and process is such that expertise should be specific (to one's discipline—or disciplines) and representative of sufficient experience(s) to allow flexibility to hone knowledge and skills upon a number of potential NELSI that arise from new (and possibly unique) applications of neuro S/T in NSID. Both TASKER and HISTORY require the capability to develop and apply neuroethical concepts in a range of (rapidly developing and shifting) conceptualizations (Rawls 1971; Lanzilao et al. 2014).

Clearly, those who would be engaged in addressing NELSI arising in and from NSID would not be neophytes or novices, but rather would be well-established professionals with considerable expertise in their respective fields (and perhaps multiple disciplines). Yet, it will be important to ensure that these individuals possess and retain the types and extent of knowledge and skills' competencies that afford greatest flexibility and agility to apprehend the NELSI that can and will likely arise (as well as those that are presently unanticipated and will only emerge as subsequent iterations of neuro S/T are developed and applied in NSID contexts). How this flexibility will be evaluated and ensured needs to be addressed if such currency is to be maintained as a cornerstone of these individuals' and groups' fluency in approaching and articulating neuroethical decision making that exerts the range of foreseeable effect(s) that neuro S/T could engender (not only in NSID situations, but on more broadly sociopolitical fronts of the world stage). Moreover, while the formation and engagement of these individuals and working groups at present is vital, it would be myopic—and therefore impractical—not to educate and train a cadre of professionals that would be (perhaps even better) suited to accept the proverbial baton as specialists dedicated to these tasks in the future. Thus, we argue that any approach to evaluation and training must also be applicable to the education of this next generation of professionals (Giordano 2012b; Anderson and Giordano 2013).

A PARADIGM FOR TRAINING AND ASSESSING NEUROETHICAL REASONING

We opine that current systems of ethics education and training, particularly those within programs of science, technology, engineering, and/or mathematics (STEM), tend to be insufficiently focused and/or applied. Thus, they do not enable meaningful apprehension and address of the kinds of "real-world" ethical issues generated within the proximate and more far-reaching aspects of current and potential uses of novel S/T (inclusive of NSID-related situations). Requirements of the National Institutes of Health (NIH; 2009) formally state that "... there are no specific curricular requirements for instruction in responsible conduct of research." As well, mandatory training in the responsible conduct of research (RCR) as upheld by NIH is inadequate, as this only affords introduction to, and basic "... awareness and application of established professional norms and ethical principles in the performance of all activities related to scientific research" (Update on the Requirement for Instruction in the Responsible Conduct of Research: NOT-OD-10-019).

According to a recent report issued by the Office of Research Integrity, "accepted practices for the responsible conduct of research can and do vary from discipline to discipline and even from laboratory to laboratory" (Steneck 2007, p. 2). Required training is not comprehensive, as not every person involved in scientific or technological activities is considered to be involved or engaged in "scientific research." Further, a reliance on "accepted practices" makes the paradigm inflexible in the face of changing and evolving ethical issues. Perhaps most worryingly, there is no statement, allusion, or implication that the ability to reason—and make decisions—under conditions of uncertainty is of any value in the responsible guidance and conduct of research.

This approach also fails to equip participants with the knowledge and skills necessary to uncover gaps in their understanding of, or reasoning around ethical challenges that may actually be encountered. Unarguably, facts are important to any ethical analysis. However, what is equally, if not more important is the ability to utilize facts, as well as a host of other variables, nuances, and experiences when assessing ethical issues, questions, and problems that can and do arise within various applications of S/T research and use. We have argued elsewhere that simple presentation of factual material about ethical issues is not sufficient to introduce, promote, and sustain the type and extent of reasoning and ethical decision making that would be of greatest value to guiding novel and provocative iterations of science (Tractenberg and FitzGerald 2012; Tractenberg 2013)—inclusive of neuro S/T in NSID. Although the NOT-OD-10-019 stipulates that "(a)ctive involvement in the issues of … research should occur throughout a scientist's career" there is no information or even suggestion that this involvement should be flexible, adaptive, grow, and/or change in any way over time. There are no requirements—nor are there incentives or guidelines—to either develop or document the capacity to reliably train others in the finely grained ethical reasoning that would meaningfully afford the type and level of expertise required. The only difference between "new" and "practicing" scientists in this paradigm is *time*, not training in ethics or experience with ethical challenges.

THE MASTERY RUBRIC FOR ETHICAL REASONING (MR-ER)

Therefore, we (re-)introduce a paradigm—the Mastery Rubric (MR)—that focuses on a learnable, improvable skillset in *ethical reasoning* (MR-ER; Tractenberg and FitzGerald 2012), as directly applicable to the evaluation, training, and fortification of multidisciplinary professionals who would constitute the working groups to address NELSI of neuro S/T in NSID. As shown in Table 17.1, this paradigm explicitly has an inherent developmental trajectory: performance of the ethical reasoning skills and steps are described in a flexible manner so that any experiences can be reflected upon in order to demonstrate either the need for additional development of a given reasoning skill, or the actual level at which that particular ethical reasoning element is possessed.

Our new paradigm does not replace formal curricula in S/T ethics (what we have referred to as *In-STEPS: In*tegrative *S*cience, *T*echnology, *E*thics and *P*olicy *S*tudies; Giordano 2012b; Anderson and Giordano 2013), but it challenges almost every feature of regnant forms of ethics training and uses performance portfolios to capture learning and development of participants. In this way, it can be seen as a valuable component (e.g., as a progression metric and/or capstone) of the type of applied ethics curriculum that has recently been called for to meet multifocal opportunities and challenges of neuro S/T (Nuffield Report 2013), as well as an evaluative and/or recurrency training tool that can be used to assess/expand the knowledge and skills of upper-level professionals. In short, the MR-ER outlines a career-spanning training trajectory emphasizing ethical reasoning (Tractenberg and FitzGerald 2012). The MR-ER paradigm assesses (and directs) the acquisition and exercise of metacognitive reasoning skills that will be used to address and discern decisions (and decision processes) that must be made about specific scenarios, while also engaging the participant to self-assess their formal reasoning skills. These metacognitive skills can be taught and practiced with ethical materials and can be used across other domains once developed (Tractenberg and FitzGerald, in preparation).

Herein, we (re-)present the MR-ER more broadly as a developmental (and assessment) tool that is both task and temporally agile and that supports knowledge bases and experiential and ethical competencies that are wholly aligned with and constituent to our TASKER approach to HISTORY (Figure 17.2). This offers a unique form of developmental and dynamic ethics training and evaluation that we conceptualize as targeting a set of six learnable, improvable types of knowledge, skills, or abilities (KSAs) that together make up ethical reasoning. These are: prerequisite knowledge; recognizing an ethical issue; identification of

TABLE 17.1

The Mastery Rubric for Ethical Reasoning

Performance Levels	Novice	Beginner	Competent	Proficient
KSAs	Limited opportunity to exhibit any of the RCR KSAs; lack of awareness of many or all of the dimensions and also of individual's own development or place in the continuum. Output focused at Bloom's levels 1–2 (understand, summarize). Uninitiated.	Increasing opportunities to exhibit RCR KSAs. Requires oversight as awareness of KSAs grows. Output focused at Bloom's levels 1–2 (understand, summarize), but at a more in-depth level than the novice exhibits. Apprentice.	Inconsistent exemplification of all RCR KSAs, but clearly emerging proficiency in RCR; reflective participation in RCR training activities—seeking additional opportunities to reinforce less well-developed RCR skills. Output represents understanding and summarization but also moving toward prediction and/or analysis. **Journeyman.**	Consistent exemplification of all RCR dimensions; proficient mentoring of less senior/proficient investigators; active and competent participation in RCR training activities, including their development and evaluation. Output represents understanding and summarization, prediction and/or analysis, as well as mechanisms by which these cognitive skills can be elicited by less-proficient scientists/RCR trainees. Master.
Prerequisite knowledge	CITI and NIH online training modules completed satisfactorily.	Participation in discussion over time on fundamental (foundational) ethical issues.		
Recognize a moral issue	No recognition or only inconsistent recognition of most issues.	Consistent recognition of only most clear-cut issues.	Increased confidence in recognition ability with respect to most, if not all, moral issues.	Identify subtle conflicts at the personal, interpersonal, institutional, or societal level. Articulate questions arising either at the level of thought or feeling. Identify moral and ethical components. Analysis of how moral/ethical question arises. Coherent synthesis of perspectives of all relevant individuals involved for full recognition of moral issues and distinction between moral and ethical issues.

Identify decision-making frameworks	List at least a few different moral points of view and their basic key characteristics; recognize within which framework an action fits.	Identify strengths and weaknesses of several different frameworks. Predict within which framework an alternative action fits.	Identification of stakeholders and relevant facts of the case; explanation of why those facts are relevant and to whom, within each framework.	Judge among frameworks for: Relevance to problem, internal consistency, and broader applicability. Capacity to create vignettes for eliciting decision-making frameworks from less-proficient RCR trainees.
Identify and evaluate alternative actions	Incomplete list of the most clear-cut options—from few frameworks. Cannot evaluate or rank well; unaware of nuances within or between alternatives.	List of clear-cut options—from limited number of frameworks, articulation of alternative actions possibly recognizing conflicts between alternatives and/or for various interests, but uncertain how/unable to reconcile these conflicts well. Incomplete evaluations and explications of alternatives.	Evaluative list and ranking of alternative actions, from most frameworks; evaluation of stakeholders for whom each alternative might be most relevant and/or compelling, and why. Incomplete explication.	Create and evaluate a relatively comprehensive list of alternative actions from the perspectives of those moral points of view that are specifically relevant to the problem. Capacity to create vignettes for eliciting lists or evaluations from less-proficient RCR trainees.
Make and justify decision	Unable to consistently justify any one decision over another adequately.	Justify at least two decisions, even if they are at odds, and predict what the outcomes would be if all (justified) decisions were taken—furthering the decision-making process but not exactly completing it.	Identify "best" alternative actions, justify the rankings. Synthesize the evidence composed for first three KSAs for a thorough analysis of what can and should be done—by whom and why.	Identify the "best" alternatives from the perspective of each stakeholder as well as overall. Critique these classifications from the perspectives of experts in the field. Capacity to create vignettes for assisting less-proficient RCR trainees in perceiving the perspectives, alternatives, and justifications.
Reflect on decision	Reflection is a very abstract, high-level cognitive activity and would not be expected for the novice or beginner; novice or beginner might make an acceptable or "correct" decision and simply be aware that they cannot justify it well.		Consideration of ramifications of earlier decisions to improve future decisions on problems that do not necessarily have a "right" answer.	Facilitating the reflection of others on ethical decision making; taking a leadership role in pursuing contextual changes that could be made to avoid, adapt, or facilitate, similar decision making in the future.

Source: Adapted with permission from Tractenberg, R. and K. FitzGerald. 2012. *Assessment and Evaluation in Higher Education* 37(7/8): 1003–1021.

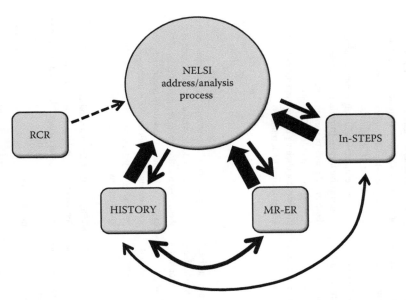

FIGURE 17.2 This figure shows how the proposed process of NELSI address and analysis (Figure 17.1) emerges from primary areas of experience, knowledge, and practice of three sources: (1) Historicity and implications of science and technology, ombudsmanship, and responsible yeomanry (HISTORY); (2) the Mastery Rubric for Ethical Reasoning (MR-ER) and participants' documented ethical reasoning knowledge, skills, and abilities; and (3) course and experiential work in integrative science and technology ethics and policy studies (i.e., the In-STEPS model) (bold arrows). Minor contributions (dotted arrow) to the neuroethical, legal, and social issues (NELSI) process from training in the responsible conduct of research (RCR) may be applicable. The reciprocal associations between these sources are represented by the curved arrows. Smaller solid arrows represent the effect of participation in the NELSI address/analysis process on the three primary sources; this process will affect participants and will feed back into the NELSI process with stronger inputs from each of those sources (for details, refer to text).

decision-making frameworks; identification and evaluation of alternative actions; making and justifying decisions; and reflecting decisions. These KSAs are focused on decision making and reasoning—attributes that have been noted to be conspicuously absent in (or to actually worsen after) existing forms of ethics training (Antes et al. 2010).

In addition to having refocused ethics training from topics to a learnable, improvable knowledge and skill set, the MR-ER also establishes a developmental schema of multilevel ethics training and competence. These levels are discussed in the subsequent paragraphs:

Novice. An uninitiated person who may have/have had limited opportunity to acquire or exhibit any of the reasoning KSAs. This would best describe an individual who has not had any formal academic ethics education, or who has only been exposed to extant (NIH-type) RCR training (given that such RCR training does not teach or support the development of ethical reasoning). At this level, ethical reasoning skills would tend to be limited to remembering, understanding, or summarizing ethical facts and concepts (Tractenberg et al. 2013).

Beginner. An apprentice who has increasing opportunities to exhibit ethical reasoning KSAs. This individual requires formative and formal oversight as their awareness of the KSAs (and how/when to employ them) grows. The beginner has had some exposure to the reasoning KSAs, but their reasoning would tend to be at a low level, and their work would be expected to reflect greater insight and depth than the novice.

Journeyman. Professionals (from any field) who exhibit all ethical reasoning KSAs, clearly show proficiency in ethical reasoning, and seek new opportunities to reinforce less well-developed skills. Their work represents understanding, summarization, and depth that permit analysis and/or prediction of ethical outcomes to the extent necessary and sufficient to function independently.

Proficient. Ethical reasoners (at the Journeyman level) who consistently exhibit all of the KSAs and who are capable and viable participants in ethical reasoning training activities, including development and evaluation. The output of the "proficient" reasoner represents functioning at the highest level of sophistication. Moreover, as originally conceptualized, the proficient individual is recognized by their documented experience and ability to capably mentor less senior/proficient reasoners, and their abilities to evaluate and remediate the reasoning skills exhibited in less-proficient reasoners and trainees.

The ethical reasoning skills in the MR-ER are foundational for engaging ethical challenges generally—and are neither topic nor discipline specific. This combination of reasoning skills and the developmental trajectory (with concretely described performance at each level) provides a new view of ethics training, competence, and assessment—as preparation for individuals to meaningfully engage evolving challenges, like NELSI in NSID, without having to be retrained for specific "fields" or "issues." Because ethical reasoning is a learnable, improvable skill set, achievement and criteria for candidacy or selection to groups or bodies charged with decision making around NELSI can be documented. Because ethical reasoning is a high level but generic skill set, the reasoner will be able to identify and analyze gaps in their own prerequisite knowledge and develop compensatory strategies and/or tactics—including, but not limited to finding individuals with the ethical reasoning skills but different HISTORY perspectives, perhaps, to form a functional (and highly functioning) team.

This paradigm not only concretely, but flexibly, describes a level of functioning that represents independent competence (journeyman) and explicitly distinguishes between the training and mentoring that individuals receive at different levels of their respective fields/disciplines. Furthermore, this paradigm supports the assessment and documentation of achieving journeyman-level ethical reasoning without requiring advanced (e.g., graduate) education in ethics, per se. Therefore, a working group could be comprised of professionals from a number of disciplines, who will have and contribute a variety of perspectives. Yet, synergy is facilitated because all possess an established level of ethical reasoning knowledge and skills (i.e., journeyman).

The MR-ER descriptions reflect the expected performance of each KSA as the individual moves from novice toward expertise and as such can be used to evaluate journeyman-level competencies (Tractenberg et al. 2010). Note that more expert-level habits of mind that are inculcated and reinforced by the MR are not the same as mere mastery of information. To the contrary, the MR-ER formalizes performance at the expert level for each of the target KSAs. As presented here, the MR-ER represents "the intended results of the cumulative contribution" of an ethics curriculum (Diamond 1998; Bresiani 2006, p. 14; Devine et al. 2007, pp. 51–52) and as such is an ideal method for identifying the functional level at which a participant must be certified before participating in activities required to engage the NELSI process represented in Figure 17.1 (see also Figure 17.2). Additionally, the MR-ER can be used to evaluate the efficiency of ethics education and training curricula (Tractenberg et al. 2010), which can highlight whether and how any given curriculum supports the development or achievement of specific and requisite KSAs. The original formulation of the MR-ER illustrated the viability of this tool in identifying, analyzing, and compensating for gaps in KSA acquisition and competence (Tractenberg and FitzGerald 2012). In this way, we propose that the MR-ER can be adopted or adapted to provide consistent evaluation of the ethics goals and capabilities necessary to address NELSI arising in and from the use of neuro S/T in NSID.

PARSIMONY, OPTIMIZATION, AND PREPAREDNESS

Attempting to develop approaches to address and resolve the neuroethical issues fostered by neuro S/T research and its possible uses and misuse in NSID may be seen as a Herculean, if not Sisyphean task. Indubitably, we appreciate the complexity and magnitude of the endeavor and effort required. At the core is that any such ethical address must begin with and proceed from fact(s). The fact is that neuro S/T, like any S/T, can and will be uptaken and used in NSID agendas worldwide. While existing treaties aim to constrain and govern the use of biochemical weapons, it is important to recall that participation in these treaties is not homogenous and often can be skirted through dual-use neuro S/T research and development. As well, a preparatory—or even precautionary—stance could establish the need for research to examine the potential ways that neuro S/T might be employed in NSID, what effects such uses might incur, and how these could be prevented, mitigated, and/or countered to protect the polis. Whether inadvertently or intentionally, this too would further neuro S/T research that could be used in and for NSID.

Simply put, we believe that it would be impossible to stop the incorporation of neuro S/T into NSID programs, and hence, related NELSI will arise and will require address, attention, and resolution. This presents a version of the proverbial "Neurath's Boat" problem, that is, one must find a way to repair the structure and function of a leaking (if not sinking) boat while underway at sea, given that it is impossible to begin from the bottom-up. The military allusion is not lost in this conundrum. What we propose herein will not make the issues any less complicated, or the task(s) any easier, but we believe that the process, methods, and paradigm offered may enable a more straightforward and effective approach—yielding Ockham's Razor (i.e., optimized parsimony) to fortify preparation and engagement of current and future NELSI generated by neuro S/T in NSID.

We recognize—and endorse—that some aspects and details of neuroscientific NSID research and development obviously need to remain classified (Giordano et al. 2010). Yet, we also advocate the need for frank, yet prudently directed and metered public communication about the broad scale and intent of such programs, so as to (1) avoid miscommunication and misapprehension and (2) gain insight into public concerns, expectations, and anxieties that undergird attendant NELSI. However, to paraphrase the U.S. President Barack Obama, the public frequently has a deficit of trust and harbors deep doubts about the workings of government (Obama 2010). Critical to any approach to NELSI of neuro S/T in NSID are public queries (and worries) of who will be involved in decisions about the scope and tenor of such research and its applications, and in what ways do these individuals' qualifications and experiences afford the capabilities demanded by both the ethical enormity of the effort, and a responsibility to public protection. It is not unreasonable to anticipate fears of "hawks in doves' feathers," and suspicions of "doves in the hawks' nest." Neither is useful, nor recommended.

CONCLUSIONS: INTEGRATION AND IMPLEMENTATION

What is required is that those involved must have knowledge and skills necessary for ethical reasoning and must also be aware of (if not experienced in) the realities, contingencies, and exigencies of NSID contexts. We maintain that the MR-ER will afford a mechanism to assess and inculcate these attributes and will also allow for more transparent and accurate public evidence (and perhaps selection) of individuals who are best prepared to guide, influence, and steer neuro S/T research and use in NSID. Additionally, we believe that the process, method, and paradigm presented herein comport well with a critical analytic approach to addressing social effects of neuro S/T, such as that proposed by Choudhoury et al. (2009), and a surety framework for guiding neuro S/T relative to identified NELSI (Shaneyfelt and Peercy 2012). Therefore, it can be employed to identify neuroethical issues, problems, and questions, ground these to past, present, and future circumstances and possibilities, and pose resolutions to effectively inform policy and law to guide use or non-use in NSID (Sarewitz and Karas 2012).

Writ large, we view the process, method, and paradigm as constituents of a three-legged platform that is required to authentically address NELSI in any context (Figure 17.3). Assessment and training of professionals (at the journeyman level) is important, but enduring commitment to guiding neuro S/T will necessitate education, both within the scientific and engineering fields, as well as the (social science and humanities') disciplines that will enable the ethically sound articulation of S/T in the social, economic, and political domains of human enterprise (Giordano 2012b; Anderson and Giordano 2013, *vide supra*).

None of these efforts in addressing NELSI are achievable or sustainable without "support." Such support is twofold and reciprocal: Extrinsically, it must provide (1) funding to study, develop, articulate, and grow programs dedicated to the education,

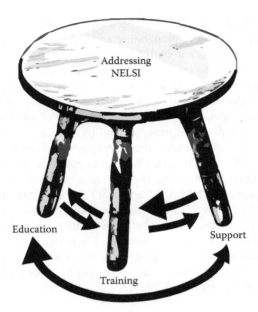

FIGURE 17.3 Diagrammatic representation of three-legged platform required to effectively address neuroethical, legal and social issues (NELSI). Arrows represent interactions; large(r) arrows indicate actual and/or potential fortitude of effect (see text for details). (Figure © S. Loveless.)

training, and activities of neuroethics professionals, and (2) policies to establish, enable, direct, and protect these programs and groups (and their resultant activities, as described subsequently). Intrinsically, support is generated by education and training endeavors that develop the individuals who will be involved in—and lead— the process, scope and conduct of neuro S/T research and use.

In conclusion, we offer that any meaningful consideration of NELSI of NSID must be based not upon a question of if neuro S/T will be used or misused (for we believe that to be a given), but rather upon questions of when, by whom, in what ways, to what extent, and to what effect. We opine that while these questions may remain uncertain at present, what is certain is the need to dedicate efforts toward developing the people, infrastructure, and modus of addressing and dealing with such inevitability and its consequences.

ACKNOWLEDGMENTS

This work was supported in part by grants from the JW Fulbright Foundation (JG); Office of Naval Research (JG); William H. and Ruth Crane Schaefer Endowment (JG); Clark Faculty Fellowship (JG); and by funding from the Edmund D. Pellegrino Center for Clinical Bioethics of Georgetown University Medical Center (KTF; JG). The authors thank Sherry Loveless for graphic artistry and assistance in the final preparation of this manuscript.

REFERENCES

Anderson, M.A. and J. Giordano. 2013. "*Aequilibrium Prudentis*: On the necessity for ethics and policy studies in the scientific and technological education of medical professionals." *BCM Education* 19(4):279–283.

Antes, A.L., X. Wang, M.D. Mumford, R.P. Brown, S. Connelly, and L.D. Devenport. 2010. "Evaluating the effects that existing instruction on responsible conduct of research has on ethical decision making." *Academic Medicine* 85(3):519–526.

Aristotle. 1966. *The Nicomachean Ethics*. (D. Ross, trans.). London: Oxford University Press.

Association of Internet Researchers Ethics Committee. 2012. "Ethical decision making and internet research." http://aoir.org/reports/ethics2.pdf. Accessed December 5, 2013.

Benedikter, R. and J. Giordano. 2012. "Neurotechnology: New frontiers for European policy." *European Journal of Government* 1(3):204–208.

Benedikter, R., J. Giordano, and K. FitzGerald. 2010. "The future of the self-image of the human being in the age of transhumanism, neurotechnology and global transition." *Futures: The Journal for Policy, Planning and Futures Studies* 42(10):1102–1109.

Bresciani, M.J. 2006. *Outcomes-Based Academic and Co-curricular Program Review: A Compilation of Institutional Good Practices*. Sterling, VA: Stylus.

Casebeer, W. 2013. "Plenary testimonial presented at US' president's commission for the study of bioethical issues." August 23. www.bioethics.gov/node/2224.

Choudbury, S., S.K. Nagel, and J. Slaby. 2009. "Critical neuroscience: Linking neuroscience and society through critical practice." *BioSocieties* 4:61–77.

Dando, M. 2007. *Preventing the Future Military Misuse of Neuroscience*. New York: Palgrave Macmillan.

Devine, S.M., K. Daly, D. Lero, and C. MacMartin. 2007. "Designing a new program in family relations and applied nutrition." *New Directions for Teaching and Learning* 112:47–57.

Diamond, R.M. 1998. *Designing and Assessing Courses and Curricula: A Practical Guide. Revised*. San Francisco, CA: Josey-Bass.

DiEuliis, D. and J. Giordano. 2014. "Neuroscience and technology (Neuro S/T) as the new dual-use frontier: Importance and necessity of neuroethical guidance and articulation." *American Journal of Bioethics Neuroscience* (in press).

Ferguson, J.R. 1999. "Biological weapons and U.S. law." In *Biological Weapons—Limiting the Threat*, ed. J. Lederberg. Cambridge, MA: MIT Press, pp. 81–91.

Giordano, J. 2010. "Neuroethics: Coming of age and facing the future." In *Scientific and Philosophical Perspectives in Neuroethics*, eds. J. Giordano and B. Gordijn. Cambridge, MA: Cambridge University Press, pp. xxv–xxix.

Giordano, J. 2011. "Neuroethics: Traditions, tasks and values." *Human Prospect* 1(1):2–8.

Giordano, J. 2012a. "Neurotechnology as deimurgical force: Avoiding Icarus' folly." In *Neurotechnology: Premises, Potential and Problems*, ed. J. Giordano. Boca Raton, FL: CRC Press, pp. 1–14.

Giordano, J. 2012b. "Keeping science and technology education in-STEP with the realities of the world stage: Inculcating responsibility for the power of STEM." *Synesis: A Journal of Science, Technology, Ethics, and Policy* 3(1):G1–5.

Giordano, J. 2013a. "Using neuroscience and neurotechnology in global influence and deterrence initiatives." Strategic Multilayer Assessment Group Report. Washington, DC.

Giordano, J. 2013b. "*Respice finem*: On the heuristics and guidance of scientific and technological advancement and use." *Synesis: A Journal of Science, Technology, Ethics, and Policy* 4:E1–4.

Giordano, J. 2014. "The human prospect(s) of neuroscience and neurotechnology: Domains of influence and the necessity-and questions-of neuroethics." *Human Prospect* 3(3):2–19.

Giordano, J. and R. Benedikter. 2012a. "An early-and necessary-flight of the Owl of Minerva: Neuroscience, neurotechnology, human socio-cultural boundaries, and the importance of neuroethics." *Journal of Evolution and Technology* 22(1):14–25.

Giordano, J. and R. Benedikter. 2012b. "Neurotechnology, culture and the need for a cosmopolitan neuroethics." In *Neurotechnology: Premises, Potential and Problems*, ed. J. Giordano. Boca Raton, FL: CRC Press, pp. 233–242.

Giordano, J. and R. Benedikter. 2013. "*HISTORY*—Historicity and implications of science and technology, ombudmanship, responsibility and yeomanry: A methodologic approach to neuroethics." International Neuroethics Society, San Diego, CA.

Giordano J., C. Forsythe, and J. Olds. 2010. "Neuroscience, neurotechnology and national security: The need for preparedness and an ethics of responsible action." *American Journal of Bioethics Neuroscience* 1(2):1–3.

Giordano, J., A. Kulkarni, and J. Farwell. 2014. "Deliver us from evil? The temptation, realities and neuroethico-legal issues of employing assessment neurotechnologies in public safety initiatives." *Theoretical Medicine and Bioethics* 35(1):73–89. doi:10.1007/S1 1017-014-9278-4.

Giordano, J. and J. Olds. 2010. "On the interfluence of neuroscience, neuroethics and legal and social issues: The need for (N)ELSI." *American Journal of Bioethics Neuroscience* 2(2):13–15.

Gross, M.L. ed. 2013. *Military Medical Ethics for the 21st Century.* Farnham, U.K.: Ashgate.

Lanzilao, E., J. Shook, R. Benedikter, and J. Giordano. 2013. "Advancing neuroscience on the 21st century world stage: The need for–and proposed structure of–an internationally-relevant neuroethics." *Ethics in Biology Engineering and Medicine* 4(3):211–229.

Levy, N. 2010. "Neuroethics: A new way of doing ethics." *American Journal of Bioethics Neuroscience* 2(2):3–9.

Lynch, Z. and C.M. McCann. 2010. "Neurotech clusters 2010: Leading regions in the global neurotechnology industry, 2010–2020." NeuroInsights Report, 2009. San Francisco, CA. http//www.neuroinsights.com. Accessed December 29, 2013.

Menzel, D.C. 1998. "To act ethically: The what, why, and how of ethics pedagogy." *Journal of Public Affairs Education* 4(1):11–18.

Messick, S. 1994. "The interplay of evidence and consequences in the validation of performance assessments." *Educational Researcher* 23(2):13–23.

Moreno, J. 2006; 2012. *Mind Wars* (1st and 2nd Editions). New York: Bellevue Press.

National Institutes of Health. 2009. Update on the requirement for instruction in the responsible conduct of research. NOT-OD-10-019. http//grants1.nih.gov/grants/guide/notice-files/NOT-OD-10-019.html. Accessed January 25, 2012.

National Institutes of Health (Funding Opportunity Announcement PAR-12-244). 2012. http://grants.nih.gov/grants/guide/pa-files/PAR-12-244.html. Accessed January 17, 2014.

Nuffield Council Report. 2013. *Novel Neurotechnologies: Intervening in the Brain.* London: Nuffield Council on Bioethics.

Obama, B. 2010. "State of the union address." http://www.huffingtonpost.com/2010/01/27/state-of-the-union-2010-full-text-transcript.n439459.html.

Racine, E. 2010. *Pragmatic Neuroethics: Improving Treatment and Understanding of the Mind-Brain.* Cambridge, MA: MIT Press.

Rasmussen, L. (ed.) 2005. *Ethics Expertise: History, Contemporary Perspectives and Applications.* Dordrecht: Springer.

Rawls, J. 1971. *A Theory of Justice.* Cambridge, MA: The Belknap Press.

Roskies, A. 2002. "Neuroethics for the new millenium." *Neuron* 35:21–23.

Sarewitz, D. and T.H. Karas. 2012. "Policy implications of technologies for cognitive enhancement." In *Neurotechnology: Premises, Potential and Problems*, ed. J. Giordano. Boca Raton, FL: CRC Press, pp. 267–286.

Selinger, E. and R.P. Crease. (eds.) 2006. *The Philosophy of Expertise*. New York: Columbia University Press.

Shaneyfelt, W. and D.E. Peercy. 2012. "A surety engineering framework and process to address ethical, legal and social issues for neurotechnologies." In *Neurotechnology: Premises, Potential and Problems*, ed. J. Giordano. Boca Raton, FL: CRC Press, pp. 213–232.

Shook, J. and J. Giordano. 2014. "A principled and cosmopolitan neuroethics: Implications for international relevance." *Philosophy Ethics and Humanities in Medicine* 9(1).

Simon, J.D. 1999. "Biological terrorism: Preparing to meet the threat." In *Biological Weapons—Limiting the Threat*, ed. J. Lederberg. Cambridge, MA: MIT Press, pp. 235–348.

Steneck, N.H. 2007. *Introduction to the Responsible Conduct of Research*. Revised Edition. Washington, DC: U.S. Department of Health and Human Services. http://ori.dhhs.gov/documents/rcrintro.pdf. Accessed January 1, 2010.

Stevens, D.D. and A.J. Levi. 2005. *Introduction to Rubrics: An Assessment Tool to Save Grading Time, Convey Effective Feedback and Promote Student Learning*. Portland, OR: Stylus.

Tractenberg, R.E. 2011. "Developing a curriculum for research in pathology residency: A pathology research mastery rubric." Northeast Group on Educational Affairs (NEGEA) Annual Meeting, Washington, DC. October 10.

Tractenberg, R.E. 2013. "Ethical reasoning for quantitative scientists: A mastery rubric for developmental trajectories, professional identity, and portfolios that document both." *Proceedings of the Joint Statistical Meetings*, Montreal, QC.

Tractenberg, R.E. and K.T. FitzGerald. 2012. "A mastery rubric for the design and evaluation of an institutional curriculum in the responsible conduct of research." *Assessment and Evaluation in Higher Education* 37(7/8):1003–1021.

Tractenberg, R.E., K.T. FitzGerald, J. Collmann, L. Vinsel, M. Steinmann, G. Morgan, A. Russell, and L.M. Dolling. (manuscript in preparation). "Ethics in and of big data: Using an ethical reasoning framework to facilitate multi-disciplinary perspectives on ethical, legal, and social issues (ELSI)."

Tractenberg, R.E., M.M. Gushta, S.E. Mulroney, and P.A. Weissinger. 2013. "Training in cognitive complexity is critical for targeting higher level thinking with multiple choice questions." *Advances in Health Sciences Education: Theory and Practice* 18(5):945–961. doi:10.1007/s10459-012-9434-4.

Tractenberg, R.E., R.J. McCarter, and J. Umans. 2010. "A mastery rubric for clinical research training: Guiding curriculum design, admissions, and development of course objectives." *Assessment and Evaluation in Higher Education* 35(1):15–32.

White, S.E. 2008. "Brave new world: Neurowarfare and the limits of international humanitarian law." *Cornell University International Law Journal* 41:177–210.

18 Postscript
A Neuroscience and National Security Normative Framework for the Twenty-First Century

William D. Casebeer

CONTENTS

INTRODUCTION

War and peace alike have both always been at least in part about the brain. As the organ of human action—and the mediator of how environments, our own cognitive processes, and our genetic inheritance shape our behavior—it must take pride of place in any comprehensive study of the causes of conflict. Despite the historical but sometimes hidden importance of the mind and brain for understanding conflict, as the essays in this volume demonstrate, we as a polity (both local and global) are only now coming to grips with the upshot of this fact for how we prevent conflict and prevail in it quickly when it is morally obligatory. In this postscript, I briefly discuss a framework for examining the practical and moral issues involved in using neuroscience to research and develop national security technologies; I also quickly consider how some of the traditional tools of moral reasoning can be used to resolve ethical issues or help us circumscribe the limits of what is required, permissible, and forbidden as we develop national security-related neuroscience technologies.

THE PRIMACY OF THE BRAIN ACROSS ALL PHASES OF CONFLICT

The U.S. military doctrine—in Joint Publication 5-0 of 2011, for instance, which outlines the military's joint planning process—identifies six phases of conflict (Joint Operation Planning 2011). These are broken out as (0—"phase zero") shape, (1) deter, (2) seize the initiative, (3) dominate, (4) stabilize, and (5) enable civil authority and then return to phase zero to "shape" yet again. Shaping consists in taking actions to affect the environment in ways that make threats to security less likely to emerge. Deterring consists in taking actions to prevent agents who desire to threaten security from doing so. Seizing the initiative consists in taking decisive actions to disable an active threat. Dominating consists in using all aspects of military power to achieve victory quickly. Stabilizing consists in taking actions to return to pre-conflict normalcy. Enabling civilian authority consists in taking actions (such as working with host governments) to reestablish nonmilitary-mediated stability and reconstitute governance. These definitions can be debated, especially by theorists and practitioners involved in military operational art, but serve as adequate entry points for considering why understanding the brain is so important for knowing how to achieve effects—and develop technologies required to do so—across all these phases.

In phases zero and one, understanding how the human brain is shaped by facts about the environment (including the social environment) and our genetic endowment is critically important. What kind of environments make it likely that conflict will break out or that peaceful means will likely not be effective in resolving disagreements about core values? Social cognitive neuroscience is especially useful here. In phase two of conflict, being able to analyze the effect of action and information on perception and decision making is very important; deterrence and influence occur at this confluence, and while rational actor models of this process have been useful, they require augmentation and bounding to be maximally useful. Neuroscience is useful here as well. In phase three, understanding how combatants make decisions and how performance is influenced by the combat environment—as well as how we treat injury and recover from and repair the stresses and wounds of war, both physical and psychological—is advanced by incorporating the latest in the neurobiology of decision making under stress. For phases four and five, understanding de-escalation, trust-building, the relationship between development and conflict resolution, and related questions about the brain and good governance are all advanced by cognitive and affective neuroscience. Theorists talk about winning hearts and minds, which in many respects is longhand for using technology to shape our brain.

While soldiers, sailors, airmen, and marines have always been concerned about achieving effects across all phases of conflict, our frameworks for the moral evaluation of the use of force—especially traditional approaches such as "just war theory"—are most thorough when it comes to phase three, the use of force in the battlefield. Questions about when the use of force is justified and against whom are paramount in this tradition and are no doubt exceedingly important. But to focus solely on these would be to neglect important questions that surface across all these phases, especially when it comes to the use of neuroscience to either tutor military action or to develop national security technologies that will be used across all phases of engagement.

NOT AT SEA IN A SIEVE: CHARACTER, CONSENT, AND CONSEQUENCE

Fortunately, the same ethical and moral frameworks that inform the development of traditional just war considerations—when is the use of force justified and in what fashion can it be permissibly applied?—can also be used to help us answer moral questions about the use of national security neuroscience technology across all phases of conflict. *Jus in bello* ("justice during the war") concerns about how non-combatants are treated, for instance, reflect praiseworthy considerations about both the rights of people not to be used as a mere means and utilitarian concerns about what practices will help minimize the harmful effects of war if it becomes morally necessary (Johnson 1999).

The three grand traditions of ethical theory can thus be of use as we think about a normative or moral framework for evaluating national security neurotechnology: these are virtue theory, deontology, and utilitarianism. The first highlights the person taking an action; the second focuses on the nature of the action being taken; and the third highlights the consequences of the action. All but the most adamant partisans of a particular approach to moral theory can agree that, at least for heuristic value, these three traditions have thrived because they focus attention on ethical aspects of a situation we might otherwise be prone to ignore. Here are thumbnail sketches of each approach.

Virtue theorists, such as the Greek philosophers Plato (427–347 BC) and Aristotle (384–322 BC), make paramount the concept of "human flourishing"; to be maximally moral is just to function as well as one can given one's nature. This involves the cultivation of virtues (such as wisdom) and the avoidance of vices (such as intemperance) and is a practical affair. Deontologists, exemplified by the Prussian philosopher Immanuel Kant (1724–1804), do not place emphasis upon the consequences of actions, as utilitarians would, nor on the character of people, as a virtue theorist would. Instead, they focus on the maxim of the action—the intent-based principle that plays itself out in an agent's mind. We must do our duty, as derived from the dictates of pure reason and the "categorical imperative," for duty's sake alone. Deontologists are particularly concerned to highlight the duties that free and reasonable creatures (paradigmatically, human beings) owe to one another. Maximizing happiness or cultivating character is not the primary goal on this scheme; instead, ensuring that we do not violate another's rights is paramount. The typical utilitarian, such as British philosopher John Stuart Mill (1806–1873), thinks one ought to take that action (or follow that "rule") that if taken (or followed) would produce the greatest amount of happiness for the largest number of sentient beings (where by happiness, Mill means the presence of pleasure or the absence of pain). The second flavor of utility we just described, "rule utilitarianism," is probably the most popular.

These three frameworks can be captured by remembering the "three C's"—"Character, Consent, and Consequence." A comprehensive evaluation of any particular neuroscience and national security technology would ask whether its development and use enables us to flourish as human beings and is conducive to the development of traits allowing us to do so ("Character"), whether the technology is being developed and used in a fashion consistent with the human right not be used as a mere means to someone else's end ("Consent"), and whether the development and use of the technology produces the best consequences all things considered ("Consequence") (Myers 1997). Given that these three frameworks cut across cultural differences and that any explicit or implicit regulatory frameworks that fall out of their consideration must speak across national boundaries, we could also speak—as Giordano has suggested—of a fourth "C," namely, Context. I see this not necessarily as being a separate axis of normative evaluation, but rather as a procedural demand that we constantly seek objectivity in our articulation of norms so that they are not parochial and can be discussed usefully and potentially be agreed upon in international fora (Rhodes 2009).

AN EXAMPLE: STRATEGIC RHETORIC, NEUROBIOLOGY, AND NORMATIVE EVALUATION

For an example of this framework in action, consider phases zero, one, and two of conflict. Shaping, influencing, and deterring involve in part as acts of communication, and acts of communication are often most effective if they are couched in terms of narratives or stories. If I am, for example, to successfully communicate my intention to provide disaster relief during a humanitarian operation that could involve the use of force (so as to reassure the victims that aid is on the way, and so as to deter organizations such as violent nonstate actors from attacking the forces provisioning relief), I will need to tell an effective story regarding our forces' involvement in an area. This is an aspect of narrative strategy. In practice, effective narrative strategies will require understanding the components and content of the story being told so we can predict how they will influence the action of a target audience. In other words, we need a sophisticated understanding of "strategic rhetoric." This is difficult to come by. Nonetheless, even well-worn and simple models of this process, such as that offered by the ancient Greek philosopher Aristotle in his *On Rhetoric*, can be very useful for structuring our thinking (Kennedy 1991).

Aristotle would have us evaluate three components of a narrative relative to a target audience: (1) what is the *ethos* of the speaker/deliverer? (2) what is the *logos* of the message being delivered? and (3) does the message contain appropriate appeals to *pathos*? Consideration of ethos would emphasize the need for us to establish credible channels of communication, fronted by actors who have the character and reputation required to ensure receipt and belief of the message. "You have bad ethos" in this context is merely another way of saying "You won't be believed by the target audience because they don't think *you* are *believable.*" Consideration of logos involves the rational elements of the narrative: is it logical? Is it consistent enough to be believed? Does it contain (from the target's perspective) nonsequiturs and forms of reasoning not normally used day-to-day? Finally, pathos deals with the emotional content of the story. Does the story cue appropriate affective and emotive systems in the human brain? Does it appeal to emotion in a way that engages the whole person and that increases the chances the story will actually motivate action? Understanding these processes is in part a matter for neurobiologists (see, e.g., the recent work of Greg Berns, Jorge Barraza, Emile Brueau, Antonio Damasio, Jonas Kaplan, Ken Kishida, Lucas Parra, Rebecca Saxe, and Paul Zak).

Some of these Aristotelian considerations will be affected by structural elements of the story (Is the story coherent? Is it simple enough to be processed? Can it be remembered? Is it easy to transmit? If believed, will it motivate appropriate action?); others will be affected by content (Does the narrative resonate with target audiences? Is the protagonist of the story a member of the target audience's in-group? Is the antagonist of the story a member of a hated out-group?). But all will be affected by the mechanism responsible for receiving and processing them to drive behavior: the brain.

Use of strategic storytelling in security contexts is interesting from a normative perspective as it does not fit neatly into traditional *jus ad bellum* (justice before conflict) and *jus in bello* (justice during conflict) considerations. However, the same overarching theories and their principles that inform just war considerations can be used—via the three C's—to map out the territory. While beyond the scope of this postscript to discuss in full, there are many circumstances where strategic storytelling in security contexts is conducive to human flourishing, can be done in a fashion which respects the rights of all involved and is accomplished with either their implicit or their explicit consent, and will produce better consequences. A world where we are able to speak truth to the power that others have to exploit the innocent in conflict environments is one where we can use reason to resolve our disagreements instead of force, and clear and effective storytelling is an important part of that process.

CONCLUSION

Given the primacy of the brain in driving human action, it is no surprise that neuroscience and its affiliated disciplines (psychology, cognitive science, biology, etc.) have an important role to play in helping us understand and channel conflict. Evaluating the moral dimensions of the use of national security neuroscience technology is an enterprise that requires subtlety of thought and respect for the limits and promise of neuroscience as a field. Fortunately, we do not have to start from scratch when thinking about a comprehensive framework for moral evaluation of national security neuro-technology: considerations of character, consent, and consequence will go a long way toward ensuring that we use our scientific knowledge and engineering expertise in such a fashion that future generations to respect our collective judgment call and thank us for them. I thank the authors and editors of this volume for continuing and enriching an important conversation about the intersections of neuroscience, technology, and morality, and am sure that posterity will pass positive judgment about their efforts.

REFERENCES

Johnson, J.T. 1999. *Morality and Contemporary Warfare*. New Haven, CT: Yale University Press.
Joint Publication 5-0. 2001, "Joint operation planning, August 11, 2011," available at www .dtic.mil/doctrine/new_pubs/jp5_0.pdf. Accessed on December 15, 2013.
Kennedy, G. trans. 1991. Aristotle's *On Rhetoric: A Theory of Civic Discourse*. New York: Oxford University Press.
Myers, C. 1997. The core values: Framing and resolving ethical issues for the air force. *Air and Space Power Journal* 11:38–52.
Rhodes, B. 2009. *An Introduction to Military Ethics: A Reference Handbook*. Westport, CT: Praeger.

Index

Note: Locators followed by "*f*" and "*t*" denote figures and tables in the text